GADSDEN PUBLIC LIBRARY
3
W9-DAO-111

AN EXHIBIT DENIED

MARTIN HARWIT

AN EXHIBIT DENIED

Lobbying the History of *Enola Gay*

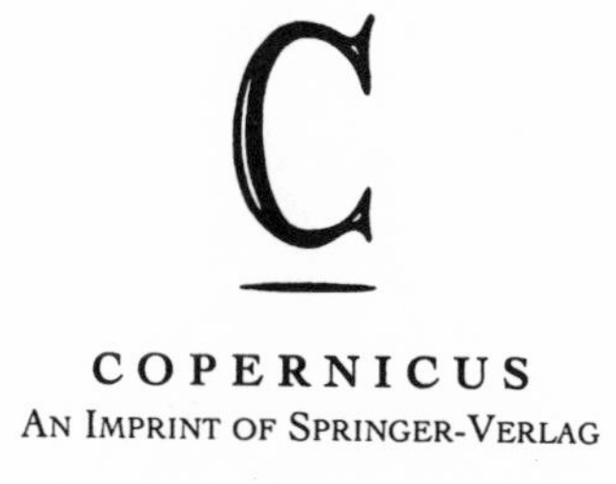

COPERNICUS
AN IMPRINT OF SPRINGER-VERLAG

Published in the United States by Copernicus, an imprint of Springer-Verlag New York, Inc.

Copernicus
Springer-Verlag New York, Inc.
175 Fifth Avenue
New York, NY 10010
USA

Library of Congress Cataloging-in-Publication Data
Harwit, Martin, 1931-
An exhibit denied : lobbying the history of *Enola Gay* / Martin Harwit
p. cm.
Includes bibliographical references and index.
ISBN 0-387-94797-3 (hardcover : alk. paper)
1. Enola Gay (Bomber)—Exhibitions. 2. National Air and Space Museum—Exhibitions. 3. Hiroshima-shi (Japan)—History—Politics and government—1993— I. Title
D787.25.H6H348 1996
940.54'25—dc20 96-18676
CIP

Printed on acid-free paper

9 8 7 6 5 4 3 2 1

ISBN 0-387-94797-3 SPIN 10539310

Contents

Preface

At 8:15 A.M., August 6, 1945, the *Enola Gay* released her load. For forty-three seconds, the world's first atomic bomb plunged through six miles of clear air to its preset detonation altitude. There it exploded, destroying Hiroshima and eighty thousand of her citizens. No war had ever seen such instant devastation. Within nine days Japan surrendered. World War II was over and a nuclear arms race had begun.

Fifty years later, the National Air and Space Museum was in the final stages of preparing an exhibition on the *Enola Gay*'s historic mission when eighty-one members of Congress angrily demanded cancellation of the planned display and the resignation or dismissal of the museum's director. The Smithsonian Institution, of which the National Air and Space Museum is a part, is heavily dependent on congressional funding. The Institution's chief executive, Smithsonian Secretary I. Michael Heyman, in office only four months at the time, scrapped the exhibit as requested, and promised to personally oversee a new display devoid of any historic context. In the wake of that decision I resigned as the museum's director and left the Smithsonian.

The losers in this drama were the American public, and most particularly the veterans of World War II, who, I am convinced, would have found the exhibition informative and inspiring. It provided insight into arguably the most important single military action of this century. Using previously secret information gradually declassified in the decades since the war, the exhibit portrayed motives, actions, and fates factually, and in the light of their times.

The exhibition's cancellation transcends normal museum affairs. The intended display had initiated a national and international debate, featured in literally thousands of media releases worldwide. Within the United States, the controversy reflected the ways our nation has begun to settle important issues—not through substantive debate, but through partisan campaigns aimed at victory by any means.

Most Americans remain unaware of the magnitude of such battles and their dominant role in formulating congressional legislation and national policy. The warring factions do their best to hide from the public the enormous influence they wield. They portray their efforts as generous works carried out for the national good, though the truth is often at odds with that picture. Win or lose, the soldiering lobbyists keep their tactics under wraps to be honed and reused in the next campaign.

A unique aspect of the *Enola Gay* exhibit was the substantial volume of privileged correspondence released by one of the lobbying organizations, the Air Force Association, even before the debate had fully subsided. These letters and memoranda dramatically reveal how much those who aggressively lobby Congress can gain for themselves.

For whatever it costs to buy influence, you can now have your own version of our nation's history displayed and opposing views suppressed at the Smithsonian Institution. Since the Smithsonian has close to thirty million visitors a year, three quarters of them American citizens, this tampering threatens widespread misapprehension about our nation's history, with potentially disastrous consequences.

Thus, when on January 30, 1995, Secretary Heyman announced the replacement of the National Air and Space Museum's planned exhibition on the mission of the *Enola Gay* by a smaller exhibit for which he alone would be responsible, he was bending to powerful forces in Congress. In turn, Congress was responding to directives from the 3.1-million-member American Legion and a number of other veterans' organizations.

In announcing his decision to mount his own display, Secretary Heyman explained,

> We made a basic error in attempting to couple an historical treatment of the use of atomic weapons with the 50th anniversary commemoration of the end of the war.... Veterans and their families were expecting, and rightly so, that the nation would honor and commemorate their valor and sacrifice. They were not looking for analysis, and frankly, we did not give enough thought to the intense feelings such an analysis would evoke.[1]

Implied in this statement, as indicated by the display he now set about to create, were two assertions, both contrary to the Smithsonian Institution's congressional charter: that a true history of the mission of the *Enola Gay* could not adequately honor the nation's veterans; and that it was more important for America to accept a largely fictitious, comforting story in this commemorative year than to recall a pivotally important twentieth-century event as revealed in trustworthy documents now at hand in the nation's archives.

In a democratic society a national museum has a particularly heavy responsibility to research the nation's history and to recount it faithfully. Our form of government is predicated on an informed citizenry; and our best guide to the future is understanding our past. For the original exhibition, *The Last Act: The Atomic Bomb and the End of World War II*, the National Air and Space Museum had gathered excerpts from the diaries of President Harry S. Truman and Secretary of War Henry Stimson, letters and memoranda exchanged between them and other wartime leaders, briefing documents, minutes of crucial meetings, and other recently declassified material on the decision to drop the atomic bomb. The new exhibit deliberately excluded this material.

I do not wish to imply that Heyman took this step of his own volition, or even that he believed it was best for the nation in the long run. I do not know. I do know that he was under intense pressure; that he could and did claim that the original exhibition had been planned before his tenure, even if he had for many years been a member of the Institution's fully informed board of regents; and that he was worried the Smithsonian's congressionally controlled budget would suffer. He chose to give in.

The battle over the *Enola Gay* did not, as Heyman and many of the newspaper accounts have intimated, suddenly emerge in the spring of 1994. It did not, as was suggested, surprise a distressingly naive museum leadership enveloping it in controversy and political machinations: Heyman's "Frankly, we did not give enough thought," is belied by a massive correspondence to the contrary that shows the seeds of the debate sown ten years earlier, even before I came to the museum, and certainly before any of the curators who eventually worked on this exhibition had arrived.

My own active involvement and responsibility for the exhibition dated back to the fall of 1987, shortly after my arrival at the Smithsonian as director of the National Air and Space Museum. In October 1987, I assembled a distinguished external advisory committee and first examined the anticipated complexities surrounding a serious exhibition of the *Enola Gay*. Our discussions were attended not only by senior museum personnel, but also by then Smithsonian Secretary Robert McCormick Adams and members of his senior staff. Following this exchange, the museum continued an intense debate on the issues throughout the winter and spring of 1988.

In the early summer of 1988, several of us met with retired Brigadier General Paul Tibbets and two veterans of the 509th Composite Group, the force Tibbets had created and commanded to carry out the atomic bomb strikes on Hiroshima and Nagasaki. These three men visited to discuss ways to help us speed restoration of the *Enola Gay*. From this meeting sprang a two-year-long fund-raising campaign, to which ultimately more than six hundred World War II Army Air Forces veterans contributed.

Almost simultaneously, the issue of how and where an exhibition should be mounted came into focus. Adams, as secretary, wrote an editorial in the *Smithsonian* magazine, with a circulation of about two million; I, as museum director, wrote one for *Air & Space*, with its circulation of about three hundred thousand. Both informed our most interested constituents that an exhibition of the *Enola Gay* was in the offing. Between them, the editorials promised that the exhibition would treat sensibilities and intensely held beliefs with respect while presenting important historical issues as forthrightly as possible.

Within days, letters began to arrive in a flow that never ceased. Particularly concerned were veterans who wished to be reassured that their sacrifices in the war would not be forgotten or even denigrated. Such assurances were easily given. The Smithsonian had every intention of honoring the nation's veterans. The best way to do that, we felt, was to offer the most informed display that we could assemble. We were sure that veterans would find such an exhibition revealing of detail they had never known and intensely interesting in mirroring the thoughts that had occupied President Truman and his advisors as they discussed the deployment of atomic bombs to win the war.

Equally obvious from the start was the faction of veterans who would never be satisfied with a dispassionate exhibit reflecting a variety of perspectives. For them, only one view could be tolerated in the nation's most popular museum. It must present the atomic bombing of Hiroshima and Nagasaki as the only alternative to a full-blown invasion of the Japanese home islands. American lives heroically sacrificed in such an assault would have to be presented as numbering half a million to a million, with many more millions of Japanese military and civilian lives lost in the ensuing slaughter. Seen in this light, President Truman's order to bomb Hiroshima and Nagasaki, and thus kill more than one hundred thousand Japanese, would be revealed as a far-sighted, courageous, perhaps even humanitarian act, designed, at minimal cost in lives, to shock Japan into surrender and end this bloodiest of all wars.

To this faction we tried to explain that an effective strategic bombing campaign and naval blockade had left Japan reeling and weakened. Our code-breaking efforts had shown the Japanese interested in a brokered peace as long as they could retain their emperor. The Soviet Union had promised to enter the war by mid-August 1945, and by early that month was massing troops on the Manchurian border. None of these factors would guarantee Japan's surrender without an invasion. But no American president, certainly not Harry Truman, who had known fighting in World War I, would have launched an invasion he knew could cost half a million, let alone a million American lives, without first exploring less costly alternatives.

Over the years, as the massive restoration of the *Enola Gay* drew to completion in time for 1995, the fiftieth anniversary of her mission, the stand-off on how the aircraft should be exhibited seemed to be settling in favor of the museum's intended broad historical presentation. We had worked long and closely with the Air Force historian, Dr. Richard H. Kohn, who headed what was then known as the Office of Air Force History and thus was the Air Force's chief representative on historical matters. We had been testing public reactions all along by mounting a number of smaller exhibitions dealing forthrightly with controversial historical matters. The Smithsonian Board of Regents, and particularly the three senators and three members of the House who served on the board, were kept informed, and they also supported the planned exhibition.

Not content with forging and testing alliances within the United States, we also turned to Japan, early in 1993. The *Enola Gay* had inextricably interwoven the histories of our two nations. Fifty years after the event, it seemed possible that the exhibition we were planning might reflect an emerging consensus on the circumstances surrounding the mission of the *Enola Gay* and on the historic consequences of the first atomic bombing. We tried to discern how we might mount such a display that included Japanese experiences, images, and artifacts recalling August 6, 1945.

These precautions were severely tested in the spring of 1994, as an increasingly vocal post–Cold War conservative mood swept over the country and the media. Ironically, the Museum's very popularity was now proving to be its greatest liability. Had this been an insignificant establishment no one would have cared. But to the veterans, the opportunity of having their own version of history displayed at the world's most popular museum was worth a costly fight.

By mid-November, with Republican victories in both houses of Congress support had given way to concerted attack. By late January 1995, a new slate of conservative Republican congressmen was installed on the Smithsonian Board of Regents; Dr. Richard Hallion, successor to Dr. Richard Kohn as Air Force historian, was attacking the exhibit; and the Smithsonian secretary was offering to comply.

Heyman's promise to erase all traces of history from his own exhibition on the *Enola Gay* suited members of Congress. In contrast, the historic approach the museum had intended to present in *The Last Act* was dubbed a violation of the museum's charter. The museum should desist, some congressmen asserted; it had not been created to discuss history. They wanted the *Enola Gay,* this most famous airplane of World War II, to be displayed simply—as a masterful piece of technology that had permitted delivery of the first atomic bomb ever dropped from an aircraft.

Forgotten was the preamble to the 1846 act of Congress that had accepted the bequest of an Englishman, James Smithson of London,

> to found, at Washington, under the name of the "Smithsonian Institution," an establishment for the increase and diffusion of knowledge among men. . . .[2]

Forgotten was the museum's own enabling legislation directing us to mount educational, historic displays:

> The national air and space museum shall memorialize the national development of aviation and space flight . . . display aeronautical and space flight equipment of historical interest and significance . . . and provide educational material for the historical study of aviation and space flight.[3]

Were those the directives to the museum to drop all aspirations to mount a historical exhibition? Where had the museum made its "basic error in attempting to couple a historical treatment of the use of atomic weapons with the fiftieth anniversary commemoration of the end of the war?"

The legislation that had set up the Smithsonian Institution and, a century later, the National Air and Space Museum, was clear enough. Only it happened to be inconvenient to the legislators who had now entered the fray. They wanted to disregard or deny it.

As soon as the museum's planned exhibition had been canceled, the newly appointed Smithsonian regent Congressman Sam Johnson, Republican of Texas, demanded the suppression of the catalogue that had been prepared for *The Last Act* exhibition.

Knowing that the controversy around the exhibition would persist for a long time, I had a year earlier asked the Smithsonian Institution Press to publish the catalogue—to reproduce, word-for-word, every piece of text in the exhibition. That way, there would be no doubt about exactly what the museum had said and what we had not. This was important because antagonists of the exhibition had made so many false allegations about the planned contents, had quoted so many phrases out of context, that I wished to set the record straight. But, in response to Congressman Johnson's request for formal assurances, the secretary complied and instructed the Smithsonian Press to withdraw the catalogue from its offerings.[4]

It is not as though the exhibition or the catalogue would have been an embarrassment to the Institution. On January 17, 1995, Heyman himself had written to the chairman of the California Department of Veterans Affairs,

> I believe the script for the exhibition now strikes the appropriate balance that provides visitors with an opportunity to learn more about this critical event while at the same time recognizing the sacrifice of those who served in the armed forces and the resoluteness of those who led our nation. . . .[5]

Confirming this assessment, the president of the Society for Military History, retired Brigadier General Roy K. Flint, former dean of the faculty at West Point, wrote to Chief Justice William H. Rehnquist, chancellor of the Smithsonian Board of Regents, on January 26, pleading that the exhibition not be canceled:

> Ending the war against Japan by employing atomic weapons was perhaps the most significant historical event of the time and therefore deserves . . . careful treatment. Moreover, the Smithsonian's prominence as the leading museum of our nation and its possession of the *Enola Gay* demand a full presentation of the context and history of those events. . . . The Smithsonian must stand publicly against the politicizing of scholarship in public discourse, and it must resist all efforts to impose conformity in the rendering of history.[6]

To this, the three most recent presidents of the Organization of American Historians, the largest organization of historians of the United States in the country, writing to the Chief Justice the following day, added:

> We are concerned about the profoundly dangerous precedent of censoring a museum exhibition in response to political pressures from special interest groups.... [It] will send a chilling... message that certain aspects of our history are "too hot to handle," so susceptible to contested points of view that they must be excluded from the public mind.... History museums should not be confined only to exhibitions about subjects for which a perfect consensus exists. Where consensus already exists, there is the least need for the presentation of information and the opportunity for members of our diverse society to be educated and formulate opinions.[7]

With this level of endorsement, what possible reason could there have been for banning the exhibition's catalogue?

James Smithson would have been appalled!

I will describe the circumstances that surrounded and led up to these events—the context within which former secretary Adams and I forged the Smithsonian's original plans for exhibiting the *Enola Gay*, where we saw eye to eye and where we differed, how interest groups sprang into being and how they gained the support of veterans' organizations that then lobbied Congress, how the Smithsonian tried to respond, how a newly installed Secretary Heyman gave in and canceled the exhibit, and how and why I subsequently resigned.

I have tried to portray how the National Air and Space Museum operated during my tenure, who its staff members were, what it took to mount an exhibition, how standards of excellence were continually monitored, and how the needs of the public were probed and respected.

I have further tried to convey the sense of the exhibition the museum was planning to mount; the approach we were taking to make understandable, in the context of their time, events that happened two generations ago, when America was at war; how that context, as expressed in previously classified but now available documents exchanged between President Truman and his closest advisors, was rejected as 'politically correct' or 'revisionist' by the veterans' organizations and their supporters in Congress; and how they marshaled their resources to see the exhibition canceled.

I have written *An Exhibition Denied* to convey what we at the museum were hoping to do. But the reader must keep in mind that an exhibition is not just words. It combines words with objects and sounds and sights. It is a medium of the senses, not just the mind. I will do my best to do the exhibition justice, to the extent that words alone can. Let those who read these pages visualize the exhibition's wartime newsreels of desperate fighting in the Pacific; let them wrestle with the difficult decisions President Truman and his advisors were weighing; let them imagine the fifty-six-foot-long, brilliantly sparkling, forward fuselage of the *Enola Gay* towering over their heads, with the *Little Boy* uranium bomb casing beneath her open bomb bay doors; let them consider the courage of a crew that for all it knew would be blowing itself out the sky with the bomb it was dropping; let them be dismayed by images of desolation and suffering on the ground in the wake of the bombings; let them feel compassion for American prisoners of war, photographed near death after release from captivity in Japan; let them ponder a huge wall, twenty-five feet long, twenty feet high, covered, wallpaper-fashion, with 70,000 warheads, one for every square inch—two of them in a different color to emphasize the enormous postwar increase in the number of nuclear warheads since the two dropped in August 1945. All this was part of the intended display.

As the museum's director throughout the time this exhibition was being prepared, I cannot expect readers to trust my personal account of our efforts, and how they eventually were denied. I have therefore adopted a style that includes the views, desires, motivations, and actions of each participant, including myself, in his own words as voiced at the time, through letters, memoranda, magazine and journal articles, Op-Ed pieces, news releases, and media interviews.

Every author brings prejudices to his work. I hope to have suppressed my own as much as possible. Others may see this story differently and write their own account some day. Until then this is the history, as clearly as I have been able to reconstruct it, of an exhibition that never took place, never was seen by anyone, and yet gave rise to the most violent dispute ever witnessed by a museum.

Acknowledgments

I am foremost indebted to my wife Marianne, who has been both my most outspoken and my most consistently loyal and constructive critic. She has read and helpfully commented on each successive draft as the book took shape.

Our son Eric, and our daughter Emily carefully read and commented on early drafts. I thank them both for their help.

The Air Force Association gave me permission to freely quote from the voluminous file of notes, memoranda, letters, telefaxes and other correspondence that they had assembled in a volume entitled *The Enola Gay Debate, August 1993–May 1995*. The information contained in these files was invaluable.

Early in 1996, Commander Luanne J. Smith (US Navy, Ret.) made available to me a body of correspondence I had not seen before, and provided answers to many questions relating to her work on the 50th Anniversary of World War II Commemoration Committee. Very kindly, she also read a version of the manuscript and provided me with comments.

Tony Capaccio, editor of *Defense Week*, similarly made available to me a collection of incisive interviews he and Uday Mohan had conducted in the spring of 1995, with historians, members of the press and spokesmen for veterans organizations.

Hubert R. Dagley II, former Director of Internal Affairs at The American Legion, also provided me with valuable background material, personal insights and comments on the script.

Additional material, was sent by Elbert L. Watson, publisher of *World War II Times*, and World War II Army Air Forces B-29 veterans Donald C. Rehl and Benjamin A. Nicks. I am pleased to acknowledge these contributions.

Mike Wallace, very kindly sent me a copy of his insightful essay "The Battle of the Enola Gay," in advance of its publication in his book *Mickey Mouse History and other Essays on American Memory*, published by Temple University Press in 1996.

Mark W. Rodgers, former Smithsonian Director of Government Affairs; Richard H. Kohn Professor of History at the University of North Carolina and former Air Force Historian; and Michael Kammen, Professor of History at Cornell University and member of the Smithsonian Council; all read early versions of the script and offered critical comments that I found valuable.

For useful telephone conversations and added insight, I also thank, Dr. Edward J. Drea, Chief of the Research and Analysis Division of the U.S.

Army Center of Military History; Jack Giese, formerly Chief of Media Relations at the Air Force Association; Dr. Alfred Goldberg, Historian for the Secretary of Defense; Bob Manhan, Assistant Director of Legislative Services at the Veterans of Foreign Wars; Dr. Oscar Rosen of the National Association of Atomic Veterans; Kenneth L. Eidnes and Frank B. Stewart , both World War II members of the 509th Composite Group; and Prof. John J. Stachel, expert on the Albert Einstein papers.

Permission to quote from letters and memoranda came from Andrew H. Anderson, Jo Becker, Brian Burke-Gaffney, Fred D. Cavinder, Robert Collings, Hubert R. Dagley, II; Russell E. Dougherty, Robert S. Faron, Thomas W. Ferebee, Roy K. Flint, Daniel S. Greenberg, William G. Hulbert, Akira Iriye, Sadao Ishizu, Michael Kammen, George C. Larson, Robert Jay Lifton, Greg Mitchell, Robert K. Musil, Carole M. P. Neves, Benjamin A. Nicks, Donald C. Rehl, Yoshio Saito, Itsuzo Shigematsu, Hugh Sidey, Luanne J. Smith, Lucius Smith III, Frank B. Stewart, Morihisa Takagi, J. W. Thiessen, Arthur Veik, Senji Yamaguchi, and Alfred A. Yee.

Permission to reproduce printed material was kindly provided by the ABC News Polling Unit, ABC News; *Air & Space/Smithsonian*; the Air Force Association; Indianapolis Newspapers, Inc.; the American Legion Auxiliary *National News*; *Asahi Shimbun*; the Associated Press; the *Atomic Veteran's Newsletter*; the *Dallas Morning News*; *Defense Week*; the Doubleday division of the Bantam Doubleday Dell Publishing Group; the *Journal of American History*; the *Kansas City Star*; Marlowe & Company; Martilla & Kiley, Inc.; the *National Journal*; the *Omaha World-Herald*; the *Philadelphia Inquirer*; Random House Vintage Books; *The Retired Officer Magazine;* the Scripps Howard News Service; *U.S. News & World Report*; the *Wall Street Journal*; Mike Wallace for *Mickey Mouse History and Other Essays on American Memory*; the *Washington Post*; the *Washington Times*; *Whittier Daily News*; and *World War II Times*.

A number of acknowledgments need to be specifically spelled out: Excerpts from "Missing the Target," by Capaccio and Mohan are reprinted by permission of *American Journalism Review*. Excerpts from *The American Legion magazine*'s "Letters to the Editor" column for October 1988; Brian D. Smith's article "Rewriting Enola Gay's History," November 1994; and "How the Legion Held Sway on Enola Gay," May 1995; are reprinted by permission, *The American Legion Magazine,* ©1988, 1994 and 1995, respectively. Excerpts from "Talk of the Town", August 13, 1990, are reprinted by permission © 1990, The New Yorker Magazine, Inc. All rights reserved. Excerpts from "The romance of technological progress: a critical review of the National Air and Space Museum," by Michal McMahon © by the Society for the History of Technology, 1981, are reprinted with permission of *Technology and Culture* and the University of Chicago Press. Permission to reproduce the August 2, 1939 letter from Albert Einstein to Franklin D. Roosevelt is granted by the Albert Einstein Archives, the

Hebrew University of Jerusalem, Israel. Excerpts from the articles "Wounds of war still color Enola Gay's place in history," by Andrea Stone, October 5, 1994, "Politics had no place in Enola Gay exhibit," January 30, 1995, and "Hiroshima Display ends in rancor," by Andrea Stone, January 31, 1995, are reprinted with permission of USA TODAY, copyright 1994 and 1995, respectively. Excerpts from Elisabeth Kastor: "At Air & Space, Ideas on the Wing," October 11, 1988; Kim Masters: "Arts Beat—The Enola Gay on the Mall," June 3, 1991; Eugene L. Meyer: "Dropping the Bomb," July 21, 1994, "Smithsonian Bows to Critics," September 30, 1994, "Smithsonian Stands Firm on A-Bomb Exhibit," January 19, 1995, "Target Smithsonian," January 30, 1995, "AU May Exhibit Artifacts from Hiroshima Bomb," April 6, 1995; Jacqueline Trescott: "Michael Heyman Airing the Nation's Attic," September 20, 1994; Coleman McCarthy, "Glory-Seekers and the Bomb," February 7, 1995; and Joel Achenbach, "Enola Gay Exhibit: Plane and Simple," June 28, 1995, are reprinted with permission of the Washington Post Writers Group, © 1988, 1991, 1994, and 1995, respectively.

My special thanks go to William Frucht, senior editor at Copernicus, whose enthusiasm and encouragement never flagged. At his side, and always ready to help was Theresa Shields. I greatly appreciated working with Lesley Poliner, senior production editor, whose care and professionalism always shone through. I also thank David Kramer for his care in copy editing the manuscript and for his perceptive comments.

No one at the Smithsonian was involved in the preparation of this book.

A Note for the Reader

Names of Principals: For ease in reading, I have included a list of frequently recurring names with a short annotation for each.
Abbreviations: Abbreviations are spelled out in full in the index.
Chronology: To help the reader with a complex eight-year history, I have included a brief chronology.
Quotations: I have chosen to correct obvious typographical errors in a quoted document; this avoids continuous interruption by designations "sic," without excessive risk of misrepresenting the document. Selectively quoting from a text to avoid excessive length carried a far greater risk for misleading the reader, but was unavoidable. Where the sense of a quotation is not altered by starting in mid-sentence, I have usually chosen to capitalize the sentence where I start the quote.

All emphases in quoted text faithfully represent the original text except in rare instances, where added emphasis is noted.

For uniformity of style, I have throughout italicized three repeatedly occurring names—*Enola Gay, Little Boy,* and *Fat Man*—even though the originally cited text sometimes had them in quotation marks, sometimes italicized them, sometimes had them in capitalized letters, and sometimes left them unannotated.

Where I have inserted an explanation or correction into a quotation it appears [in square brackets].
Notes: I have designated the location of letters and other documents not generally available with a simple code:

MH indicates that the document is in my possession.

NASM indicates that the document has been archived at the National Air and Space Museum or, in a few instances, might be found in the files of the Museum Director's office.

NASM/MH indicates that the document can be found in both these locations.

Carmen Turner File is a file available from both *NASM* and *MH*. It is a loose-leaf binder that had been prepared for Under Secretary Carmen Turner in August 1991, and comprises ten tabulated sections containing letters, memoranda, newspaper clippings, minutes of meetings, articles published by Smithsonian officials, and other materials. It represents a capsule history of preparations for the exhibition of the *Enola Gay,* as understood at the time.

log book indicates entries that I made in a log in which I recorded commitments that I would have to honor or proceedings of meetings at which significant decisions were to be made.

AFA The Enola Gay Debate refers to a compilation of several hundred letters, memoranda, resolutions, and other documents reprinted and bound by the Air Force Association, Arlington, Virginia, in May 1995. The volume is divided into ten sections: 1. Chronology; 2. Key Documents; 3. March Report; 4. Script Analyses; 5. Tiger Team; 6. Releases; 7. AFA Articles; 8. Congress; 9. Other Groups; 10. Historians. In citing a particular document, I identify the section in which it can be found.

AFA Enola Gay Coverage refers to two volumes the Air Force Association reprinted in May 1995. They were issued simultaneously with *The Enola Gay Debate*. These two volumes, respectively identified by designations "1994" and "1995," contain copies of several hundred news clippings each, published between March 1994 and May 1995.

For completeness, readers might wish to note that the Air Force Association issued *The Enola Gay Debate* and *Enola Gay Coverage* as part of a four-volume set. The fourth volume is designated *The Crossroads: The End of World War II, the Atomic Bomb and the Origins of the Cold War.* This is the first draft of the label script for the *Enola Gay* that the National Air and Space Museum completed. It is dated January 12, 1994. The script underwent major changes in the twelve months that followed, but interested readers may wish to refer to this early version.

List of Principals

Adams, Robert McCormick	Secretary of the Smithsonian Institution, 1984–1994
Alison, Thomas	Curator in the Museum's Department of Aeronautics
Aubin, Stephen P.	Director of Communications, Air Force Association
Bearss, Edwin	Chief Historian, National Park Service
Bennett, W. Burr, Jr.	World War II veteran, B-29 reconnaissance flight photographer
Bernstein, Barton J.	Professor of History, Stanford University
Blackburn, Paul P.	Director, U.S. Information Service, United States Embassy, Japan
Blute, Peter I.	Congressman from Massachusetts, Republican
Burke-Gaffney, Brian	International advisor to the Nagasaki municipal government
Collins, Michael	Astronaut, National Air and Space Museum Director, 1971–1978
Conable, Barber	Smithsonian regent, former President of the World Bank
Constantine, William M.	Brigadier (USAF, Ret.), Museum docent
Cooper, Charles D.	Editor, *The Retired Officer* magazine
Correll, John T.	Editor in Chief, of the Air Force Association's *Air Force* magazine
Crider, Gwendolyn ("Gwen") K.	National Air and Space Museum Deputy after August 1993
Crouch, Tom D.	Chairman of the Museum's Department of Aeronautics
Dagley, Hubert R., II	Director of the American Legion's Internal Affairs Commission
Dear, John	Jesuit Father, peace activist
Detweiler, William M.	American Legion National Commander, Sept. 1994–Sept. 1995
Dietz, Thomas	Museum curator, Department of Aeronautics
Dornan, Robert K.	Congressman from California, Republican
Ezell, Linda Neuman	Museum's Assistant Director for Collections Management
Ferebee, Thomas W.	Bombardier on the *Enola Gay*'s Hiroshima mission

Fetters, J. Michael	Museum's Assistant Director for Public Affairs
Freudenheim, Thomas L.	Smithsonian Assistant Secretary for Museums
Gallagher, Ray	Assistant flight engineer on the *Bockscar* mission to Nagasaki
Gayler, Noel	Admiral (USN, Ret.)
Gaynor, Margaret C.	Smithsonian office of Government Relations, director until 1992
Gernstein, Joanne M.	Museum curator, Department of Aeronautics
Giese, Jack	Air Force Association Chief of Media Relations
Gingrich, Newt	Republican, Speaker of the House after 1995
Goldberg, Alfred	Historian for the Secretary of Defense
Goldberg, Stanley	Historian and biographer of General Leslie R. Groves
Gore, Albert, Jr.	Vice President of the United States, Vice Chancellor of the Smithsonian after 1993
Goss, Kenneth A.	Air Force Association Director of National Defense Issues
Govier, Victor M.	Project Coordinator in the Museum's Department of Exhibits
Gray, Hanna Holborn	Smithsonian Regent, former President, University of Chicago
Hallion, Richard P.	Air Force Historian after 1991
Harada, Hiroshi	Director of the Hiroshima Peace Memorial Museum after April 1993
Harrington, Herman G.	Chairman of the American Legion's Internal Affairs Commission
Hatch, Monroe W., Jr.	Executive Director of the Air Force Association
Herken, Gregg F.	Chairman of the Museum's Department of Space History
Heyman, I. Michael	Smithsonian Institution Secretary after September 19, 1994
Hiraoka, Takashi	Mayor of Hiroshima
Hobbins, James M.	Executive Assistant to the Smithsonian Secretary
Hoffmann, Robert S.	Smithsonian Institution, Acting Provost, from September 1994
Horigan, Richard D., Jr.	Foreman of the Museum's Preservation and Restoration Division until 1993
Hunter, Duncan L.	Congressman from California, Republican
Hutchison, Kay Bailey	Senator from Texas, Republican
Iriye, Akira	Professor of History, Harvard University

Itoh, Tatsuya	Director, Nagasaki International Culture Hall
Jacobs, William "Jake"	Chief designer of *The Last Act* exhibition
Johnson, Sam	Congressman from Texas, Republican, Smithsonian Regent after January 1995
Kassebaum, Nancy Landon	Senator from Kansas, Republican
Kicklighter, C.M.	Lt. Gen. (USA, Ret.), Executive Director, 50th Anniversary of WWII Commemoration Committee
Kilcline, Thomas J.	Vice Admiral (USN, Ret.), President, The Retired Officers Association
Kohn, Richard H.	Air Force Historian, 1981–1991, Professor of History, University of North Carolina
Kuriyama, Takakazu	Ambassador to the United States from Japan
Lewis, Tom	Congressman from Florida, Republican
Linenthal, Edward T.	Professor of Religious Studies, University of Wisconsin, Oshkosh
Lord, Winston	Assistant Secretary of State for Pacific and East Asian Affairs
Makovenyi, Nadya	The Museum's Assistant Director for Exhibits
Malenfant, Richard E.	Director's staff, Los Alamos National Laboratory
Manhan, Bob	Veterans of Foreign Wars, Assistant Director of Legislative Services
Matano, Hanako	Smithsonian Institution special representative for Japan
Mathews, Rusty	Senate Appropriations Committee, Interior and Related Agencies staff
McInerney, Tom	Lt. General, U.S. Air Force, Assistant Vice Chief of Staff
McPeak, Merrill A. ("Tony")	General, U.S. Air Force, Chief of Staff until October 1994
Mikesh, Robert	Senior Curator in the Museum's Department of Aeronautics
Mims, Bradley A.	Smithsonian Congressional liaison officer
Motoshima, Hitoshi	Mayor of Nagasaki
Neal, Homer A.	Vice President for Research, University of Michigan, Ann Arbor, Smithsonian Regent
Nelson, Richard H.	Radio operator on the *Enola Gay* mission to Hiroshima
Neufeld, Michael J.	Lead curator, *The Last Act* exhibition
Newman, Constance ("Connie") Berry	Smithsonian Institution Under Secretary after July 1992

Nicks, Benjamin A.	Veteran of 35 combat missions as a B-29 pilot in the Pacific
Pisano, Dominick A.	Curator in the Museum's Aeronautics Department
Powers, Peter G.	Smithsonian General Counsel
Rabbitt, Frank	Museum docent
Reese, William C.	Foreman of the Museum's Restoration and Preservation Division after 1993
Rehl, Donald C.	B-29 pilot with the 509th Composite Group
Rehnquist, William H.	Chief Justice of the United States, Chancellor of the Smithsonian Institution
Rodgers, Mark W.	Director, Smithsonian Office of Government Relations, 1993–1995
Rooney, William A.	Veteran, with B-29s as intelligence officer in World War II
Shigematsu, Itsuzo	Chairman, Radiation Effects Research Foundation
Skaggs, David E.	Congressman from Colorado, Democrat
Smith, Luanne J.	Commander (USN), member, 50th Anniversary of WWII Commemoration Committee
Smith, Lucius, III	President, Atomic Library and Technology Foundation
Solomon, Gerald B.	Chairman, House Rules Committee after 1995, Republican from New York
Soter, Steven L.	Special Assistant to the National Air and Space Museum Director
Stephens, Wendy	Associate Director, later Deputy Director of the Museum
Stevens, Ted	Senator from Alaska, Republican
Stewart, Frank B.	509th Composite Group Veteran; President, *Enola Gay* Restoration Association
Stolz, Preble	Professor Emeritus of Law, University of California, Berkeley
Thiessen, J.W. "Joop,"	Vice Chairman, Radiation Effects Research Foundation
Tibbets, Paul W., Jr.	Commander of the 509th Composite Group; *Enola Gay* pilot at Hiroshima
Turner, Carmen	Smithsonian Under Secretary, 1990–1992
Van Kirk, Theodore ("Dutch")	Navigator of the *Enola Gay* on her Hiroshima mission
Von Hardesty, D.	Chairman of the Museum's Aeronautics Department until 1988
Widnall, Sheila E.	Secretary of the Air Force

Woodside, Patricia ("Patti") A.	Chief, Film & Video Production in the Museum's Department of Exhibits
Wolk, Herman	Historian at the U.S. Air Force Center for Air Force History
Yates, Sidney R.	Congressman from Illinois, Democrat

1

Remembrances

Late in the summer of 1980 a small band of men approaching retirement age convened in Washington. At five-year intervals in the previous thirty-five years they had met in other cities to reminisce and exchange news.[1]

On this occasion they also had an additional attraction. They would be able to visit their beloved *Enola Gay.* With great expectations, they drove to Silver Hill, in Suitland, Maryland, just outside Washington's city limits, where the National Air and Space Museum has its Paul E. Garber Restoration, Preservation, and Storage Facility.

When they had last seen her, she was a proud, brilliantly shiny, beautifully sleek B-29 Superfortress—the most powerful bomber the Army Air Forces flew in World War II. In 1945, the *Enola Gay* and the men who were now visiting her had ended the war. Others could also claim to have contributed. But the *Enola Gay* and the men of the 509th had, some would argue, actually ended the war all by themselves.

Fifteen years later, largely inspired by these veterans' visit that day, the National Air and Space Museum would be preparing an exhibition on the mission of the *Enola Gay*. Prominently featured in that gallery would be the restored fifty-six-foot-long forward fuselage of the aircraft, memorabilia donated by the men of the 509th, and a video-film the museum had produced, in which crew members of the *Enola Gay* and her sister ship *Bockscar* recalled their missions.

Here is their story.

THE 509TH COMPOSITE GROUP

The 509th Composite Group had been created in September 1944 when Major General Leslie Groves, the man in charge of the Manhattan Project to construct the atomic bomb, foresaw the need for a dedicated corps of men trained to drop the bomb on targets in Japan. He chose twenty-nine-year-old Lieutenant Colonel Paul Tibbets to assemble and command the group.

Groves provided Tibbets with fifteen Boeing Superfortresses and eighteen hundred men, and ordered him to shape them into a self-contained, secret outfit. Tibbets was to control his own maintenance, engineering, ordnance, medical, radiological, and technical units, and his own set of troop transport aircraft and military police. These provided the required self-sufficiency and with it, the urgently demanded secrecy. If Tibbets ran into any bureaucratic problems, he needed only to mention the code word "Silverplate," which revealed nothing about the group's mission, but magically cut through red tape. If thwarted nevertheless, he had direct access to Groves and if need be, to H. H. "Hap" Arnold, commanding general of the Army Air Forces.[2]

Tibbets's top secret mission was to forge a group to deliver an atomic bomb to Japan and survive. For this he had to devise the means and train his crews to drop this incredibly powerful bomb and escape before its terrible blast could consume them. For months, he did not know when the bomb would be ready or exactly how much it would weigh. But he knew it was going to be hard even to get the loaded Superfortress off the ground. Whether she would be able to struggle to an altitude of 30,000 feet with that bomb in her belly was anyone's guess.

Altitude was important. Tibbets and his crew would enter a hairpin turn immediately after releasing the bomb and beat a retreat to gain added distance from the point of explosion. If the B-29 could gain a few thousand feet of added altitude, that would add time to the forty or forty-five seconds for the bomb to fall to its detonation height. Every second gained meant added distance from the blast and greater safety for aircraft and crew; the Army Air Forces had no intention of making this a suicide mission.[3]

The B-29s in which the 509th trained were not yet those they would need to carry out their mission. New Superfortresses would have to be acquired and modified for the task. On May 18, 1945, the Martin Aircraft factory in Omaha, Nebraska, delivered aircraft No. 44-86292 to the U.S. Army Air Forces (USAAF). She was one of 536 Boeing-designed B-29s to be assembled by the Omaha plant, and one of four thousand Superfortresses to be built and delivered by Boeing, Martin, and other companies for the war. This particular Superfortress had been designed as a Model B-29-45-M0. Her wingspan of 141 feet 8 inches, length of 99 feet, and four Wright 2,200 hp, R-3350-57 Cyclone engines permitted her to take off weighing sixty-seven tons fully loaded—about twice her weight empty of fuel, crew, and bombs. With this takeoff weight, she could cruise at 190 to 200 miles per hour. The aircraft's ceiling, the maximum altitude she could reach, was 35,000 feet, nearly seven miles above sea level.

For four weeks after leaving the factory, Superfortress 44-86292 was modified to make her a "Special Mission" aircraft. Then, on June 14, she was picked up by one of Tibbets's right-hand men, Capt. Robert A. Lewis, and ferried to Wendover Army Air Force Base (AAFB), Utah, where the 509th had been training in isolation for the past nine months. By June 27 Lewis and his crew were ready to head for Tinian Island in the Marianas. Along the way, they stopped at Mather AAFB in California and in Hawaii before reaching Guam on July 2. There, the aircraft's bomb bay was further modified. Leaving Guam on July 6, Lewis first headed to Kwajalein before finally taking off for Tinian, where the 509th Composite Group was now assembling.

PRACTICING FOR PERFECTION

The next few weeks were spent on practice runs. On July 12 the aircraft participated in a raid on Marcus Island, fully loaded with seven thousand gallons of fuel and twenty 500-pound bombs, weighing the maximum sixty-seven tons on takeoff. Like all the other 509th heavy bombardment air crews, Lewis and his men were required to fly half a dozen missions to prepare for battle conditions and their ultimate mission.[4]

Though occupied with the many pressing problems of commanding the entire group, Tibbets made time to fly on some of these practice missions. But he was under strict orders not to go along on the flights over Japan. He knew too much for the United States to risk his capture. Other members of the 509th, who had been told little, could and did fly practice runs over the Japanese home islands. They knew their mission was special; they knew the maneuvers they would have to carry out; but they knew little else.

THE MISSION

By Sunday, August 5, 1945, everything was ready. The clouds that had hung over Japan's home islands for a week were clearing. Tomorrow would be the day.

Aircraft 44-86292 had not yet been named. Tibbets got a painter to brush "ENOLA GAY" in bold black capitals just below the pilot's window on the aircraft's port side. It was his mother's maiden name and Paul Tibbets's way of honoring her for standing by his side in an often rocky early career.[5]

Just after noon that Sunday, the Mk-1 atomic bomb—nicknamed *Little Boy*, in spite of its ten-foot-length, two-foot-diameter, and four-and-a-half-ton weight—was removed from its heavily guarded assembly hut on Tinian's North Field and loaded into the modified bomb bay. Tibbets watched every move and recalls thinking in disbelief that this single bomb was claimed to have the explosive power of two hundred thousand of the 200-pound bombs he had dropped over Europe and Africa three years before. But so far, Tibbets was the only member of the 509th to know that secret. A few of the others would have to be told before day's end.[6]

Besides the *Enola Gay*, six other aircraft were to participate in the mission. Three were weather planes to be dispatched ahead of the others. *Straight Flush*, commanded by Claude Eatherly, would be on her way to Hiroshima; *Jabbitt III*, with John Wilson in charge, would fly to Kokura; and *Full House*, piloted by Ralph Taylor, would head for Nagasaki. Hiroshima was the prime target, but if clouds prevented visual sighting of landmarks, Kokura and Nagasaki were potential alternate targets. Charles Sweeney's *The Great Artiste* and George Marquart's unnamed aircraft No. 91 carried cameras and special instrumentation and were to escort the *Enola Gay* to her target. The seventh and final aircraft in order of takeoff would be *Top Secret*, piloted by Chuck McKnight. He was to fly only as far as Iwo Jima, to stand by as needed.[7]

That evening the seven crews taking part in the mission gathered for the preflight briefing shortly after supper. Later, at 11:00 P.M., the crews of the *Enola Gay* and the two planes that would accompany her to her target received a final briefing. This was the first time they were told the expected power of the bomb they would drop. They were stunned; but the enormity of the explosion explained those violent escape maneuvers they had been practicing immediately after bomb release, procedures they had all practiced to perfection.

Tibbets had chosen Theodore "Dutch" Van Kirk as his navigator and Thomas W. Ferebee as his bombardier. Both had flown with Tibbets on bombing missions over Europe during the early years of the war. Other members of *Enola Gay*'s crew were Robert A. Lewis, copilot; Wyatt E. Duzenbury, aircraft flight engineer; George R. "Bob" Caron, tail gunner; Joseph A. Stiborik, radar operator; Richard H. "Junior" Nelson, radio operator; Robert Shumard, assistant aircraft flight engineer; Jacob Beser, "Raven" operator / radar officer; Navy captain William S. "Deak" Parsons, weapons officer on loan from the Manhattan Project, which had built the bomb; and his assistant, Morris R. Jeppson, proximity fuse specialist.[8] The names of all except the last three, whose functions were largely related to the bomb, would later be stencilled on the aircraft's side to chronicle their participation on this historic flight.

Of the mission on which they were about to embark, Dick Nelson, a twenty-year-old kid on the crew at the time, today recalls, "You knew it was big, you just didn't want to mess anything up. . . . When we were in the air somebody said . . . this bomb cost as much as an aircraft carrier. . . . Well, . . . then you really get the monkey on your back."[9]

Just after 1:00 A.M., the crews drove a Jeep out to the flight line. When Dutch Van Kirk, then aged twenty-four, remembers that late-night scene with the aircraft lit up by spotlights, he thinks of a Hollywood premiere. Dick Nelson likens it to a supermarket opening, "Klieg lights and all kinds of photographers. . . . You're almost embarrassed." But as Dutch is quick to add, this scene had not been staged by news reporters. The Manhattan Project needed to document the event for history.[10] For that purpose, *New York Times* science writer William L. Laurence, who had been given a leave of absence from his newspaper to write the official history of the atomic bomb effort, had just flown in that morning, though arriving too late to be included on the mission.[11]

Some of the men were excited or disturbed by all the attention. For Tom Ferebee, who seems imperturbable, "The only difference between that and other missions I'd flown was that there's an awful lot of people around the airplane and floodlights . . . which I didn't expect. . . . There wasn't much excitement as far as I was concerned. It was just another mission."[12]

The most famous photo coming out of this session is that of a smiling Paul Tibbets, sticking his head out the pilot's window, just above the "O" in "ENOLA GAY." He is waving with his right hand at the nighttime crowd surrounding the plane—none of whom he can presumably see against the glare of the lights focused on him and brilliantly reflected off his aircraft's polished surface. The picture taking continued until close to 2:00 A.M., when Tibbets called a halt so they could go ahead with their preflight preparations.

Van Kirk recalls the mission as, "Rather routine, really. And one of the reasons for it was it went exactly according to plan. . . . Any time you have a good plan and everything goes according to it, . . . things do appear to be routine, and that's how our mission went." Takeoff took place as scheduled at 2:45 A.M. Tibbets held the aircraft at low altitude while Captain Parsons crawled back to arm the bomb. When they reached Iwo Jima, Tibbets circled the island to let the other two airplanes catch up, and with them on his wings, he gradually climbed to altitude. They had a seventeen-hundred-mile trip ahead to Hiroshima and the crew took turns napping. This would be a thirteen-hour-long round-trip mission, and for a while there was little to do.

Claude Eatherly, whose plane had preceded them, reached Hiroshima, found the weather clear, and radioed back. Then he turned home. Hiroshima now was the target.

As the *Enola Gay* approached the city, the crew could clearly see it from more than fifty miles away. Van Kirk recalls Ferebee making a long bomb run, probably eight or nine minutes, eventually setting his sights on the target, the landmark T-shaped Aioi bridge. He remembers thinking, "If we'd ever sat on a bombing heading like this over Europe for this long a time, they'd have really blasted us out of the sky. But there was no opposition."

As the bomb dropped, the aircraft jumped, relieved of its weight, and Tibbets went into his sharp turn. Forty-three seconds later, as the bomb reached detonation altitude preset at 1,890 feet above ground, the sky lit up. Even with dark goggles over the crew's eyes, they felt as though someone had sparked a flashbulb in their eyes. The shock wave arrived another forty-five seconds later. This was the moment of truth. The aircraft rocked, but withstood the blast. The immediate danger was over. Meanwhile, the mushroom cloud was rising faster than anything any of them had ever seen, soon reaching an altitude of nine miles, three miles above their own cruising altitude. Down below, Nelson recalls, "The town was just a big mess of flame and dust." Van Kirk says it "looked like a pot of bubbling tar." There wasn't much to be discerned, so the three aircraft turned and headed for home. Nelson recalls their saying, on the way back, "that the war was over.... We couldn't see how they could possibly go on any longer ... with this device."

Looking back, Ferebee says, "We don't think we should be glorified or the airplane be glorified. Just ... show what it did in that period of time. Things are much different now, and people look at things much different. But it's got to be considered in the time period that it happened."

The *Enola Gay*'s copilot, Bob Lewis, did consider it at the time. As the last entry in the log he had been asked to keep of the mission, he wrote, "My God!"

The Hiroshima mission did not end the war. Despite the scale of destruction, the Japanese did not immediately give up. Nelson thinks, "It required something else, and that something was the second bomb.... The second bomb was a necessity.... It did show that we had more than one weapon." He feels that the bomb dropped on Nagasaki "worked—because three days later they did concede defeat."

Where the Hiroshima mission's *Little Boy*, like most bombs, was cigar-shaped, the plutonium bomb dropped on Nagasaki looked more like a giant egg, and was aptly nicknamed *Fat Man*. *Little Boy* was a uranium bomb. It was also the only one of its kind in existence. The uranium isotope of atomic mass 235 was arduous to isolate from its sister isotope of mass 238, occurring in far greater abundance in ordinary uranium ores.

Material for a second bomb of this kind would not have been available for several more months. The explosive for the bomb dropped on Nagasaki was plutonium, an element that does not exist naturally on Earth. It had to be manufactured from uranium 238 in giant nuclear reactors in a huge new plant built for the Manhattan Project by the Army Corps of Engineers at Hanford, Washington. By August 1945, Hanford was producing enough plutonium for two or three bombs a month.

Tibbets decided not to go on the second bombing mission himself, and assigned the command to Chuck Sweeney. Since Sweeney's *The Great Artiste* had been outfitted for instrumentation, he was assigned to fly *Bockscar*, while Fred Bock, who normally piloted the plane that bore his name, would command the instrumentation aircraft on this flight.

The Nagasaki mission was plagued with troubles. A faulty fuel pump prevented complete use of the fuel on board. Kokura, the primary target for the mission, was so clouded in that Sweeney had to give up after circling the city for some time and go for the alternate target, Nagasaki. That city also was heavily overcast, but Sweeney did drop the bomb. Then, heading home he was so low on fuel that he had to land on Okinawa to refuel before proceeding back to Tinian.

THE 509TH VETERANS

The men of the 509th who visited the museum's Garber facility that summer day in 1980 shared many feelings: pride in having served their country on missions that were hazardous in the extreme and could have threatened their lives; pride in having contributed to the rapid conclusion of a long and terrible war; horror at the magnitude of the destruction and loss of life on the ground; awe at the power of the bombs they had delivered; and hope that the awesomeness of nuclear weapons will make total war unthinkable from now to eternity.

Many of the men also believe that, though two hundred thousand may have died in Hiroshima and Nagasaki, the quick end to the war ultimately saved more lives than it took. That certainly is Paul Tibbets's view: "You hate to weigh one thing against another, but sometimes you save more lives than you take."[13]

Ray Gallagher, assistant flight engineer on the Nagasaki mission, may have summarized his comrades' feelings best, years later, in the summer of 1994, when he, Tom Ferebee, Dick Nelson, and Dutch Van Kirk were recording their stories with National Air and Space Museum's filmmaker, Patricia (Patti) Woodside. Woodside was preparing a fifteen-minute film for visitors to see as a final commemorative feature in the museum's planned exhibition, *The Last Act: The Atomic Bomb and the End of World War II.*[14]

Ray Gallagher's recollections seemed to us to provide the most powerful and perhaps also the most honest summation for the film and the exhibition:

He recalls a lovely summer day in 1985 when a Japanese television crew came to interview him on the fortieth anniversary of the bombings and set up their equipment in his yard. The sun was shining, the flowers were in bloom—a lovely, peaceful, perhaps unlikely setting to talk about war. He felt anxious. What would he say if they asked him, as he was sure they would, "Are you sorry you dropped the bomb?"

The interview started; the reporter wanted to know everything; Ray recalls, "Oh man, he went here and there and all over."

Ray stops a moment and then continues, "Last question he had, he says, 'Are you sorry you dropped the bomb?' And—I don't know—God must have been with me that day, because I says, 'You know Mister, at that era in our life there was a monster loose. That monster was war. It was killing people, destroying homes, mothers, fathers—oh gosh, so many heartaches. And here this comes along, and it stopped all that! If you'd had it, you would have used it. We had it and we used it. And we stopped it! Many, many, many, many people got to go home.' "

Gallagher stops, then continues: "I says, 'At your years I doubt whether it means anything to you—the term "you got to go home." But if you're a soldier and you're just a civilian that went to war, and that's what we were, your first thought is, "When am I going to get home?" And that's what Truman said. He says "Bring the boys home." And that's what he did. He brought the boys home.' "

Ray halts again. Recalling this has shaken him. "I know we caused," he stops not quite knowing how to get it out, "a lot of heartaches, I am sure." Another pause. How can he say it when it is so awful? "But," his voice breaks, he swallows, "we did all right."

He turns away.

2

A Solemn Vow

The *Enola Gay* had played a pivotal role in the lives of the 509th veterans visiting the National Air and Space Museum's Paul E. Garber Restoration Facility that day in 1980. In their youth, on Tinian Island in the Pacific, they had worked on the aircraft, flown in her, or walked past her a hundred times with pride. Ken Eidnes, twenty-two years old at the time, had taken some of the first color pictures of her in an era when color photos were rare. Driving to Garber that day, he and his comrades could still recall this powerful aircraft gleaming under Tinian's tropical sun.

Imagine their disappointment when they were ushered into a gloomy shed. On the concrete floor next to the wall lay a dull-gray hulk, severed in half. Forward and aft sections were propped up side by side on makeshift steel yokes painted a garish yellow. Where the wings had been removed, the gashed fuselage gaped. Where the engines had been removed from the wings, disheveled tubes protruded from the hollows. The rear gunner's turret was smashed; years earlier, vandals had intruded. Birds had followed and made the turret their home, tearing apart webbing to construct their nests. The engines also had become home to birds, whose corrosive droppings gutted their once smoothly moving parts.

The *Enola Gay* was a wreck!

Who could have allowed this to happen? Who was responsible for this outrage?

To fully understand, we must return to the days following the Allies' victory.

A POSTWAR ODYSSEY

Once the war had ended with the signing of the armistice on the battleship Missouri on September 2, 1945, the *Enola Gay* remained on Tinian until November 6. She was then flown to Roswell AAFB in New Mexico to serve with a squadron that would remain operational despite the massive reduction in force that had returned many of the men to civilian life.

After the war, in the summer of 1946, the United States conducted Operations Crossroads at Bikini Atoll in the Marshall Islands to test the effects of nuclear explosions. Bikini's isolation in mid–Pacific Ocean, just north of the equator, made the atoll an ideal site. It provided adequate secrecy and was sufficiently remote from habitation to prevent a threat from radioactive fallout—or so it was hoped.

After undergoing modifications, the *Enola Gay* was ready to join the tests. Tibbets flew her to Kwajalein Island in the Marshalls but then found himself and his aircraft largely shunted aside. A different crew was selected to conduct the first test, whose prime task was to see the effect of an atomic explosion on seventy obsolete naval vessels assembled in Bikini's lagoon.

Bikini is an extinct volcano that juts up from the ocean floor. The irregular crater rim here and there protrudes from the water in a necklace of small islands enclosing a shallow, central lagoon, only two hundred feet at its deepest point, but stretching twenty-five miles east-west and fifteen miles north-south. Most of the little islands are barely a hundred yards wide, a few hundred yards long, and quite flat. Only a few rise more than twenty feet above high tide. Two of the islands, Bikini and Eneu, occupy as much area as the two dozen others combined. Even Bikini, the largest of them all, claims less than a square mile of land.[1]

What these little islands lacked in size they made up in beauty. In 1946 they were covered with lush tropical vegetation and tall palms. Wonderfully colored fish swam in the lagoon. Multicolored seashells adorned the beaches. Bikini Atoll had been home to 162 Malayo-Polynesian natives, who lived on coconuts, the fish they caught, and local vegetation.[2] Within months of the war's end, however, the atoll's entire population was evacuated so that tests could be conducted. These Bikinians, their children, and their grandchildren were not to see their homeland again for over four decades. When they finally returned, their paradise had been transformed into a wasteland that the United States was attempting to clean up and restore.

The bomb dropped on July 1, 1946, at Bikini was identical to the plutonium bomb dropped on Nagasaki. As Tibbets recalls it, the Air Force had decided to replace him and his men with a new, less-experienced team. At dawn on July 1, they set out to drop their bomb on the USS *Nevada*, right at the center of the cluster of ships and painted a bright orange to make her easily visible from 30,000 feet. But because of an apparent error in the crew's computations, they missed the target ships by a third of a mile.

General Curtis LeMay, in charge of the entire operation, was furious. He asked Tibbets to return to Washington as his personal emissary to report the unfortunate incident to General Carl Spaatz. Tibbets left at once, returning in the *Enola Gay* without her having taken part in the tests.[2] Later that month, the Army Air Forces decided to mothball the *Enola Gay*, and on July 24, 1946, she was flown to Davis-Monthan AAFB, at Tucson, Arizona, for storage. There she stayed for three years.

REDWING

Ten years later, I was at Bikini and witnessed the first hydrogen bomb drop from an aircraft. That bomb missed its target by four miles—ten times worse than the 1946 miss.

I was at Bikini because in early 1955 I had been drafted into the U.S. Army as a private. I was then twenty-four years old with a master's degree in physics, and had been assigned to the Chemical Corps' radiological warfare unit to participate in the 1956 series of tests code-named "Redwing."

For the hydrogen bomb drop, a group of us staked out our neutron detectors on tiny Namu Islet. One could stroll its length in a few minutes. Unlike the main island of Bikini with its tall palms, Namu was barren, open to the sky and the bright tropical sun.

Placing the neutron detectors was not difficult. But we took our time, clambering around to familiarize ourselves with their location. We knew we'd be in an immense hurry to find them when we returned later. The entire island would then be radioactive and completely transformed.

Far overhead an aircraft was flying patterns. We watched the pilot's maneuvers for a while. He was too high for us to be sure, but we figured he was practicing his run, finding the best way to safely escape the hell he'd create.

The bomb Tibbets had dropped over Hiroshima, and escaped by steeply banking and diving his B-29, was a tiny firecracker compared to what would be set loose here. The pilot of the B-52 intercontinental bomber, who for the first time in history would drop a hydrogen bomb, would have to escape a blast 250 times more powerful, in an aircraft that, for all its jet engines, still could not fly that much faster than the *Enola Gay*.[3]

Having laid out our detectors we got back into our landing craft, whose skipper returned us to Bikini Island, across the lagoon. There we awaited the right winds and the wholesale evacuation of the atoll to the fleet at sea.

On the afternoon before the shot, a handful of us who were to helicopter to ground zero just after the burst were taken aboard the helicopter carrier USS *Badoeng Strait,* from which we would take off right after daybreak. We couldn't predict what levels the radiation on Namu would reach. The plan was to helicopter in to check that the levels were tolerable. If they were, we'd swoop down, pick up our detectors and beat the fastest retreat we could.

We turned in early that night. The next morning we'd be getting up well before dawn to get ready.

Just before daybreak we were on deck listening to the countdown. At 5:50 A.M. on that morning of May 21, 1956, the United States dropped a hydrogen bomb from an aircraft for the first time. The bomb took about a minute to fall from an altitude of around 50,000 feet down to one tenth that height. There it exploded, showing the world that our country had the might to deliver such weapons on any enemy.

We were thirty miles out at sea. The heat of the bomb that lit up the sky struck at once. Though wearing thick goggles, we had faced away to protect our eyes. Now we turned around and saw the enormous fireball that stretched miles across the sky, partly hidden behind a cloud, but filling the entire field of vision and rising incredibly fast followed by a giant pillar that rose from below. The entire macabre evolution to this point, and further as the fireball gave way to an enormous cloud that kept endlessly spreading and approaching, proceeded in total, eerie silence. For two minutes we stood riveted by this ghastly scene. Then came the startling crack that shook us awake. The blast of pressure had reached our ship.

We went below then, to get ready for the helicopter ride to ground zero. But as the minutes went by, we learned that the bomb had missed the island by four miles. No seriously high neutron fluxes were expected. We might as well wait a few days and then go in more leisurely to retrieve our detectors by ship.

NIGHT TURNS TO DAY

Other shots followed, both at Eniwetok and at Bikini, as we shuttled back and forth laying out our detectors and then retrieving them. For one predawn explosion at Bikini we remained at our home base on Eniwetok. But we knew it would take place at 5:30 A.M. We got up early to see whether at the distance of 220 miles we would even notice anything. Foolishly, we walked to the edge of the island nearest to Bikini, as though coming a hundred yards closer would make a difference. Then, at the appointed hour, the night suddenly turned into day, undulating in brightness for several seconds before relapsing into darkness. It was as if a bomb exploding over Philadelphia had lit daylight in Boston.

We returned to our tent and went back to sleep. Twenty minutes later, the whole tent shook as a rumbling sound wave arrived, rattling our bunks, as though some huge truck was lumbering by. The sound had taken that long to reach us. A few minutes later, another rumble, as another sound wave reached us more slowly.

In preparation for yet another bomb burst, we had emplaced our detectors on a pretty little palm-crested island. The carcass of a Japanese fighter plane was still sitting there, eleven years after the war's end. We climbed all over it, of course, curious to explore its cockpit. We were all too young to have seen a real Japanese fighter plane in combat. Then we left to await the explosion.

When the burst occurred a few days later, we did helicopter in to retrieve our detectors. The bomb had exploded at ground level near one end of the island. As we approached we saw a huge azure bowl of water that had not been there before. Coming in closer, we saw its origins. The bomb had vaporized half the island down to well below sea level. The other half also was unrecognizably changed. The tall palms were gone. All signs of vegetation had been blasted away. The Japanese fighter had disappeared. There was no soil. Only fist-size rubble remained. We circled around without landing. There were no neutron detectors to be retrieved.

We were all young and gripped by the excitement of being there. When we were not working, we went skin diving, fascinated by the colorful tropical fish that populated the atolls. We collected seashells more beautiful than any we'd ever seen before. We would talk about our plans when we'd get out of the army; most of us wanted to go back to school and earn a higher degree. The GI bill would help out.

The Pacific Proving Grounds was a weird world. There were long lulls in our work as we waited for favorable weather permitting further explosions. To relieve boredom there were always new films, shown in a large outdoor theater that was pleasant on those balmy Pacific nights. But first on the huge screen each evening was Marilyn Monroe. She would lean out at us voluptuously and breathe, "I hate men who talk."

We got the message: Secrecy was crucial. The Cold War and arms race were deadly serious. We had better watch what we said or wrote home. The cost of loose talk was incalculable.

But the recollection that stuck longest was different. It was the memory of an island, half of which had been vaporized from the face of the earth, and of the sky that morning when we had stood on our little island at Eniwetok, staring expectantly toward Bikini, only to see night turned into day.

HAP ARNOLD AND PAUL GARBER

At the end of World War II, the Army Air Forces of the United States were engaged in a struggle to create a new United States Air Force independent of the U.S. Army. Chief strategist in this political battle was General H. H. "Hap" Arnold, who had commanded the Army Air Forces throughout the war. In 1946 he took two steps to assure the future of American air power.

He masterminded the establishment of an Air Force Association, a conglomerate of individuals and industries passionately dedicated to military aviation. Its purpose was to make the Air Force strong, and to keep it strong. In time, the AFA would grow to two hundred thousand members and become the most powerful voice for military aviation in Congress.

Arnold also set out to preserve as much of World War II Air Force history as possible, and to keep it in the public eye in the nation's capital. It was in Washington that favorable publicity would be most valuable in guarding against imminent postwar cutbacks. He saw to it that a large and varied collection of Allied and enemy war planes were spared trips to the scrap pile in the general postwar reduction of armed forces and equipment. He wanted these aircraft saved for a national museum dedicated to flight.

To strengthen his hand, Arnold solicited and received supporting petitions from 267 military aviation enthusiasts, core members of the AFA. They lobbied the Congress of the United States to authorize the creation of a National Air Museum, to be administered by the Smithsonian Institution.[4] Located on the National Mall in the heart of Washington, the museum would literally be in full view of Congress at all times, a constant reminder of the primacy of aviation.

The new museum's Congressional charter directed it to

> memorialize the national development of aviation; collect, preserve, and display aeronautical equipment of historical interest and significance; serve as a repository for scientific equipment and data pertaining to the development of aviation; and provide educational material for the historical study of aviation.[5]

To assure that the museum would have no lack of significant aircraft for display, Arnold directed the Air Force to transfer to the Smithsonian the collection he had saved.

In assembling this collection, Arnold had sought the advice of Paul E. Garber, the Smithsonian's first curator of aviation. Garber, born in 1899, had joined the Smithsonian in 1920, to remain there till the end of his life. When he died in 1992, he had served the Institution in one capacity or another for seventy-two years. During that period, he had personally collected more than half of the 350 aircraft in the museum's collection, including the original *Flyer,* in which the Wright brothers had carried out their first controlled flights in 1903, and Charles Lindbergh's *Spirit of St. Louis.*

Certainly the most historic of the military aircraft selected by Arnold and Garber for the National Air Museum's collection was the *Enola Gay.*

STORAGE PROBLEMS

Selecting airplanes for a collection was one thing. Figuring out where to put them was another. The United States Congress legislates through two different procedures. It first authorizes an action, and then may appropriate funding. The Congress had authorized but not funded the museum's construction, and so no museum could be built. Smaller airplanes could be housed in various sheds that Paul Garber might find, but the *Enola Gay*, 99 feet (30 meters) long, 141 feet (43 meters) in wingspan, and 28 feet (8.4 meters) in tail height, was far too large to be thus tucked away.

Since the Smithsonian had no airport of its own, a suitable location to gather this new collection needed to be found. On July 3, 1949, the National Air Museum and the Air Force Association jointly sponsored the National Air Fair, the country's largest air show to date, at Park Ridge, Illinois—the site of today's O'Hare Airport at Chicago. Befitting an occasion of this magnitude, the two organizations, twin offspring of Hap Arnold's vision, arranged for Paul Tibbets to fly the *Enola Gay* from Davis-Monthan to Park Ridge for a ceremony to officially hand the aircraft over to the Smithsonian's care.[6]

When the Korean War broke out, the Air Force suddenly needed Park Ridge. The Smithsonian's collection was in the way. On January 12, 1952, the *Enola Gay* was once more shuttled away, this time to be put in temporary storage at Pyote Air Force Base, Texas. There she remained for nearly two years, until retrieved and flown to Andrews AFB, Maryland, at the outskirts of Washington, D.C.

Andrews is a base with restricted access; the aircraft assigned to the president of the United States are hangared there. The Smithsonian would have encountered difficulties in bringing a separate security force to this military base, and instead relied on the United States Air Force to guard this aircraft, just as it guarded its own planes. But the Air Force apparently had no available hangar space, and kept the *Enola Gay* outdoors, unguarded, at a remote site on the airfield. Curiosity seekers soon found her, broke in, and took out a number of small, readily removed parts as souvenirs. Where intruders had gained entrance, birds, insects, and the elements also found access to inflict damage.

Paul Garber became increasingly concerned. If kept at Andrews any longer, where the Air Force was unable or unwilling to provide better security and care, the *Enola Gay* would be irreparably damaged through neglect. Since no other airfield would do any better, the only alternative was to take her to safe storage. This meant taking the entire aircraft apart, because she was far too big to be transported along available roads.

The complex disassembly was begun on August 10, 1960, and required nearly a year. The fuselage had to be separated into two halves. The wings had to be removed and dismantled into inboard and outboard portions, and the engines had to be taken out of the wings. The enormous empennage had to be removed from the rear section of the fuselage, propeller blades were dismounted from the engines, and much more. All this took until July 21, 1961, when the disassembled components were finally moved to the museum's storage facility in Suitland, Maryland. Here at least they were safe from vandals and further deterioration through weathering or infestation. The temporary buildings erected at Silver Hill, in Suitland, were not much more than primitive unheated sheds; but Paul Garber figured that some day the museum would have the resources to undertake a restoration, and then the parts would be there, better preserved than had the aircraft remained outdoors.

THE VETERANS

Nineteen years later, in 1980, that is where and how the men of the 509th found their *Enola Gay*. Some were stunned; others merely depressed. A few were outraged, accusing the Smithsonian of wanton neglect. They didn't know the aircraft's history, and they had no tolerance for excuses. The Smithsonian had accepted the aircraft from the Air Force and was responsible for its care. It was that simple!

Donald C. Rehl, of Fountaintown, Indiana, who thirty-five years earlier had piloted a B-29 as a first lieutenant with the 509th, recalls, "That's when we realized the condition and decided to do something about it."[7] Several of the men began to seek ways to get the aircraft restored. At that time, most of them were still working, but five years later many would be retired and would have time to press for its restoration. They were either going to make the Smithsonian care for their aircraft and its history or have it transferred, perhaps to Offutt Air Force Base in Omaha, Nebraska, where the Strategic Air Command was headquartered and had a museum.

Their concerns were shared by some on the museum's staff. In mid-1983, Walter Boyne, a former B-52 pilot with the Strategic Air Command, was made director of the National Air and Space Museum. By December 1984, he had assessed the magnitude of the task and given instructions for the restoration to start.

In the 1980s, members of the 509th began to meet more frequently. At their 1984 gathering in Philadelphia, they started serious discussions. Don Rehl and his former navigator in the 509th, Frank B. Stewart, of Indianapolis, began to write dozens of officials urging the restoration of the *Enola Gay*. Interviewed in 1985, Stewart said, "It ticked me off. We figured, what the hell!"[8] Both men visited Silver Hill shortly before the fortieth anniversary on August 6 to see the aircraft and the ongoing restoration. To a reporter Stewart said, "They are starting to move on it, but they don't have the money." Rehl added, "The timetable is five to seven years away.

We feel that it's been forty years and it's time." Stewart thought he could get the donations to restore the plane faster.[9]

For Stewart, "The reasons to restore the aircraft [were] considerable." No other aircraft so unalterably changed the course of human events. He wanted it "to serve as a symbol so that this ... will never happen again. We're not hawks—nobody wants nuclear war. We think it could be a good teaching tool."[10] Rehl's motivations in 1985 were somewhat different. He cited an invasion of Japan and the potential participation of the Soviet Union in such an effort as another reason. "Millions of lives would have been lost if an invasion would have been necessary, and Russia would now occupy part of Japan just as they [did] Germany and other countries." For him that was another reason why "*Enola Gay* should be placed on public display where people can see it along with the story to justify its use."[11]

Stewart, who was only nineteen at the time of the bombings, sold aeronautical equipment after the war, and then went to Indianapolis where he worked as a trouble-shooter for the Department of Housing and Urban Development. Rehl, who was twenty-two at the time of the Hiroshima raid, had stayed in the service for a while after the war, had later entered the car business in Indianapolis, and eventually started his own insurance agency. He now was retired, and had some time to devote to the cause.[12]

Neither Rehl nor Stewart had taken part in the Hiroshima or Nagasaki missions. They had remained in New Mexico, ready to fly a third atomic bomb to Tinian when it was ready. Rehl thinks they would have participated in the next raid if the war had not ended.[13]

Forty years later, Rehl's and Stewart's efforts to have the aircraft restored had the strong support of Paul Tibbets. Tibbets, by then seventy, had not seen the *Enola Gay* since the day he had personally delivered her to the Smithsonian at Park Ridge, Illinois. He went to Silver Hill twice in the weeks before the fortieth anniversary. He recalled the first of these visits as "A sad meeting. [My] fond memories, and I don't mean the dropping of the bomb, were the numerous occasions I flew the airplane.... I pushed it very, very hard and it never failed me.... It was probably the most beautiful piece of machinery that any pilot ever flew."[14]

Tibbets should know. His entire career had centered on flying. He stayed on in the U.S. Air Force until 1966, when he retired as a brigadier general. He had spent nearly thirty years in the service, much of them testing advanced aircraft. Leaving the Air Force, he joined Executive Jet Aviation, an all-jet air taxi company based in Columbus, Ohio, where he became the company's president in 1976. He stayed on until late 1986, when the company, by this time solidly established, was sold. He was then seventy-one.

THE VOW

Tibbets, Rehl, and Stewart were not alone in trying to get the Smithsonian to move faster in the mid-1980s. Early in 1986, William A. Rooney, of Wilmette, Illinois, launched a first missive to Smithsonian secretary Robert McCormick Adams, sending it through the office of Senator Barry Goldwater to make certain it received due attention. In the late 1960s, Goldwater had been one of the strongest voices in Congress in support of providing the funds needed to build a National Air and Space Museum on the National Mall. And he still was seen as a very powerful friend by the Institution.

Rooney was a former Army Air Forces intelligence officer with wartime service in B-29s as a member of the 40th Bomb Group and in the 20th Bomber Command, headquartered in India and China. Later, he had been an advertising executive in Chicago, and was now retired. Not only was Rooney upset at the Smithsonian's lack of care for the *Enola Gay,* he wanted the aircraft rapidly restored. He also had very definite views on how she was to be displayed.

On June 13, 1986, Adams replied. He challenged Rooney's assertion that the Institution had for four decades neglected the *Enola Gay.* He explained that the restoration of the aircraft had begun and that Rooney would be welcome to inspect it. "Walter Boyne . . . will be happy to arrange a visit for you if you are ever in the Washington area."[15]

Turning to the question of display, Adams noted that the aircraft was too large to fit into the present museum. He then mused, "You may well ask, how will we ever exhibit it? Here the story gets more complicated and uncertain, although I am confident it will have a happy ending." He referred to the planned Museum Extension at Washington's Dulles International Airport, and added, "Such a facility has had a place in the formally-approved, future construction program of the Smithsonian regents for some time. . . . I can also tell you confidentially that, when we do manage to complete the new extension, the *Enola Gay* will have a place in it." Adams then embarked on a final passage that infuriated Rooney.

> Having said all this, let me also say—although here you may part company with me—that we have to give great care to precisely *how* we exhibit the *Enola Gay.* As you are aware, it is not simply another piece of hardware but something of great symbolic importance. Having participated not long ago in a major effort to introduce the Smithsonian to Japan, for example, I know we will need to think through carefully how our manner of presentation will go over with a friendly and important nation that has a powerful set of its own symbolic associations with the *Enola Gay*—and that sends many thousands of visitors through the National Air and Space Museum every year. I think it would be wrong, in short, to deal with the *Enola Gay* merely as a significant step in the development of several vital aerospace technologies. "A decent respect for the opinions of mankind," to use a

famous old phrase, requires us also to touch on the demonstrated horror and yawning future risk of the Age that the *Enola Gay* helped to inaugurate.

Yours was a very thoughtful letter, and I am grateful for the interest and concern that it reflects.

Rooney took seven months to think this over, and then exploded. On January 22, 1987 he wrote to Adams,

> I have read [our exchange of letters] over many times trying to find some part of it which could place us in accord. I find none.

After again accusing the Smithsonian of neglect in caring for the aircraft, he referred to Adams's last paragraph and wrote,

> You offer me the quote "A decent respect for the opinions of mankind." I offer you one in response. I can give you the author of this quote. It was Harry Truman. He said, "I haven't heard anyone apologize for Pearl Harbor."
>
> This develops the issue further on two fronts. Does your "Decent respect for the opinions of mankind," give respect to the millions of Americans who fought in WWII? Do you read the Japanese correctly or does your opinion take its lead from the State Department plus your own determination to punish America for having dropped the atomic bomb? . . . [B]y burying the *Enola Gay*, do you feel you have assisted the peace lovers of the world in achieving their ultimate triumph? If you wish me to do so, I believe I can produce a considerable body of opinion in Japan which will show Japan was grateful that we did drop the atomic bomb for, they say, if we had not, the war would have gone on.
>
> I see in this exchange a no win situation on my part for you to sit in your chair in the Smithsonian with all of your authority while I am just an angry citizen. I shall tell you this however: Accompanying this letter is a picture of Paul Tibbets leaning out of the pilot's window of the *Enola Gay* prior to its famous flight. I am dedicated to seeing the *Enola Gay* restored and on display and Paul Tibbets putting his head out of the window of the *Enola Gay* once again. Only this time the *Enola Gay* will be properly restored and displayed whether or not it is in the Smithsonian. I intend for this to take place within my lifetime and that of Gen. Tibbets, and I am in my 70th year. . . .
>
> In summary, Mr. Adams, I see this situation as a contest between a Washington Satrap with all of the infrastructure and Washington social, political and government connections on his side. On the other side is one old American citizen with a conviction.[16]

3

The National Air and Space Museum

THE MUSEUM ON THE MALL

Hap Arnold's charisma had persuaded Congress to authorize a National Air Museum. His dowry of aircraft provided by the United States Air Force had also made this the best-endowed museum of aviation in the world. For the Smithsonian he had created a quagmire of problems.

Authorization permitted the Institution to begin planning to build a museum. That did not count for much unless Congress was also willing to appropriate funds for construction, which they were not. Even in 1966, when Congress authorized the museum to change its name, to the National Air and Space Museum, no appropriation for a building was in sight. The museum existed in name only; but it possessed a huge collection of airplanes and spacecraft for which it could not properly care.

Finally, with the war in Vietnam dragging on and the country in a deep malaise, Arizona's influential Senator Barry Goldwater, the Republican contender for the presidency in 1964 and now also board chairman of the Air Force Association's Aerospace Education Foundation, persuaded the Congress to appropriate funds for a museum on the National Mall. Goldwater, a pilot in World War II and a general in the Air Force Reserve, told his fellow legislators that this was "a cause that is right ... a cause that deserves a fight." He wanted a museum that would celebrate aviation and spaceflight and inspire the nation's youth. This vision excluded the *Enola Gay.* "What we are interested in here are the truly historic aircraft," he declared at hearings. "I wouldn't consider the one that dropped the bomb on Japan as belonging to that category."[1] When the funds that could be appropriated sufficed only to build a museum not quite adequate to house the large bomber, Goldwater was not disturbed.

The new National Air and Space Museum that began to rise on the National Mall in the early 1970s, became the nation's birthday present to itself on the two hundredth anniversary of the republic. Readied on time and under cost, the museum opened its doors to the public two days ahead of schedule, on July 2, 1976. Within a month one million visitors had crowded through its doors. Overnight, it became the most visited museum in the world, averaging between eight and nine million visitors a year over the next two decades.

A visitor entering the building from the National Mall immediately comes upon several of the most significant air and spacecraft in the history of flight: the first airplane ever to fly, the *Wright Flyer*, in which the two Wright brothers made four flights on December 17, 1903, the longest lasting just fifty-nine seconds; the *Spirit of St. Louis,* in which Charles Lindbergh dodged rain squalls and clouds, fighting off sleep, on his lonely, thirty-three-hour, 1927 transatlantic flight from New York to Paris; and the tiny, bright orange Bell X-1 *Glamorous Glennis,* in which Air Force pilot Chuck Yeager became the first man to fly faster than the speed of sound. He named the airplane after his wife—a nice touch for this museum thronged by families with children. Also there is the *Apollo 11* capsule that returned Neil Armstrong, Buzz Aldrin, and Michael Collins safely to Earth after man's first successful landing on another celestial body—the Moon. Michael Collins later became the museum's first director, the man responsible for bringing the building to life.

These all are real: the actual historic craft! Nothing is fake!

None are models, though for the spacecraft still in orbit, or on Mars, the museum can only display engineering backup units that could have been flown but were kept in reserve.

Kings and presidents, astronauts and cosmonauts, billionaires and school groups, and young couples with six-week-old infants pass each other here, their eyes not on each other, but on the amazing machines inspired by dreams of flying.

ENOLA GAY AMONG THE MISSING

For many years, the most truly epoch-making craft absent from the museum was the *Enola Gay*. Original plans for a building envisaged spaces considerably larger than those seen now, including one sufficiently large to accommodate even this enormous airplane. But funds were tight, plans were redrawn, and the designed spaces became smaller to make the museum affordable. For lack of funds, the fully assembled *Enola Gay* would have to be displayed elsewhere—at a larger museum extension.

This museum extension was in any case needed. The museum on the National Mall lacks any storage space to house the aircraft, spacecraft, and aerospace memorabilia not on display. With 350 airplanes and a similar number of spacecraft, the museum is responsible for an air force larger than that of many small nations. Thirty thousand other, mostly smaller items round out this vast collection of aerospace artifacts. Only a fraction of this collection is on exhibit at any given time.

If not here, where are all those objects?

Some of the larger airplanes have for decades been stored at Davis-Monthan AFB just outside Tucson, Arizona. There, the dry desert air retards metal corrosion and maintains the aircraft in reasonable shape, though the intense summer heat deteriorates rubber, leather, and plastic components. Smaller aircraft have found their way over the years into temporary metal sheds that Paul Garber, in his inimitable way, had managed to get erected at Silver Hill, in Suitland, Maryland. This storage and restoration facility now bears his name.

Unlike the National Air and Space Museum building on the National Mall, Silver Hill boasts no soaring glass and marble structures. Only a very few of the metal sheds even have heating, air conditioning, or humidity controls of any kind. Roofs leak. Corrosion is everywhere. Infestation is a constant hazard. The collections management crew, responsible for caring for the museum's treasures, struggles valiantly, but is losing the fight. Unless better buildings are made available, most of this superb collection of the nation's—of man's—early flying machines will steadily deteriorate into oblivion. I estimate we are irrevocably losing three percent of the collection each year.

THE MUSEUM EXTENSION

Realizing this impending loss, as early as 1981, Noel Hinners, Michael Collins's successor as the new museum's director, initiated a search for an additional facility. The museum extension he proposed would replace the deteriorating temporary buildings where many critical functions were carried out. These included preserving and restoring the aircraft and spacecraft in the collection; preparing, maintaining, and updating the exhibits in the museum on the National Mall; housing and preserving the museum's archives; and storing the artifacts not on display at the museum or elsewhere on loan.

The museum extension would also be essential for eventually displaying the restored, fully reassembled *Enola Gay*. The structure would have to be designed with indoor spaces sufficiently large to house and display this and even larger aircraft and spacecraft in the museum's collections. Among these were the space shuttle *Enterprise*; the prototype for the Boeing 707, the first commercially successful jet; a World War II Mitchell B-25; and many other historic craft.

Once a sizable aircraft lands, it cannot be transported farther, except again by air. Highways are too narrow, overpasses too low. The museum extension would have to be built on airport grounds, with access to an active runway capable of handling even the largest jet aircraft: a space shuttle, flown to an airport piggyback on its Boeing 747 mother craft, required a runway at least 8,500 feet long.

The extension would also have to be located within an hour's drive of the Mall to facilitate daily visits, in either direction, by curators and other staff members. A facility at a greater distance would be less accessible and would require greater autonomy and added staffing. This would unnecessarily duplicate capabilities already available in the main building on the Mall, inevitably raising operating costs.

Many airports were investigated. Some were too far away; others, including the Baltimore Washington International Airport (BWI), at the time were reluctant to serve as a site. Only the Washington Dulles International Airport met all the criteria and was willing to accept the extension on its grounds.

Given the circumstances, the Smithsonian Board of Regents had an easy choice. At their January 1984 meeting they agreed to make Dulles the site of the extension and asked the six congressional members of the board to introduce legislation authorizing its planning and construction.

THE SMITHSONIAN INSTITUTION

The National Air and Space Museum is only one of a large family of museums and research facilities that together constitute the Smithsonian Institution. To show how this assemblage operates, I need to briefly note its purpose and structure.

The Smithsonian was established by Congress in 1846 in response to a surprise bequest by an Englishman, James Smithson, who, having no heirs, left his entire wealth to the United States. He stipulated that it was to be used

> to found at Washington, under the name of the 'Smithsonian Institution,' an establishment for the increase and diffusion of knowledge among men.[2]

These twenty-two words have, in the course of a century and a half, fashioned the Smithsonian as it appears today. Its operations are carried out in a dozen museums and half a dozen additional research centers devoted to science, history, and art. While much of the activity takes place in the nation's capital, research centers and outposts can be found as far away as Kenya, Panama, and Hawaii.

The Smithsonian is governed by a board of regents, chaired by its chancellor, the Chief Justice of the United States. The nation's vice president serves as vice chancellor. Three senators, three members of the House of Representatives, and nine citizens appointed by the Congress complete the board's membership.

The board meets three times a year—at the end of January, early in May, and in the third week of September—always on a Monday morning. Its meetings typically last two hours. A three-member executive committee of the board convenes at intervals between board meetings and works with the secretary on pressing matters that cannot await a session of the full board.

The secretary of the Smithsonian is the Institution's chief executive officer, appointed by the regents and responsible to them. While the secretary, working with the board, provides an overall direction to the Institution's efforts, an under secretary, working as the Institution's next highest official, acts as chief operating officer, looking after day-to-day operations.

The largest of the museums, each operated by its own director, line Washington's National Mall, that stretches from Capitol Hill to the Washington Monument and beyond. Anyone can walk in. There is no entrance fee. The Air and Space Museum is one of the most modern in this family of buildings. The Smithsonian "Castle," where the secretary, the under secretary, and their staffs have their offices and where the regents always hold their meetings, is the oldest. It is a dark reddish-brown, with quaint spires. For the Smithsonian staff, the Castle represents the seat of authority.

CONGRESSIONAL LEGISLATION

Even with the regents' inside track to Congress, through the six congressional members on their board, progress on major issues can be slow or even confrontational.

While the board had decided in January 1984 that the Air and Space Museum Extension should be built at Dulles International Airport, the House of Representatives was not yet ready to proceed on this major project. In March 1985, the Institution was asked not to proceed with master planning on the Dulles extension without further authorization. This put a stop to all activity.

Two years later, in the summer of 1987, Governor William Donald Schaefer of Maryland reversed his state's original decision and announced that Maryland would now be interested in making a strong bid to house the extension on BWI airport grounds and to help the Smithsonian defray construction expenditures.

Having received this offer, Secretary Adams decided to seek a meeting with Congressman Sidney Yates, chairman of the House Appropriations Committee's Subcommittee on Interior and Related Agencies, whose purview included the Smithsonian Institution. On January 4, 1988, he obtained an appointment on short notice, and the two of us headed up Capitol Hill to appeal to Yates to release the Institution from the moratorium on planning.

Adams and I cannot have spent more than half an hour in Yates's office before emerging with his approval to go ahead. The way to start master planning for the Extension now seemed clear.

Unfortunately, this turned out to be no more than a mirage.

First, however, I have to turn to my own appointment to the Smithsonian.

4

A New Director

ADAMS'S CHOICE

Walter Boyne, who had been director of the National Air and Space Museum since the spring of 1983, abruptly left the Smithsonian in midsummer 1986, leaving the National Air and Space Museum directorship vacant for a year. Within weeks, Secretary Adams began to look for a successor.

In 1984, the Board of Regents had brought Adams to the Smithsonian to raise the Institution's standards of scholarship. Under his predecessor, Secretary Dillon Ripley, the Smithsonian had rapidly grown, adding museum after new museum. The public loved the new buildings and their exhibitions and came by the tens of milllions each year. But in the rush to produce displays, curators had little time to conduct research and to infuse the exhibits with thoughtful scholarship. Given James Smithson's wish to create an insitution "for the increase and diffusion of knowledge," the Smithsonian was doing remarkably well in diffusing knowledge through exhibitions, but was falling behind on its "increase" through scholarly research.

Adams wanted to get started, especially at the National Air and Space Museum, where he felt reforms were most needed. He appointed Dr. James Tyler, a marine biologist at the National Museum of Natural History, to serve as the museum's interim director.

Tyler was a no-nonsense administrator. He wanted the staff to do its job. That meant examining whether people were doing the work implied by their job titles and earning the salaries they were paid. His aim was to make it easier for his successor, whoever the new permanent director would be, to decide how to fill these positions in the long run.

After a search that culminated in the spring of 1987, Adams had a choice of two finalists. He invited both to meet the senior museum staff and then met with the staff himself to hear their preferences. One candidate was a retired four-star general who had served in the U.S. Air Force with great distinction and integrity. I only learned many years later that he had been the museum staff's choice recommended to Adams.[1]

Adams was reluctant. Walter Boyne had come from the Air Force and Adams thought the museum now needed to pursue new directions. For him, aviation and spaceflight were not just the domain of the military or the aerospace industry. He wanted the museum to address itself also to broader issues of public interest. He decided to offer the position to the other candidate.

I was an MIT-trained astrophysicist, had been a professor of astronomy at Cornell University for twenty-five years, and had chaired its astronomy department for a five-year term. As a space scientist, I had designed and constructed sensitive equipment that my students and I placed aboard rockets, high-flying NASA aircraft, or satellites to obtain a clear view out into the universe. I had also become interested in work that previous generations of astronomers and engineers had carried out. With my Cornell colleague, the historian Pearce Williams, I initiated, cofounded, and codirected the University's program in the history and philosophy of science and technology. This brought together scientists, engineers, historians, philosophers, and sociologists. We wanted to gain a deeper understanding of technology and give our students better insight into its origins and influence on everyday life.

I had applied for the directorship of the National Air and Space Museum because with so many million visitors coming to the museum each year, where else could one find such a great opportunity to inform people about the amazing changes aviation and spaceflight had brought about in this century?

Even before I accepted the position, Adams and I had briefly talked about several major issues. I knew that the restoration of the *Enola Gay* had begun, and we discussed how she might best be exhibited. We agreed that she symbolized a pivotal change in twentieth-century warfare and doubtlessly in world history. Her Hiroshima mission had introduced a nuclear weapon with which the human race for the first time in its existence could demonstrably annihilate itself. The display of the aircraft in a thoughtful exhibition that commemorated her mission, soberly and without histrionics or fanfare, seemed like the way to proceed. This was no idle discussion. With the restoration progressing, it was high time to prepare for an exhibition, one that would unavoidably raise passions.

On June 24, 1987, I was officially named the museum's director. I was to take over in mid-August.

FIRST THOUGHTS ON EXHIBITING THE *ENOLA GAY*

For the next eight weeks I was busy phasing out my work at Cornell and transferring my space research to the museum. The Smithsonian had agreed to my setting up a laboratory whose researchers would bring informed scientific insight to the museum's exhibitions on space exploration.

For two years I had been a member of a science team designing a new space telescope that would be able to measure faint infrared radiation—tiny amounts of heat—reaching us from distant stars and galaxies out in the universe. The ten-year project to build this billion-dollar Infrared Space Observatory (ISO) was an effort spearheaded by the European Space Agency (ESA). With the support of the National Aeronautics and Space Administration (NASA), I served on the otherwise all-European team of scientists. NASA and ESA were working to establish agreements that would give American astronomers rights to use this advanced observatory.

On June 26, I left for one of the ISO science team's quarterly meetings at the European Space Research and Technology Centre (ESTEC) at Noordwijk, the Netherlands. From there I was to go on to London for an international conference on infrared astronomy. In between meetings I was thinking ahead to my new job and the many different tasks that would need attention as soon as I started. I had talked with Jim Tyler, who had filled me in on the main problems, and I was beginning to think about the exhibition of the *Enola Gay.*

Late one of these nights, I jotted down notes on how an exhibition might be mounted in a historical context that would show how the *Enola Gay*'s mission was a natural step to follow in the progressive escalation of bombing in World War II. I listed different topics and items that might be displayed, and still have the page on which I wrote down my thoughts that night. Their intent is still clear today:

> This is not an exhibit about the rights and wrongs of war, about who started what, and who were the bad guys and who the good. It is about the impact and effects of bombing on people and on the strategic outcome of conflicts. Is bombing strategically effective? Are the costs worth the strategic gains? How great is human error? How predictable are enemy casualties or one's own losses? What are the losses to humans who become the victims—civilian or military, it doesn't matter.[2]

A suitable title for this exhibition, I thought, might be "From Guernica to Hiroshima—Bombing in World War II." At Guernica, in 1937, during the Spanish Civil War, the Germans had practiced the bombing they would later perfect early in World War II. At Hiroshima, in 1945, the war had all but ended.

This approach to a display fitted in with the mood of the country that summer. While we were witnessing an escalation of technology in President Ronald Reagan's Strategic Defense Initiative, which the media had dubbed "Star Wars," Reagan was also discussing with the Soviet Union's Communist Party General Secretary Mikhail Gorbachev how to wind down the Cold War.

Also in this vein, David C. Beeder, of the World-Herald Bureau, wrote in an August 6, 1987, article,

> The *Enola Gay*, the made-in-Nebraska bomber that dropped the first atomic bomb 42 years ago today, attracted little attention on the anniversary . . . in suburban Silver Hill, Md., where it is being stored.
>
> Two groups, one from Japan where the bomb was dropped, obtained permission to see the restored front section of the plane. Gary Mummert, a leader of a suburban peace organization, said the two groups plan to ask the government to use the plane as a memorial. "For a symbol as important as this, we think it should be used as a teaching tool for peace," he said. "It was a political liability for a while."
>
> Mummert, 41, a former Air Force navigator, led a group of six people to the warehouse where the plane is being restored. A group of 10 Japanese tourists visited the plane later in the day, joining Mummert's group to present their proposal that the plane be converted into a peace memorial.[3]

Three or four other articles with the same message also commemorated this gathering. The group, which included two Japanese survivors of the Hiroshima raid, also handed out a proclamation ending in the words

> we propose that the everlasting symbolism of the *Enola Gay* be used to educate humanity to the horrors of a nuclear war. That it not be made a monument to war, but a memorial of peace, a constant beacon throughout the land lighting for us the necessity of nonviolent conflict resolution in a nuclear-armed world. . . . "Lest we forget; we must never allow it to happen again."[4]

STARTING WORK

Monday, August 17, 1987, was my first day at the museum. By September 23, the same Gary Mummert and a number of colleagues had obtained an appointment to see me. They came armed with their proclamation, and asked that the *Enola Gay* be made part of a national shrine to peace. I listened to the request, but expressed doubts that the Smithsonian would even have the authority to participate in such an effort without a presidential or congressional directive reflecting the will of the country. I saw no reason to believe that national sentiment would back such a move. I told them I would make some inquiries with the Smithsonian's legal counsel and then respond to them.

On December 2, 1987, I sent a response to one of the leading members of the group. I noted that the museum was restoring the aircraft now and "I hope that we will then be able to exhibit the airplane as part of a serious look at strategic bombing in World War II—with the destruction these bombs inflicted and their indelible imprint on history ever since."[5]

At the suggestion of Smithsonian legal counsel I also added, "The National Museum holds all of its collections, by statute, as trustee 'for the increase and diffusion of knowledge among men,' and cannot dispose of objects unless they are duplicative or of demonstrably inferior quality."[6] In other words, given the importance of the *Enola Gay*, it would be illegal for us to hand the aircraft over to some other organization.

I never heard from the group again. But there were others who also wanted the museum to give them the aircraft. A fairly typical request had come earlier, in a letter dated August 25, 1987, from Arthur Veik, of Humphrey, Nebraska. It was forwarded by Nebraska congresswoman Virginia Smith:

> I am saddened to learn that the *Enola Gay*, after restoration, is destined to be exhibited at a museum [extension] at Dulles International Airport. Since she was built at Bellevue it seems to me she ought to be brought back home to Nebraska.... In your recent constituency letter, Mrs. Smith, you indicated that expanded tourism could reap huge economic benefits for Nebraska. I couldn't agree with you more. I believe this made-in-Nebraska bomber could be a help.... During World War II I flew in 35 bombing missions attacking targets from Singapore to Manchuria and all over the Japanese homeland. My memories remain sharp and clear....[7]

On October 6, Margaret Gaynor, director of government relations for the Institution, sent me a draft of a polite reply to Congresswoman Smith, explaining that "exhibition planning is now underway as a component of a larger exhibit on strategic bombing."[8]

THE RESEARCH ADVISORY COMMITTEE DEBATES

Just how this planning was to be done was under intense discussion at the museum. I was still trying to get to know my new colleagues, and knew too little about their background, professional expertise, or possible biases. Accordingly, I decided to establish a Research Advisory Committee, reflecting a wide range of views and bringing broad expertise to its task. I asked the staff to nominate people they considered outstanding, added a few names of my own, and called a meeting of the committee for two and a half days beginning on the morning of October 26, 1987. Chairing the group was Dr. Herbert Friedman, an early pioneer of space research and member of the National Academy of Sciences. One of the topics discussed at length was whether and how the *Enola Gay* might be exhibited.

Secretary Adams attended the committee's closing session on the morning of October 28, where it delivered its findings and recommendations. The minutes of the meeting illuminate the discussion on an *Enola Gay* exhibition that had taken place during the preceding two days. On one side of the debate was Admiral Noel Gayler, a World War II naval aviator who later had risen to four-star rank and served as commander in chief of Pacific Forces. He had also headed the National Security Agency—for a time so secret that it didn't even officially exist. Opposing Gayler were the other members of the committee. The minutes have Admiral Gayler saying,

> Strategies of air warfare are certainly a most important historical subject, but this is quite a different one. This has to do with—and I am looking for a word stronger than "propriety"— . . . the validity . . . of exhibiting the *Enola Gay* in this institution. As I see it, she has a noble distinction as an aircraft, like any one of 50 others. The mission over Japan was not in any tactical or operational sense distinctive.
>
> The Japanese were essentially defeated. We were flying airplanes all over the empire, at will. I was the operations officer of the task force at that time—with Japan and defined ports for us to strike. And except for accidents, we didn't lose any airplanes.
>
> So, there was nothing aeronautical about it. The thing that made the mission distinctive was . . . that we used the nuclear weapon for the first time against human beings. . . . [I]f we put that thing on exhibit, we cannot fail to give the impression that we somehow are glorifying that mission or taking pride in it.
>
> We can't reverse history; we can't change it, but we don't have to make it the forefront in the most important museum in the capital of our country. . . . I think that if we do that, [it] will tarnish the reputation of the Museum.

To this Adams responded,

> This is an issue of great importance to us . . . a sputtering fuse, in effect. . . . So far a fairly low level of agitation has continued on that matter, for certainly as long as I have been here. . . . I suspect it goes back to the beginning of the Museum itself. . . . I am delighted that it has come up . . . because I would like to get the widest possible views on what we do . . . in the context of our not knowing what vagaries of opinions and coalitions will appear in the Congress on the matter. . . . It is a question that is just fraught with complexities, any way in which we go.
>
> My own view would be that if we did move to exhibit the *Enola Gay*, . . . it would have to be done with extraordinary sensitivity [and] that it could not be done without taking account of what happened as a result of that bombing, in other words, dealing with the devastation that at least that

atomic bomb caused, rather than simply putting the aircraft itself on exhibit.

I don't think the latter is adequate, but as soon as I say that, that becomes a political act of its own kind and of quite an extraordinary kind.

As the museum's director, my opinion was that

it will take us several years to finish the restoration of the airplane. . . . I would like to use that time to speak with as many people in Congress and outside as possible to have a dialogue or a multisided exchange that precedes any actions, so that people know what we are about to do, . . . know what our viewpoint will be, and can exercise some degree of criticism. . . .

And Prof. James Hansen, a historian of technology at Auburn University and the youngest member of the group, commented,

The first law of the history of technology is that technology is neither good nor bad, nor is it neutral. . . . I teach over 1,000 freshmen a year at Auburn in large classes, and I know that the overwhelming majority of these young people would want to see the *Enola Gay*. They would enjoy and benefit from discussing all of the issues embodied in that airplane. I think that those who are going to criticize the showing of the *Enola Gay* are going to be those who don't believe in this first law of the history of technology, who really want American technology presented as good. The overwhelming majority of American people would want to see the *Enola Gay*, for whatever reason.

But Gayler would have none of that:

There is a technology that, I think we would all agree around this table, can only be described as bad, and that is the technology of nuclear weapons.

To which Alex Roland, also a historian of technology and a professor at Duke, countered,

One of the reasons that this is so controversial . . . is because we don't know the end of the story. . . . [W]e know what crossbows did; and we can make more or less final judgments on them. We don't know what the final judgments are on man's space flight, nuclear weapons, military armaments in the air; and because of that and because these are continuing debates . . . I think that the appropriate criterion for decisions on whether to display or not to display is whether the artifact is as significant as we understand it at the present time. . . . I don't think that there is any question that this is "good" or "bad." You gave the thumbnail sketch of why we shouldn't put up [the *Enola Gay*] because of the unmitigated horror of nuclear weapons. . . . Let me give you the thumbnail sketch of the opposite view. . . . This . . . plane . . . that dropped the weapon . . . was engaged in an historical event.

But Gayler again objected:

> I want to suggest that we do know the end of the story with respect to nuclear weapons. They will destroy us unless we do something about them. I will tell you why I know it; because there is a non-zero probability that in any given year there will be a nuclear war. And if you sum those up, the probability will become one some day. So, you know, there we are. [Exhibiting the *Enola Gay* would be] an act with incalculable consequences, no matter how you do it, no matter whether you surround it with symposia on what the effects were . . . and future nuclear wars might be. . . . It could become a pilgrimage for the most radical right, or it could lead to a serious and constructive discussion of what is involved, and anything in between. . . . [I]t is an extraordinarily sensitive and complex and important question for us.

At this point, Dr. David Challinor, the Smithsonian Institution's assistant secretary for science, and about to retire after many years, added a startling note:

> I agree with the Admiral. Since I've had this job—I was even quoted in *Newsweek* about 15 years ago—[I've been] saying, "the *Enola Gay* will never be on exhibit while I'm here." Well, I have another two months to go, so I guess I am still safe; but the perception of this plane I think is, in part, generational. . . . I do know that when we first exhibited . . . the first atomic bomb in this museum there was a lot of discussion about how it should be displayed. It was shown against a backdrop of a large photomural of the bomb exploding; and other photos of the resulting devastation. We are locked in both generationally and philosophically and our feelings will probably change with time. . . .
>
> I don't think there is a precise answer to a question like this.[9]

Noel Gayler's views cannot be considered to represent those shared by most World War II veterans in the late 1980s. To many they may even sound distorted. But they are quite similar to the opinion of Fleet Admiral William D. Leahy, chief of staff to Presidents Roosevelt and Truman throughout the war. In his memoir *I Was There*, published in 1950, Leahy wrote,

> It is my opinion that the use of this barbarous weapon at Hiroshima and Nagasaki was of no material assistance in our war against Japan. . . .
>
> My own feeling was that in being the first to use it we had adopted an ethical standard common to the barbarians of the dark ages. I was not taught to make war in that fashion . . . by destroying women and children.[10]

Even earlier, in his 1948 book *Crusade in Europe*, General Dwight D. Eisenhower recalled a July 1945 conversation with Secretary of War Henry L. Stimson at Potsdam, where Stimson had just informed him of the successful atomic bomb test-explosion at Alamogordo, New Mexico. Eisenhower wrote,

> I expressed the hope that we would never have to use such a thing against any enemy because I disliked seeing the United States take the lead in

> introducing into war something as horrible and destructive as this new weapon was described to be. Moreover, I mistakenly had some faint hope that if we never used the weapon in war other nations might remain ignorant of the fact that the problem of nuclear fission had been solved. I did not then know, of course, that an army of scientists had been engaged in the production of the weapon and that secrecy in this vital matter could not have been maintained. My views were merely personal and immediate reactions; they were not based upon any analysis of the subject.[11]

I have quoted the advisory committee discussion at such length because most of the arguments the museum would encounter in the seven years to follow were already raised within weeks of my arrival at the museum.

Only the way forward was not clear.

5

A Reluctant Start

A MUSEUM MALAISE

Three weeks after the advisory committee meeting, Von Hardesty, chair of the museum's aeronautics department, sent me a long memorandum on "Strategic Bombing and the proposed display of the B-29 *Enola Gay*." It reported on two meetings he had held, respectively, with members of the museum's aeronautics department and senior staff from all parts of the museum. These had been prompted by my noting that there were two spaces in the museum that might just barely be sufficiently large to house the *Enola Gay*, despite the aircraft's enormous wingspan.[1]

For many years the standard reply to those urging the museum to display the *Enola Gay* had been that the aircraft was too big to fit into the museum. I suppose that as a scientist I tend to be skeptical, and so I asked to see the actual dimensions of the spaces. These showed that the largest of the open galleries on the main floor might be able to accommodate the *Enola Gay* with her wings stretched diagonally from corner to opposite corner.

Nadya Makovenyi, who headed the museum's exhibits department, had a dollhouse-sized model of the museum in which aircraft scaled to the same dimensions could be placed. This provided an impression of what visitors would see when they came to the museum. When she placed an exact scale model of the *Enola Gay* in the dollhouse, we found two major galleries in which the aircraft would fit. This terrified the curators, because it also showed that the only way the airplane could be accommodated in either gallery was to clear out everything currently on display—an upheaval none of them was ready to face.

Hardesty's memorandum reported that the museum staff was already busy doing other exhibitions and restorations, which they were reluctant to relegate to something of lesser priority. Further, the museum had told petitioners who wanted the early naval airplane NC-4 or the pre–World War II Junkers 52 exhibited that they were too large to be displayed. The *Enola Gay* was even larger. If we were to display it, "This could be a potentially embarrassing situation to explain away." Also, "It would upset the balance, and it would dwarf our other aircraft and exhibits." Referring to the planned museum extension at Washington's Dulles Airport, Hardesty's memorandum went on,

> The most reasonable compromise for this difficulty would be to exhibit [the aircraft] at Dulles where it could be seen in conjunction with other World War II bombers, rather than to set it apart here at NASM. Its exhibit could certainly cause security problems.
>
> One final item: if a strategic bombing exhibit is done, we could consider the possibility of using a piece of the *Enola Gay* in it. Suggestions involved using the nose section or another representative piece. . . .

Referring to a narrow balcony overlooking the museum's gallery on World War II, Hardesty followed these thoughts with a three-point conclusion:

> 1. Yes, an exhibit on strategic bombing is needed and could be very effective. The balcony in World War II would be the best location.
> 2. We will need to consider the overall impact on various people and schedules before a final decision can be made.
> 3. The exhibit of the entire *Enola Gay* at NASM is definitely not a good idea.[2]

At the time, I was annoyed by these recommendations, but several years later, I would find myself reluctantly facing the choice of displaying only part of the *Enola Gay* or not displaying her at all.

I was annoyed by Hardesty's advice because it added to my mounting concerns. In the three months I had been at the museum, I had gained the impression that a year without a permanent director had left a deep malaise. Every proposal I made was met with a string of objections in favor of inactivity. Nobody wanted to do exhibitions. Curators complained that Secretary Adams wouldn't give them promotions unless they buckled down to do research. Adams, they claimed, was "not a museum person." He had been provost at the University of Chicago. He wanted to change the Smithsonian into a university. There were no rewards in producing exhibits, so why do them.

My response to this was to elicit a written agreement from the assistant secretaries and from Secretary Adams that excellence in scholarship would be rewarded, but so would excellence in producing exhibits, in the acquisition and care of collections, and in public service. Such activities were all part of a curatorial department's work. Not every curator would have to be active in all of these aspects of the museum's responsibilities, as long as the museum as a whole satisfied all these demands. Curators would need only to demonstrate excellence in the performance of those duties that were assigned to them.

The museum had wonderful exhibits showing the history of aviation and spaceflight. But it was remarkably devoid of explanations on what kept an airplane from crashing, allowed a balloon to rise high in the atmosphere, or kept a spacecraft from tumbling down to earth. Why did some aircraft have propellers when others didn't? What was the difference between a jet engine and a rocket? Why does an aircraft have to bank when it wants to turn to change direction? Both Adams and I wanted to have at least one gallery in which visitors, especially children, could learn how the amazing machines on display actually worked.

I also wanted to encourage youngsters to take an interest in aviation and spaceflight. The public complained that the United States was losing its international lead in technology. By getting the most gifted youngsters interested at an early age, I felt we would have better odds to remain competitive. I wanted to do a futuristic exhibition on space exploration. It would not be a science fiction story. Instead, I wanted youngsters to see what the nation's best scientific and technological minds predicted about future possibilities and directions, primarily to show that there were interesting and exciting careers that would be open to those willing to go into space exploration and technology.

The curators objected to such approaches on the grounds that the National Air and Space Museum was a "historical museum." But who had said that? There was nothing in the legislation that specified such a narrow restriction. To be sure, most curators had been trained as historians, but we also had scientists on board, and we should make the best use of them to go ahead with exhibits that would be of strong public interest.

THE HEART OF THE MUSEUM

The National Air and Space Museum has an impressive, skilled restoration staff. At our restoration facility in Suitland, Maryland, I would find not only tools, but also books and old manuals on the restorers' workbenches. Here was a new breed of restorers who not only worked with their hands but with their minds. Many would go through the museum's archives looking for original drawings and parts lists for the aircraft they were to restore. Some frequented antiquarian bookshops and had their own private collections of books on older airplanes. Often the restoration specialists knew a great deal more about the aircraft they were restoring than the curators who were supposed to instruct them on how the restoration was to be undertaken.

I felt the restorers were just as much at the heart of the museum as the curators. And so were the designers and members of the exhibits production staff. They too had imagination, verve, and skill. My task was to see how all of us could work best together, to take advantage of each others' talents and skills.

Each group, however, had its own dynamics and traditions. For curators, I had to make certain that they were provided sufficient time and freedom to pursue research on topics that conformed to the museum's congressional charter. I would not interfere with that research, except to make certain that it had merit. If it did, it would find its way into journals and books. If it was not of sufficiently high quality, the editors of those publications would reject the work. This meant that I would only need to assess the curators' research from time to time, through a careful professional accomplishment review that solicited opinions from peers both inside the museum and at other research establishments. In short, curators would be responsible for pursuing their own research. But its quality would be periodically checked and would influence their advancement.

For exhibitions, the rules had to be different. The public would never view an exhibition at a national museum as reflecting solely the views of a single curator. Visitors to the National Air and Space Museum saw an exhibition as representing a museum, if not a Smithsonian, or even a national consensus. That made the museum and the Institution ultimately responsible for every exhibition we mounted. As director, I would have to take responsibility for defining the directions of exhibitions, because ultimately I would be held accountable. Moreover, where particularly sensitive or potentially controversial exhibits were concerned, I would have to keep the secretary of the Institution apprised, since the public and the Congress might also hold him responsible. These were my concerns as I read Hardesty's report.

Parts of his considerations were certainly valid. Any exhibit of the *Enola Gay* would raise security problems. We would have to expect demonstrations, some of which might pose the threat of violence. And even if the *Enola Gay* could physically fit into the museum, the disruption of removing so many other artifacts to make room for just one enormous airplane needed to be carefully considered and might indeed prove to be a very bad idea. Hardesty was clearly right in voicing that concern.

On the other hand, mounting an exhibit on strategic bombing without the *Enola Gay* and placing it on the balcony of the World War II gallery was tantamount to burial. The balcony is a tiny area, which only a few people ever visit. A set of narrow stairs restricts access. The area certainly was not geared to handling the hundred thousand visitors a month that a really interesting exhibit would attract.

Finally, given that many considered the *Enola Gay* an icon, I insisted at the time that we had to keep looking for ways to exhibit the entire aircraft. We could not exhibit just "the nose section or another representative piece." I told the staff that dismembering an icon was equivalent to desecration. Too many people would be dismayed.

Years later, given no other options, I did reluctantly embrace the idea of exhibiting just the aircraft's fifty-six-foot-long forward fuselage. But that did produce the predicted resentment. General Paul Tibbets vocally objected to our displaying his plane crippled, and others also attacked the decision.

For now, however, I instructed that we keep on looking for ways to exhibit the entire aircraft.

THE MUSEUM'S FLOORS MIGHT CAVE IN

On December 7, 1987, Linda (Lin) Neuman Ezell, charged with managing all the museum's collections, distributed a memorandum asking for clear directions on the *Enola Gay* restoration: "If we do exhibit the B-29 at NASM in the early 1990s, it will have a decided impact on the restoration schedule, and we would prefer to have that decision firmed up before we issue any more draft plans."[3]

Four days later, she followed up with a second memorandum in which she stated that her current schedule estimated completion of the restoration by 1994. She continued, "If we pull everyone off their planned assignments, except for those artifacts that have already been slated for an exhibition, we could complete the *Enola Gay* restoration much sooner, by 1992. However, we would be forced to ignore other artifacts that both the Collections Management staff and the curators believe should be cared for as soon as possible."[4] Outlining the impact of exhibiting the *Enola Gay* within only five years, in 1992, she listed the manpower requirements and alluded to the possible use of volunteers. She also expressed her belief that restorations should not be driven by the requirements of exhibitions, as they had been in the past, but rather by needs to preserve the invaluable artifacts the nation had placed in the museum's care. Turning to work that would need to be done to install the aircraft in the museum, she also estimated the time required to clear out the air transportation gallery, if that was the gallery chosen for exhibiting the *Enola Gay*. All this, she pointed out, could be done, but she preferred to do it at a more relaxed pace than that required for a 1992 opening.

A TEMPORARY STRUCTURE TO EXHIBIT THE *ENOLA GAY*

Perhaps the most significant part of Ezell's letter was her mentioning that the floors in the museum were too weak and would break under the *Enola*

Gay's weight unless they were heavily jacked up from the garage immediately below.

While these were serious concerns that would constrain our undertaking, I was glad at least to obtain well-founded considerations that I had to take seriously. I instructed Ezell to proceed with the restoration at a reasonable pace. At no time should she have fewer than two of her restoration staff assigned to the *Enola Gay*. If she had opportunities to assign more people, all the better, but two should be the minimum.

Given the floor-loading problem, I also started searching for alternatives. One thought was to erect a temporary building in which to display the *Enola Gay*. Preferably this shelter should be on the National Mall, as close as possible to the museum. I discussed the issue with several members of the staff, and within weeks they had identified seven potential vendors who might be able to erect such a structure on the ground-level, paved terrace at the museum's west end or on the National Mall.[5] For the most suitable structures, prices ranged from $2 million to $2.5 million.[6] That was a pretty steep price to pay.

Mentioning that the military services often cooperated with the museum on arrangements that could have mutual benefits, someone came up with the idea that the Air Force might be able to help with mobile hangars that they employed for some of their aircraft.

I wrote to Air Force Lieutenant General Charles McDonald on March 14, 1988.[7] Two weeks later we had our answer. Unfortunately, the general could not help us in this instance: He wrote back that the Air Force had no movable hangar large enough to house the *Enola Gay*. He ended on a cheerful note, wishing us "Good luck with your exhibit."[8]

We would certainly need it!

THE WASHINGTON FINE ARTS COMMISSION

We began to look around again to see whether we might be able to find less expensive temporary structures that could satisfy our needs for exhibiting the *Enola Gay* on the Mall. The first step was to see Carter Brown, director of the National Gallery, who also served as chairman of the Washington Fine Arts Commission, one of the organizations that would have to agree to our erecting any structure on the Mall.

Each summer a whole variety of major displays were mounted on a plot of land immediately adjacent to the museum, across 4th Street. Sometimes a massive fleet of helicopters, tents, tanks, and other military vehicles would appear there overnight and stay a while, only to be replaced by a circus tent and show, or by a fair of some kind. While such events usually lasted no more than a couple of weeks, and our request would ask for the use of the site for at least a year, and preferably for as much as three, it did not seem to me that we were asking for a vast deviation from usage routinely sanctioned in the past.

I obtained an appointment with Brown for April 15, 1988, and outlined our ideas to him in his office atop the National Gallery's East Wing. We had a cordial conversation. Brown pointed out that the Mall is considered sacred ground. He thought the Fine Arts Commission would probably be opposed to the idea of putting up a temporary building sufficiently large to house the *Enola Gay* and keeping it on the Mall for three years. He could foresee public reactions to the proposal at the open meetings that the commission would have to hold. Given these difficulties, I said I'd try for now to see what we might be able to do within the museum or later with a fully assembled aircraft at the museum extension. However, I might be back with a formal request if none of those other options worked out.

That was April 1988. The problem of finding a suitable site to exhibit the *Enola Gay* would be with us for another four and a half years.

HARDESTY WRITES AGAIN

In the meantime, Von Hardesty had written to me again on December 28, 1987:

> When the whole matter has been discussed in the Aeronautics Department, a consensus viewpoint has emerged: The strategic bombing exhibit has been warmly endorsed and the proposed exhibition of the *Enola Gay* at NASM has been opposed for numerous reasons. . . .
>
> I strongly recommend that the *Enola Gay* be designated as a future exhibition (perhaps the core exhibition) for the Dulles facility and we turn instead to a broader consideration of how, over the next three years, we could develop a new exhibit on strategic bombing. There may be alternative exhibit ideas as well. In making this proposal I feel I am echoing a consensus that exists within the department.

This letter of Hardesty's also raised a more specific and more serious problem than his previous memorandum.

> The *Enola Gay* project is problematical as well for the curatorial staff in another way: what do we want to say about strategic bombing and who would provide intellectual leadership for such a complex undertaking? Any brief discussion of the theme of bombing, Guernica to Hiroshima, exposes a set of complex (and controversial) historical problems. Our concept of this exhibit remains vague and ill-defined, the assumption being that the exhibitry would be relatively easy to fashion once we found a place for the *Enola Gay*. The research and preparation for this essential phase of the exhibit should not be underestimated. As a department, we are anxious to take on this challenge, but we feel we can do it without the *Enola Gay* at NASM.[9]

Hardesty was implying that I might not be able to count on the museum's existing curatorial staff to produce some of the more challenging exhibitions that I felt our visitors would find interesting; we might need to hire additional staff with the requisite expertise.

I resolved that we would have to pursue a two-pronged strategy. I would try to bring to the museum a number of outstanding senior people to work closely with the museum's excellent exhibition and restoration staff to produce the exhibitions I hoped we could mount. In addition, I would obtain the Institution's consent to use a part of the museum's budget to support the tuition costs of young curators so that they could accelerate their studies and more quickly take their places as fully proficient members of our staff.

TIMING OF THE EXHIBITION

We had many discussions in the fall of 1987 and the winter of 1988 about the best time to mount an exhibition on the *Enola Gay*. Some felt that this was too controversial a topic, and we should wait until all World War II veterans had passed away. This was at odds with the veterans' own wishes. They wanted to see the *Enola Gay* once more, before they were too old to visit the display.

I felt that we should have the exhibition as soon as possible. My concern was that if we waited too many decades, we would be able to gain information solely from archival documents. While these were more reliable than human memories, they also were readily misinterpreted if one waited too long. Language evolves appreciably over a period of fifty to a hundred years. Words assume different meanings and connotations; assumptions embodied in them change. By mounting an exhibition fairly soon, I felt we would still be able to work with people who had played key roles during the war, and who could point out to us any misinterpretation in our reading of archival documents.

Lin Ezell had said that the earliest the museum's craftspeople could complete restoration was 1992, but that she would prefer not to have them work in such haste and solely on one aircraft. That suggested a somewhat later target date, with the exhibition opening in the fiftieth anniversary year, 1995. An added benefit, I thought, would be the heightened popular interest in the mission of the *Enola Gay* during the anniversary year of the bombing.

6

Searching for a Home to Display the *Enola Gay*

OFFUTT AFB

While we at the museum were searching for a site best suited to displaying the *Enola Gay*, there were plenty of others who were willing to find such a site for us—usually with the aim of wresting ownership of the aircraft from the Smithsonian. Most of the veterans who were so inclined favored Offutt Air Force Base, at Omaha, Nebraska.

Offutt AFB, headquarters to the Strategic Air Command (SAC), maintained an aviation museum largely dedicated to strategic bombing. In the late 1970s, the National Air and Space Museum had written Offutt and actually offered to transfer the *Enola Gay*, provided the base museum would pledge to house her adequately in perpetuity. After considering the offer, the Air Force decided it could not give such a promise. Hangar space was valuable and might be needed for more urgent needs in a crisis. Obtaining no guarantees, the Smithsonian withdrew the offer.

Others were less willing to give up. While former 509th pilot Don Rehl was trying, in 1985, to get the museum to restore the aircraft on a faster schedule, he was also seeking ways to take the *Enola Gay* away from the Smithsonian. To a reporter from the *Indianapolis Star* he conjectured that the Confederate Air Force, a group of enthusiasts that restores World War II era planes and flies them in air shows, could quickly restore the *Enola*

Gay, and that they could ferry restored pieces to Offutt for assembly there.[1] As far as Rehl was concerned, that would place the aircraft in more responsive Air Force hands. He approached Jack L. Allen, the SAC museum's director. According to the reporter, Allen replied,

> The impact of the *Enola Gay*'s mission upon war strategies was a major reason the Strategic Air Command was created. . . . While the SAC Museum would jump at a chance to house the *Enola Gay*—and has about $100,000 in escrow which could be used as 'seed money' to raise the additional funds needed to construct a new building—the SAC Museum is not in a position to wage battle with the National Air and Space Museum over possession of the aircraft.[2]

Many others, however, repeatedly brought up the same idea. On January 22, 1988, a citizen, Paul Filipowski of Gainesville, Florida, wrote Congressman Buddy MacKay,

> The Smithsonian National Air and Space Museum is waiting for a new [extension] to be built at Dulles Airport. But it is unlikely that this expensive museum will be built in the near future due to its cost. A better solution would be for the Smithsonian to loan the *Enola Gay* to the Strategic Air Command Museum at Bellevue, NE. They would be glad to display the plane.
>
> Could you inquire of the Smithsonian the reason for taking so many years to renovate the *Enola Gay*? . . . The brave airmen who flew in the 509th Composite Group are all getting up there in age. It would be fitting for this B-29 to be renovated as soon as possible so that General Paul Tibbets & his men could witness the *Enola Gay* on display in a national air museum.[3]

Secretary Adams replied to Filipowski and MacKay on March 14, to tell them about the museum's progress on the *Enola Gay*.[4] But Filipowski wrote to Adams again on March 26:

> Efforts of an informal restoration committee of WWII 509th members have been made to ensure the restoration & display of the B-29 before Gen. Paul Tibbets passes on. . . .
>
> These 509th veterans feel that the pace of the work has been held in check due to political rather than budgetary constraints. Whether one agrees with the decision to drop an atomic bomb on Hiroshima or not, the *Enola Gay* is doubtless the most important WWII plane in your collection. I can't understand why only two restoration specialists are working on the plane. . . . These combat veterans deserve the thanks of this nation. . . . Can nothing be done to speed up the process?[5]

This time I responded. My letter of April 27, 1988, stated in part,

> The *Enola Gay*, as you pointed out in your letter, is an aircraft of historical significance. . . . Our long-range goal is to restore it, to display it, and to use the aircraft as part of a larger exhibition on strategic bombing. These

> are our plans. The process of restoration is complex and the labor intensive, but this work is proceeding at a steady rate. A restoration team is permanently assigned to the project to assure regular progress. . . .
>
> As you probably know, the *Enola Gay* is too large to fit into the National Air and Space Museum on the Mall. Consequently, we are exploring alternative places for exhibition with our new extension . . . its ultimate home.[6]

This elicited a quick reply:

> I am pleased that your plans are to exhibit the plane at some future date. . . . As you are doubtless aware, there is a growing interest in the restoration of the "*Enola Gay*." An "*Enola Gay* Restoration Committee" has been formed (1010 East 86th Street, Indianapolis IN 46240, (317) 848-9361). I'm not a member of the Committee but I do support them in their efforts.[7]

I was delighted! I called the Indianapolis number Filipowski had given, and talked with Frank Stewart, about whom I knew nothing since I had not been at the museum in 1985 when he and Don Rehl had visited the Garber facility.

FRANK STEWART

Stewart and I had a congenial conversation and a few days later, on May 15, he wrote to me on stationery whose letterhead spelled out "*Enola Gay* Association," showed a picture of the aircraft, and listed seven names, among them Don Rehl, Elbert Watson, Rolland Nail, and Bill Rooney.[8]

I have already introduced Rehl and Rooney. Watson had corresponded with Bob Adams a couple of times in the previous several months. He had written on April 14 to express his concern about the slow pace of restoring the *Enola Gay*; he was impatient to see her displayed.[9] While he mentioned that he was acting as media representative for the "*Enola Gay* Restoration Committee," his letterhead identified him as publisher of the *World War II Times*. In April, I had not yet become aware of the existence of the committee, and Watson's letter did not state what it did. That was to become clearer in the weeks ahead.

Stewart's letter now also introduced Rolland Nail, who would be visiting us with him.

> Just a quick note to confirm our telephone conversation May 10th relative to our forthcoming meeting in your office 9:00 A.M., June 9th with yourself and Linda Ezell.
>
> We will have Gen. Paul Tibbets (Ret), Mr. Rolland "Tack" Nail, and myself. We are confident that you will be pleased with all the ideas and assistance we have to offer. Our ultimate goal is to see the *Enola Gay* restored and displayed in the shortest time practical.
>
> I would like to express my personal gratitude for the most courteous and positive reception extended to our group, and look forward to an enjoy-

> able relationship and, hopefully, a firm program with reasonable timetables.[10]

He added that it would be helpful to have a man-hour and cost estimate for completion of the restoration, as well as an idea of the resources currently assigned to the project, adding, "We simply want to get the job done that will benefit everyone." The letter was signed "Frank B. Stewart, President, *Enola Gay* Restoration Assn., Inc."

On receiving Stewart's letter I wrote back,

> Currently we have two people working on the *Enola Gay* restoration. The total estimated amount of work remaining to complete the restoration is ten man-years. We have, however, had a recent offer from the aviation museum in San Diego to restore the engines, and that would accelerate the process quite a bit.... We have had our staff visit them and they have quality workmen—predominantly volunteers—working for them....
>
> As you will understand, we can only assign a fraction of our restoration staff to the *Enola Gay* at any given time. If we had a bigger staff, or could add manpower by bringing in workmen on contract or as skilled volunteers ... the project could move ahead faster....
>
> [W]e still face the problem of finding a suitable place to exhibit the airplane. The current Museum is too small.

I concluded by mentioning our plans for ultimately exhibiting the *Enola Gay* at the hoped-for Dulles extension. "Both houses of Congress have bills in front of them now asking for authorization.... Discussions with an architectural firm are starting next week." I closed, saying,

> Any aid we could receive from you would be appreciated, as will any suggestions you might have to offer. The restoration of this airplane is clearly long overdue, as we approach the half-century mark since the end of the war.[11]

Before the planned meeting I received a short note Stewart had written June 2, thanking me for my May 20 letter and saying, "The estimation you gave in reference to the remaining man-hours needed for restoration is encouraging. The best estimate we have been able to put together is almost twice that.... We are prepared to address all of the major problems that you mentioned in your letter, and look forward to our meeting."[12]

PAUL TIBBETS VISITS

At 9:00 A.M. on June 10, 1988, General Paul Tibbets, Frank Stewart, and Rolland Nail came to my office. We were joined by several members of the staff.

I was struck by Tibbets's appearance. Given that he was nearly seventy-three years old he looked remarkably fit. His hair was still quite dark and tousled, and he walked with an easy gait.

Our visitors began by telling us that the *Enola Gay* Restoration Committee, which they represented, recognized that the museum was short of funds. They didn't want to take the aircraft away from us, but wished to have it displayed for its historical significance. Recognizing that the extension could be many years away, they suggested the Strategic Air Command (SAC) museum as a suitable interim site. They said that General Jack Chain, commander of SAC, would like to have it. They also said they had a list of people with experience in working on B-29s in the service, as well as people at the Confederate Air Force, who would be able to work on the plane.

Turning to the question of funds, they said that the people who would like to help us with money also would need a proposal from us. We said we would be pleased to provide one.

Tibbets added that he liked beautiful machinery, and hoped to have Frank Stewart's Indianapolis group restore it "in a timely manner." Where the aircraft would ultimately be displayed should not matter now. He thought we needed to make certain that X (a fixed number of) people were assigned to stay on the job of restoration every day.

Sitting there talking with Tibbets felt unreal. Having been at the museum ten months by now, I was accustomed to the world leaders and historic figures of aviation and spaceflight who were always coming through. But they came mostly as VIPs or for commemorative events. To actually sit down to negotiate a significant arrangement with a personage who already was in the history books was different—as though Odysseus had come to my office with two trusted lieutenants forty-three years after the sack of Troy to see how the museum was getting along in restoring their Trojan Horse.

We all shook hands before they left my office to proceed to the Garber facility to view the restoration

Ten days later, Stewart wrote,

> Rolland Nail and I wish to express our gratitude to you and your staff for the courteous reception we received, together with Gen. Paul Tibbets, in your offices June 10th. The personal tour of the Garber facility, and of course the *Enola Gay*, was an unexpected bonus.
>
> We were impressed with the dedication and professionalism of your staff, and compliment everyone for the open, positive approach to our common goals.
>
> The *Enola Gay* Restoration Assn. will begin redrafting our plan and proposal for assistance. Later in the year we plan to draft a "Letter of Intent" to ensure agreement of commitments of mutual benefit.
>
> We appreciate the high priority assigned to the long overdue restoration of the *Enola Gay*.[13]

The letter was cosigned by Rolland Nail.

A few weeks later, I sent Stewart a letter with a two-page proposal which outlined "how the *Enola Gay* Restoration Committee might assist the National Air and Space Museum with the restoration of its B-29.... This will ensure that the many visitors who come to this city will have the opportunity to see the *Enola Gay* exhibited in a context that will allow them to understand its importance as a technical object and as a symbol of the global conflict in which it played such a critical role."[14]

The proposal listed various ways in which they could help us, including public relations support, specifically "by keeping the various organizations that have an interest in the project abreast of progress being made. Such positive press would counter the many articles and editorials that contend that the Smithsonian is not restoring the aircraft." The committee could also help with technical knowledge and expertise, work force support, other restoration support, and video support, all of which were enumerated in the two-page memorandum.

Work force support was to cover the cost of hiring three temporary employees over a three-year period to help speed up the restoration and was estimated at $235,000. Video support would help us produce a short five-minute film on the *Enola Gay* restoration project to be shown in the museum. "The film will briefly consider the importance of the aircraft and its history, but concentrate on its treatment as a museum artifact—its restoration and exhibition.... The committee could play a valuable—and visible—role as the underwriter of such a film. The budget should not exceed $15,000."

On July 25, Frank Stewart called me to say that his committee would like to help us with the restoration of the *Enola Gay*. His board had agreed, but they would like to have the airplane in Omaha, Nebraska. I replied that we would not like their commitment to hinge on that.

A few weeks later, on August 18, 1988, he wrote again

> Members of the *Enola Gay* Restoration Assn. are reconsidering the goals and degree of direct involvement in the restoration and display process. For this reason and other events that have surfaced since our June 10th meeting in Washington, there will be a delay in presenting our proposal.... However, it is the desire of EGRA to prepare a realistic plan that will not only reward both organizations but also the citizens of the United States, who deserve the benefits of an accelerated restoration and display.[15]

Then, on September 1, Lin Ezell wrote to Stewart to let him know the breakdown of work on the *Enola Gay* in terms of time, expenditure, and types of required skills. For more information, she suggested that Stewart might call Richard Horigan, the museum's restoration shop foreman, whose phone number she included.[16]

Stewart submitted a friendly note for the October issue of *American Legion* magazine for inclusion in its "Letters" section. In part it stated,

> Several of us on the *Enola Gay* Restoration Committee met recently with Dr. Martin Harwit, NASM director, to determine the status of the restora-

> tion and to offer our staff and financial assistance to speed up the restoration.
>
> Although NASM staff enthusiasm runs high and the work is now further along, about 25,000 man-hours of work remain, and still no firm date has been established for the completion of the first building of the proposed NASM extension.
>
> The widely held belief that the Smithsonian staff is insensitive and inflexible is not true. The staff is making progress and the workmanship is superb. We would like the restoration to be complete by 1990 so that the plane can be displayed at an interim site until the NASM extension is built.
>
> Readers who wish to assist in the restoration should send their donations to the *Enola Gay* Restoration Association, 1010 E. 86th St. Suite 61J, Indianapolis, IN 46240.[17]

I was pleased. Working with the veterans and forging closer relations could only help us. I thought we were off to a reasonably good start with Stewart's group and with Paul Tibbets. Unfortunately, both the *Enola Gay* Restoration Association and others still were inclined to fight us for possession of the aircraft.

7

Planning an Exhibition

STAFFING THE MUSEUM

I now need to introduce several new members of the museum staff who had joined after my arrival. In one way or another each came to play an important role in formulating the exhibition of the *Enola Gay*.

Among the first people to arrive was Steven Soter, who had been trained as a planetary astronomer but had an enormously broad range of interests in both history and science, and came in as a special advisor to me. One of the many roles I assigned him was to independently critique every major exhibition or film the museum undertook. That provided me with an added internal check, called my attention to any problems that might be developing, and alerted me to the needs for resolving potential difficulties.

Another newcomer was Gregg Herken, a historian of science and technology who had taught at Yale and was working at the California Institute of Technology when we asked him to join as the new chairman of the Department of Space History. He soon became lead curator on an exhibition on the Intermediate-Range Nuclear Forces (INF) treaty, which saw the elimination of thousands of rockets with nuclear warheads that had been facing each other across Europe. Since Gregg had written *The Winning Weapon*, a highly praised study of President Truman's postwar diplomacy on the atomic bomb, I often sought his advice on the *Enola Gay* exhibition.

Tom Crouch, a prolific historian of aviation, particularly of the early years of flight, had been at the National Air and Space Museum for many years before moving to the Smithsonian's National Museum of American History, where he rounded out his interests by immersing himself in social history. I asked him to return to the Air and Space Museum because I felt we needed to have curators who not only knew the history of the technical aspects of aviation and spaceflight, but also could exhibit the impact that the technology has had on life in the twentieth century. Crouch returned to the museum as its new chair of the Aeronautics Department late in 1989, in time to help his colleagues working on a new gallery on World War I.

Michael Neufeld, originally a social historian, first came to the museum in 1988 as a fellow, to write a book on the development of the German V-2 rocket, arguably the most influential advance in aeronautics and astronautics of the mid-twentieth century. The development of powerful rockets dramatically changed the face of modern warfare and introduced an era of space exploration that saw men going to the Moon, unmanned spacecraft exploring the outer reaches of the solar system, and Earth-orbiting telescopes providing a new view of the universe never available before. When the museum sought in 1990 to hire a lead curator for the exhibition of the *Enola Gay* we followed federal procedures and first approached numerous senior American scholars, but none of them were willing or available to take on this complex task. Finding none, we offered the position to Mike Neufeld, a Canadian citizen who clearly had the required credentials. Later, the museum would come under attack for having a curator who was Canadian taking the lead on this exhibition—as though Canada had not fought shoulder to shoulder with the United States throughout World War II.

Aviation and spaceflight keep evolving. They are not just historical topics from a distant past. To stay abreast of the times, the museum needed always to maintain a staff familiar with front-line developments. On the retirement of Robert Mikesh, curator of military aircraft, we were fortunate to attract Thomas Alison, who had just retired as a colonel in the U.S. Air Force. He had flown nearly a thousand hours in the long-secret SR-71, a spy plane that could cruise at three times the speed of sound and fly at altitudes above 70,000 feet. The museum needed curators who at one time had been privy to classified information on important military advances. They might not be permitted to divulge those secrets, but they would know what kind of still-classified airplanes and aerospace artifacts of interest might become available to the museum's collections in the future. Their inside knowledge would permit them to plan accordingly. Having worked so closely with operational aircraft and aircraft maintenance crews all his military life, Alison, who joined the staff in 1993, soon gained the respect and admiration of the restoration staff with whom he closely worked.

MUSEUMS TELL STORIES

Exhibitions at a museum are not just a jumble of arbitrarily selected items each with its own identifying label. The choice of objects to display or not to display begins to set a direction the exhibition will take. The ordering or patterns in which they are arrayed evoke particular insights. The selection of lighting levels establishes a mood. The emphases of words or film footage introducing the exhibition or describing objects further influence the visitor. Each of these choices, whether deliberate or unconscious, shades the visitors' perceptions, the story they will bring away from the exhibition, the way they will later describe the display to friends.

Museums cannot help but tell stories. Curators and exhibition designers are storytellers. Histories are particular types of stories museums tell. They are stories about events that actually happened, and museums are responsible for portraying those events accurately and truthfully.

In recounting history, the museum must take into account sound scholarship; but it must also gauge its audience: What does the visiting public already know about the subject? What is the appropriate vocabulary level that will enable both adults and children to enjoy the exhibit? How much needs to be explained? Will the typical visitor bring some common misconceptions to the exhibition, and do those need to be dispelled? If a topic is sensitive or controversial, should it be presented at all? If so, how should the museum broach it?

These are the questions to be answered, and they need to be tackled in stages. After working as far as possible without turning to others for help, the museum staff needs to assemble critics, advisors, and focus groups to view these first attempts at a finished product and comment on them. Their help is needed to address problems of accuracy, balance, and perceptions. These three words—accuracy, balance, and perceptions—need to be defined, because exhibitions have a different character from other forms of publication.

Accuracy concerns factual information: Are the facts, figures, names, ages, dates, weights, measures, all correct? Assuring accuracy throughout a large exhibition is no trivial matter. At the National Air and Space Museum, many members of the staff will independently sift through an entire script, since it is notoriously difficult to weed out all factual errors. The museum seeks to establish a level of accuracy that will make our exhibits a reference source on aviation and spaceflight as reliable as the very best texts and encyclopedias.

Balance refers to the selection of facts and objects included in the exhibition. For a balanced presentation of a complex topic, this choice is crucial. It can also be excruciatingly difficult. One major source of contention in mounting an exhibition on the *Enola Gay* was the existence of two communities that saw the aircraft quite differently. On the one hand, the museum had in 1987 been approached by Gary Mummert, representing a group that wanted to build a national shrine around the *Enola Gay* as a warning against nuclear war. On the other, we had the letters of Bill Rooney and statements from Don Rehl, who saw the aircraft as a saver of lives that would have been lost had the bomb not been dropped.

Some might wish to dismiss such popular feelings as based on anecdotal evidence that careful scholarship and research could show to be right or wrong. But this is a mirage. As the exhibition of the *Enola Gay* began to take shape, we found two entirely distinct communities of respected historians in the field who, as far as I could tell, saw military history through very different eyes and seldom agreed—historians attached to the universities and historians attached to the military services. The university historians concentrated on the social and diplomatic aspects of military history, while those attached to the services paid heed more to logistics, strategy, and tactics. The service historians tended to see the views promoted by their university counterparts as politically motivated, while the university scholars felt the military people lacked broad perspective. Debates on "balance" seldom ended in consensus.

Perceptions, in contrast to accuracy and balance, are not what the curator puts into an exhibition, but rather what the visitor takes away. If proper balance is hard to achieve, discovering what visitors with widely different backgrounds will perceive on seeing the exhibit can be even more difficult. Here advisors alone are not sufficient. The museum staff must seek out as many representative focus groups as possible to obtain their reactions and to make sure that the perception of the visitors—what the visitors *learn* on seeing the exhibition—is also what the exhibition was meant to *teach*.

The discussions with the Research Advisory Committee had already focused on one of the most challenging problems the museum would face—to shape the exhibition in its entirety so that visitors would leave with reasonable impressions and perceptions. Admiral Noel Gayler in particular had worried that the huge, gleaming *Enola Gay* would convey a celebration of raw power. Was this the impression of the United States the country wished its national museum to project? We did not think so. And yet the *Enola Gay*, displayed by herself, could hardly do otherwise.

To achieve balance, the text would have to be worked into the overall structure of the exhibition in such a way that imbalances evident elsewhere would be counterbalanced by juxtaposed text or images. To avoid the perception of cold, heartless militarism, we needed to infuse awareness that the bomb had caused damage and suffering.

The Research Advisory Committee had already pointed out that the *Enola Gay* could not be exhibited simply to illustrate an advance in aviation technology, as could Chuck Yeager's Bell X-1 *Glamorous Glennis*, the first aircraft to break the speed of sound. The *Enola Gay* was not just any bomber, nor just any B-29 bomber. She was the first bomber to have dropped an atomic bomb. That bomb had instantaneously killed close to a hundred thousand people. This was a fact that could not be suppressed without distorting history. On the other hand, this historic fact must also be placed in context, by displaying the history of strategic bombing, first by the Axis powers and later by the Allies, that had preceded Hiroshima.

STRATEGIC BOMBING

I had originally proposed that the *Enola Gay* be displayed in just this historic context. But in his December 28, 1987, memorandum, Von Hardesty had argued that the museum should separate strategic bombing from the *Enola Gay* and first mount an exhibition that concentrated on strategic bombing alone.[1] This memorandum, as well as his earlier one of November 18, had emphasized the need for us to thoroughly inform ourselves before tackling an exhibit on the more controversial history of the *Enola Gay*.[2] This learning process soon started, but I was not ready to divorce strategic bombing from the *Enola Gay*.

On March 16, 1988, only days after Steven Soter had joined the museum, I called a strategy meeting with Hardesty, Soter, and aeronautics curator Dom Pisano. We were joined by Tom Crouch and Art Molella, both of whom at the time were curators at the National Museum of American History. Stanley Goldberg, an independent historian, who several years earlier had curated an exhibition on the atomic bomb at the American History Museum participated as well. We decided we needed to focus our thoughts more sharply, and Soter volunteered to assemble our ideas in a document he would circulate to the rest of us during the weeks ahead.

Dated April 18, 1988, and titled "Strategic Bombing Exhibit," Soter's memorandum began by citing a review of the museum by historian of technology Michal McMahon, written in 1981.[3] A key passage from the review, which described the museum as it appeared to McMahon a few years after it had opened in 1976, labeled it " ... largely a giant advertisement for air and space technology ... the unmixed blessings of continued technical advance." McMahon's critique continued,

> The omission of the *Enola Gay* from the halls of the new Air and Space Museum—or rather, omission of the "offensive" themes in 20th-century history represented by the World War II bomber—can be seen as the first crisis of the new museum.[4]

Soter expressed the group's consensus that a display of strategic bombing and the *Enola Gay* should not convey the message of "unmixed blessings." He wrote, "The proposed exhibition on strategic bombing will deal honestly and forthrightly with what might be called "the dark side of aviation. Its centerpiece will be the fully restored *Enola Gay*."

"Strategic bombing" needed to be defined, and Soter summarized it succinctly:

> Strategic bombing is aerial bombardment of an enemy's homeland to destroy his war production capacity. In contrast, tactical bombing provides direct support for military forces engaged in land or sea battles. Strategic bombing first emerged on a massive scale during World War II, and there it took two forms: daylight bombing of specified targets chosen for their importance in war production. . . . and the so-called "area bombing" of entire cities, usually at night, intended to destroy the housing (and lives) of workers and thereby to lower enemy morale. . . . Area bombing ("city busting") has remained highly controversial, both with respect to its morality and to its claimed military effectiveness. . . . From the end of the Thirty Years War until the area bombing campaigns of World War II, the deliberate targeting of civilian populations had been the exception rather than the rule in military practice.[5]

Soter then traced the military doctrine that foresaw the Japanese bombing of Chinese cities and German bombing in Spain just before World War II broke out. Strategic bombing had started long before Guernica and gone on far past Hiroshima: "After 1945, strategic bombing continued to evolve, through the Korean and Vietnam Wars, leading to the elaboration of nuclear arsenals. It is this aspect which makes the subject most relevant today."

Soter also proposed an organizational outline for the exhibition; listed some interesting historic lessons that might be emphasized, such as the consequences of "excessive reliance on the supposed infallibility of technology"; discussed the mood to be created—"somber"; mentioned examples of visual art, film, and literature that might be used in the exhibition; and pointed out the need for additional people the museum would have to either bring on board or assign to the project, especially a senior curator for the gallery and one or two staff members for historical research and video interviews with members of the Strategic Bombing Survey. He proposed that two series of colloquia be held, one to educate ourselves internally, and one for the public around the time of the exhibition's opening.

One way of organizing all this, Soter thought, could be first to have an exhibition on strategic bombing in one of the museum's galleries, perhaps as early as 1990. This could later be incorporated into an exhibit in a temporary building on the Mall of the fully assembled *Enola Gay*, after she had been restored. Finally, the entire strategic bombing exhibit would be displayed at Dulles when the museum extension was ready. The second of these three steps was optional.

Soter also noted that financial support for these exhibitions might be difficult to obtain from industry. Industrial subventions had become an increasingly important source of funding for the museum's major exhibitions. He conjectured that "Japanese sources may be interested in helping to restore the B-29."

This last suggestion was quickly dispelled: In the late 1980s, Japanese industrialists were continually coming to the museum, promising us tens of millions of dollars if we were to lend the Wright brothers' *Flyer*, the *Spirit of St. Louis*, or the Space Shuttle *Enterprise* for an exhibition in Japan. These were not objects we were willing to lend; they were simply too valuable to risk loss or damage in transport. On one occasion, however, when one of the major Japanese firms, Nippon Television, again approached us with such a request, I asked whether they might have an interest in paying part of the costs of restoring the *Enola Gay*, in exchange for the possibility of later displaying the aircraft in Japan. After all, the *Enola* Gay was also part of Japan's history. The reply we received from their president, Morihisa Takagi, was unambiguous.

> [(W]e are not able to enthusiastically endorse bringing the *Enola Gay* to Japan. The Hiroshima and Nagasaki bombings remain firmly imprinted in the Japanese consciousness, much as the Holocaust does with the Jewish people.
>
> Although I am sure that the Air & Space Museum's exhibition will be a very well done, dispassionate display, it is an area of great sensitivity to the Japanese and would not have the impact we want to create. [We are] firmly committed to developing an exhibition that would be an unsullied symbol of what man can achieve—a[n] enterprise that will lift our spirits and focus on what technology can accomplish for the future of all of us.[6]

Michal McMahon would have noted the implied "romance of technological progress" in this letter, but that romance was just as prevalent in the United States as in Japan. In the years that followed we would receive countless letters in this spirit, from the most diverse sources.

Mr. Takagi's letter, written in late 1988, did, however, have a great influence. It showed how careful we would need to be to respect Japanese sensitivities, and it also started me thinking about mounting this exhibition without private funding support.

GIVING NOTICE

In the summer of 1988, Secretary Adams and I decided that we needed to let the general public know of our intentions to exhibit the *Enola Gay*. That might stop the inflow of letters accusing us of "deliberately hiding" the aircraft. It would also announce how we intended to display her. In early June, Adams sent me a draft of his monthly *Smithsonian* magazine column and asked me to comment. His article, which would appear in the July 1988 issue, read in part,

> The significance of most human artifacts found in museums lies safely in the past. . . . There are, however, certain classes of objects that erase the possibilities of detachment. Icons of the atomic age are one of these. . . . Even now . . . the outlook for the *Enola Gay*'s early exhibit (at least in fully reconstructed form) is still clouded. It is understandable, in light of this record, that some have asked whether the Smithsonian once made a private decision never to place the aircraft on public exhibition. If not, how did things go so wrong?
>
> The answer to the first question is that we are in the business of confronting and learning from history, not suppressing it. It *will* be exhibited. Neither in the surviving records nor in the memories of staff members of those years is there any hint of ambivalence on that score. There is none now. . . . The answer to the second question is less clear. The National Air Museum . . . had a minuscule staff and an equally small budget . . . And while NASM was planned to contain a gallery big enough to house [the aircraft] when the Museum was authorized in 1964, a substantially smaller structure had to be built when funding was finally approved in 1971. . . .
>
> An indissoluble part of an exhibit of the *Enola Gay* should be some account of what happened at Hiroshima—then and afterward. Probably, the somewhat doubtful overall effectiveness of earlier and subsequent non-nuclear bombing—in Germany during World War II, and in Vietnam—also should be looked at to provide a comparative perspective.
>
> In taking advantage of a rare moment of absolute technical superiority to bring to an end one war, the coupling of the *Enola Gay* with Hiroshima placed before us limitless horrors of another in which the grossly destructive powers would be more evenly divided. Beyond this, it must cause us to reflect on how much of the extraordinary human achievement of ascending so far, and so abruptly, from the Earth has been funded and energized by the general scramble for superiority in ways of killing one another.[7]

In response, I sent Adams a draft of my intended column for the August/September issue of *Air & Space Smithsonian*, with a note saying that I had taken a somewhat different thrust from his, and that

> It is intended to be a brief progress report; but I am also hoping it will evoke a response from any people violently opposed to us putting the airplane on exhibit. That might help me to assess the amount of opposition we might face. If I receive a flood of objecting letters, I will need to worry about security far more than if our plans for the *Enola Gay* appear to meet general approval.[8]

I attached a copy of the draft, which differed little from the article as it appeared. The last three paragraphs from the article contained the key message I wanted to convey:

> The *Enola Gay* will be displayed in a setting that will recall the history of strategic bombing in World War II. As distinct from tactical bombing, which was designed to destroy specific military targets, strategic bombing was meant to break an enemy's overall ability to respond militarily. It was

> aimed at eliminating critical resources, such as ball bearings or gasoline, thereby paralyzing the enemy, or at wreaking enough havoc to break the will of a nation. The B-29 has been called the ultimate realization of the strategic bomber in World War II.
>
> The practice of strategic bombing has raised many questions that have been debated ever since the end of World War II. These issues are critical, because the threat of war has never entirely left us. We need to ask: How effective were the raids militarily? Did the cost to the enemy actually exceed the cost of losses to the bomber command? And, above all, how high were the losses in civilian lives?
>
> The vocabulary of war is now different. No longer do we talk of "thousand-bomber raids" and "carpet bombing." Instead, we debate "mutually assured destruction," "nuclear winter," and "megadeaths." Otherwise little has changed.[9]

These two articles laid out our intentions for the public. Soon we were to receive responses, both pro and con.

THE STRATEGIC BOMBING SERIES

The Research Advisory Committee met again on the days October 24–26, 1988, and once again discussed the *Enola Gay*. A short excerpt from their final report conveys the sense of their deliberations:

> *Enola Gay* and the Strategic Bombing Exhibit
>
> At its 1987 meeting, the Committee engaged in a somewhat contentious discussion of the appropriateness of exhibiting the *Enola Gay* and of undertaking a major exhibit on strategic bombing in WWII. Again at this 1988 meeting the committee could not reach unanimity on the issues. Admiral Noel Gayler remained staunch in his opposition to displaying the *Enola Gay*; other members of the Committee endorsed the exhibit plans now progressing in the Museum. . . .
>
> The consensus of the committee was to proceed with planning to exhibit the *Enola Gay*, but with great caution. It was agreed that a colloquium devoted to strategic bombing, past, present, and future should be organized by the museum with participation by experts from the scientific, political, and educational communities. The colloquium should help to develop a broader consensus on how to deal with such controversial issues.[10]

The advisory committee's recommendation followed a set of ideas that had been advanced a week earlier, when Steven Soter, Tom Crouch, Dom Pisano, Gregg Herken, and I had met at 3:00 P.M., on October 18, 1988. We wished to organize a series of talks, panel discussions, symposia, and films at the museum that would provide us with various perspectives on strategic bombing from the point of view of historians, military leaders, air crews, civilians, scientists, and writers. We came up with lists of potential participants, and decided the series should be videotaped and result in a book highlighting the most significant presentations.[11]

We thought that many of the presentations would be of interest not only to us at the museum, as we tried to better inform ourselves, but also to a larger public. Public events could be held in the museum's Langley theater, which seats nearly five hundred people, and announced in our regular calendar of events. Others, of a more detailed historical nature, might be held primarily for the benefit of interested experts in one of the Institution's smaller conference halls.

Funding this series might be a problem, but I promised to contact the MacArthur Foundation to seek their support. On November 1, I wrote to them,

> The Air and Space Museum has begun planning for a major exhibition, tentatively entitled *From Guernica to Hiroshima: Strategic Bombing in World War II.* In preparation for this exhibition, we would like to hold a colloquium series on strategic bombing, which would culminate in a retrospective symposium with surviving members of the 1946 Strategic Bombing Survey. We plan to videotape the proceedings (and a number of ancillary interviews) in order to document vividly for future generations the ways in which participants such as John Kenneth Galbraith, Paul Nitze, and George Ball came to understand the role and the legacy of strategic bombing after more than 40 years of reflection. We would also seek to bring in Sir Solly Zuckerman and Freeman Dyson to present a British point of view and, if possible, one or two surviving representatives from the Axis side. We expect to publish the proceedings, together with an introductory essay to set them in perspective.
>
> The Museum is seeking a sponsor. . . .[12]

Four months later, in the last week of March 1989, I received the good news that the MacArthur Foundation would fund the series.[13] We went into action at once, and by the fall of that year, the museum began mounting a sixteen-month-long series of talks, panel discussions, symposia, and films on "Strategic Bombing in World War II." It was designed to provide better insight on World War II bombing, but was meant also to put us in touch with the community of experts who could ultimately help us both intellectually and politically in successfully mounting an exhibition of the *Enola Gay*.

Fifty prominent people were invited. Incredibly, forty-nine responded positively. (I do not remember who declined.) Included among these were former Secretary of Defense Robert McNamara, former National Security Advisor McGeorge Bundy, Generals Curtis LeMay and Bernard Schriever, physicists Philip Morrison and Freeman Dyson, economist John Kenneth Galbraith, author Kurt Vonnegut Jr., as well as many prominent historians and participants in World War II bombing, intelligence, and other activities.

Public response to this series was positive, and the military and scholarly communities also appeared pleased. A book of proceedings from this series is under consideration by Professor Tami Davis-Biddle of Duke University.

TESTING PUBLIC OPINION

Between 1989 and 1991, the museum was also engaged in a variety of other projects involving air and space warfare. We consulted widely to better inform ourselves on historical issues, and to more carefully grasp public attitudes and political realities surrounding difficult exhibitions and public events. We felt we needed to proceed carefully with a project as difficult as exhibiting the *Enola Gay.*

In the spring of 1990, curator of space history David DeVorkin opened a revised exhibit of the museum's V-2 rocket. Initially this World War II rocket had been displayed with a simple label, announcing it as the first operational rocket. DeVorkin now fashioned a small display around the missile's base. On the far left, where the story began, were pictures of Wernher von Braun in discussions with high-ranking, wartime German officers. On the far right were pictures of Wernher von Braun in discussions with high-ranking, postwar American officers. In between, reading from left to right, were pictures showing the rocket's provenance, including pictures of the underground factories where slave labor had produced these missiles. About five thousand people had died where the V-2s had hit. About four times as many had died working under horrible, starvation conditions in German subterranean factories. In the middle of the display was a photograph of the destruction a V-2 had wrought in Antwerp. A corpse lay on the ground. Among the many warplanes, military rockets, and armaments on exhibit, this was the first picture of death in the entire history of the museum.

Nobody objected. The only media review was Daniel S. Greenberg's column in the *Houston Post*, stating,

> Truth in labeling has achieved a rare breakthrough in an exhibit of military rocketry at the Smithsonian National Air and Space Museum, a temple of aerospace lore where tradition has called for bland captions on horrifying instruments of war. . . . However, a recent visit revealed that on one exhibit a striking change has been quietly installed. It's the exhibit of the terror weapon Hitler unleashed in the finale to World War II, the V-2 rocket, which killed thousands of civilians in France, Belgium, and Britain.[14]

Another exhibition that spring was spearheaded by Gregg Herken, chairman of the Department of Space History, and Cathleen Lewis, curator of Soviet spacecraft. The Intermediate-Range Nuclear Forces (INF) Treaty had specified that each side could retain, for static display, fifteen of the nuclear missiles that had for some years faced each other across the European continent. The treaty did not specify whether each side could retain only rockets from its own arsenal or whether an exchange might be possible. We decided it might be nice to try for such an exchange, and began working with the Department of Defense, the Department of State, and the Embassy of the Soviet Union. We were successful in petitioning Secretary of Defense Carlucci; Chairman of the Joint Chiefs of Staff Admiral Crowe; Soviet ambassador Dubinin; and on the occasions of their visits to Washington, Soviet defense minister General Yazov and chief military advisor to President Gorbachev, Marshal Akhromeyev.

Eventually, after talking for two years with anyone willing to listen, they all said yes. The U.S. Army provided us with two Pershing-II missiles, one of which the U.S. Air Force flew to Moscow, to return with a Soviet SS-20. Soon both missiles were standing tall, next to each other in the Museum's "Milestones" gallery. Present at the opening of this exhibit were the new Chairman of the Joint Chiefs of Staff General Colin Powell, arms negotiator Paul Nitze, who had been involved in the treaty negotiations, and the new Soviet ambassador Bessmertnykh, who previously had also been an arms negotiator.

Again, the public was pleased. I had been somewhat worried about visitor response, since the Cold War was only just drawing to a close, but nobody seemed upset at bringing a Soviet missile into the national museum. An August 1990 *New Yorker* article, musing on the dramatic pace of disarmament at the end of the 1980s noted,

> In December, 1987, during Mikhail Gorbachev's first visit to Washington, the American flag and the red flag of the Soviet Union flapped side by side across the portico of the Old Executive Office Building, just next door to the White House. The casual passerby on Pennsylvania Avenue could look up and wonder what a time traveler from the fifties would feel if he saw those two flags there. *The Russians are coming! They're here! They won!* Yet a time traveler from the winter of 1987 would be almost as startled today if he entered the Smithsonian Institution's National Air and Space Museum, a few blocks from the Capitol, for there, side by side, are an American Pershing II and a Soviet SS-20 missile.[15]

We shot a video film of the restoration of the *Enola Gay* specifically to inform the public. Produced by the museum's filmmaker, Patti Woodside, it included footage of the atomic blasts and the devastation and injury on the ground. For more than four years, starting in the fall of 1990, this five-minute film clip played continuously at the entrance of the World War II gallery, from morning to night, 364 days a year—not including Christmas Day, when the museum is closed. With so many millions visiting the museum each year, and with those most interested in World War II drawn to this gallery, it is remarkable that neither the museum nor the Institution ever got a single letter or phone call, either positive or negative, regarding this clip. Since this film displayed many of the same features that we planned to exhibit in the gallery on the *Enola Gay*, we saw this as a good omen.

When the American offensive of the Gulf War started in the winter of 1991, we thought we should try to see what we could do on very short notice to inform the public about the new forms of air power now coming to the fore. These included "stealth fighters" that were largely invisible to radar and "smart bombs" that could be guided to a target with almost incredible precision. We were fortunate to have Dr. Richard Hallion visiting the museum as a Lindbergh fellow that year. This project was of direct interest to his line of research.

Years earlier, Hallion had been a curator at the museum. It was natural for him, therefore, to take a lead on this project. We decided we would not be able to cover all aspects of the air war, and instead concentrated on demonstrating the amazing improvement in targeting accuracy that this war had introduced. Bombs and missiles could now land within about twenty feet of their designated impact point, whereas the best figures for World War II showed that even in daylight, bombs tended to scatter in a three-thousand-foot radius around their intended target.

This was an important development, because it might enable warring nations to destroy each others' military facilities without having to attack civilian populations. The area bombing practices of World War II, culminating in the use of the atomic bombs, were largely a consequence of an inability to hit designated military or industrial targets precisely. Now that great accuracies were available, I wrote in an article in *Air & Space* magazine that spring,

> Military analysts will need to think through all the ramifications of the Gulf war before deciding on a new nuclear strategy. But if that strategy were to involve a dramatic reduction of the world's nuclear arsenals, all of us would breathe a little easier for our children and grandchildren.
>
> Wars inevitably claim innocent victims. If the lessons learned from the Gulf war could enable us to lower the nuclear threat that has hung over the world now for nearly half a century, we would at least have gained something from all these "smart" bombs besides just one more set of sophisticated weapons.[16]

The small Gulf War exhibit, placed in the museum's central "Milestones" gallery, was timely and aroused great popular interest. Behind the scenes, however, it had also raised unfortunate tensions. Hallion had wanted the exhibit to make a number of claims for Air Force capabilities that none of us at the museum were able to verify. Working for the Air Force, but on leave at the time, Hallion may have had inside information not accessible to us; but ultimately I had to decide to stick to just the information that we were certain was correct. We owed that to the public. Hallion was quite unhappy, and in later years, recalled this incident to a number of us on different occasions.

Finally, in 1991, curators of aeronautics Dominic Pisano, Thomas Dietz, and Joanne Gernstein opened a new gallery on World War I. They commented on the carnage of the air campaign, which saw life expectancies for newly arriving pilots as short as three weeks. They also emphasized the rapid industrial expansion that had produced 219,000 airplanes, worldwide, in only four short years, beginning in 1914, just eleven years after the Wright brothers' first precarious flight. Most of those aircraft were destroyed almost at once, usually with a young man inside. The gallery also showed a film clip on the birth of strategic bombing in World War I and its consequences for the remainder of the century, including World War II, the atomic bombings, and finally the Gulf War. Newspapers nationwide, among them the *Washington Post* and the *Chicago Tribune,* gave us strong reviews.[17,18] We were pleased.

In its way, this gallery also was a test of public opinion that could guide us in mounting an exhibition on the *Enola Gay.*

CORDIAL TIES TO THE AIR FORCE

The success of the strategic bombing series of 1989–90 and other public offerings we undertook at the time owed a great deal to then Air Force historian Richard H. Kohn. When the museum had first assembled a list of potential participants in the series, Kohn approached me to let me know that the Air Force was upset. Many leading experts of air power familiar with evolving Air Force doctrine had not been invited.

I immediately asked Kohn to sit down with us to propose names we might add to our list. As a result we were able to arrange for a far more informative and interesting, if sometimes contentious, series that included a broader range of views.

Many in the Air Force apparently felt that our original list of invitees reflected a bias of a museum staff with a narrow political agenda. I do not believe that there was any deliberate bias or attempt at exclusion. A deeper problem was at play and would continue to plague us.

I have already mentioned the existence in the United States of two communities of military historians working with little awareness of each others' contributions. In organizing a symposium or an exhibition, the museum's academically trained curators tended to invite contributors, advisors, and critics who came from the same academic community to which they themselves belonged. We were rather dependent on the Air Force historian and a small number of other scholars we knew in the services to suggest experts we needed to invite to present the military community's views.

Dick Kohn often advised us on such matters. I was sorry when after many years of service to the Air Force he decided to return to academic life, as professor of history and Director of the Curriculum in Peace, War and Defense at the University of North Carolina at Chapel Hill. After Kohn's departure, the museum was never able to reestablish the same high level of rapport with the Air Force that we had enjoyed before. Scholars attached to the military began to feel deliberately excluded. Relations eroded just when they were greatly needed to help the museum on the most difficult exhibition it had ever undertaken—the display of the *Enola Gay.*

OTHER EXHIBITIONS AND FILMS

I do not want to leave the impression that the museum was engaged solely in exhibitions on the military aspects of aviation and spaceflight. These were only a small part of our offerings.

One of the first projects we undertook after my arrival at the museum was a wide-screen "IMAX" film showing our planet viewed from space. This award-winning film, *Blue Planet,* released in November 1990, soon was playing in more than one hundred museum theaters worldwide. Within a few years, more than fifteen million people had seen it. It showed a variety of natural forces that shape the earth—ice ages, volcanic eruptions, hurricanes, and asteroid impacts—and ended by showing that human beings by now also have the ability to change the nature of our planet.

In late February 1992, we opened the most-visited exhibition ever to be displayed in Smithsonian history—at least until my departure. *Star Trek* dealt with the popular television series and showed how this series, which had first aired in the late 1960s, had looked at then-current issues of war and peace, gender, race, and democracy and dictatorship. By placing the episodes in the twenty-fifth century, the program had been able to discuss these issues openly.

Star Trek, like so many other science fiction stories, had encouraged many children to take up careers in space exploration. We felt that by mounting this exhibition we might achieve a similar aim. By the time the exhibit closed, after eleven months, nearly nine hundred thousand visitors had passed through it.

In December 1992, for the five hundredth anniversary of the Columbian voyages, we opened a futuristic gallery in which, working with some of the finest space scientists and engineers in the country, we looked ahead to discern along what lines space exploration might develop in the decades and centuries ahead. We pointed out that the laws of physics will constrain our ambitions for space travel, and that physiological effects may be another important limitation. We explained the different roles that humans and robots can play, and tried to show youngsters what kinds of careers they might expect to find if they chose to become space explorers. I felt the museum had an important educational role to play in going beyond the science fiction that children might read or see in the movies, and to be as precise as possible about our current knowledge of space exploration, while explicitly acknowledging the vast scope of the still unknown.

In winter 1994, we opened a small exhibition on Patty Wagstaff, three-time winner of the national aerobatics championship, open to both men and women. Central to the display was the Extra 260 she had flown in winning the championship three times in a row, from 1991 through 1993. We wanted to encourage young girls to think seriously about flying, and show them there were no goals that women could not attain just as well as men. Flying is accessible to anyone.

In June 1994, we opened another IMAX space film, *Destiny in Space*. It showed spectacular close-up views of Mars and Venus, and made theater visitors feel they were flying over the surfaces of these planets, seeing the landscape with the clarity enjoyed by airline passengers flying thirty thousand feet over, say, Colorado or New Mexico. The raw data for this film had been gathered by unmanned space probes sent by NASA to orbit the two planets. We had asked Jet Propulsion Laboratory, in Pasadena, California, to process the data with a computer to give the impression of flying over the planets' surfaces. Not even leading space scientists had seen planetary surfaces with such clarity before we made this film available to the public. It, too, became popular and was seen by millions of theatergoers in museums worldwide.

By May 1995, two major projects were approaching completion: *How Things Fly* is a major gallery explaining the principles of aviation and space flight in understandable terms—what makes a balloon rise, keeps an airplane from crashing, makes a rocket go, and keeps satellites from tumbling down to earth. A new IMAX film, tentatively titled *Cosmic Voyage* would explain how nature, from subatomic particles to the largest structures in the universe, is constituted, and how everything we find in nature originated in an initial explosion, followed by gradual evolution to the state of the universe we now perceive. Finally, we also were making plans for an exhibition that would display and contrast spacecraft built in the United States and in the former Soviet Union. It would distinguish different technical approaches, and show youngsters that this is a highly competitive field, where America cannot afford to lag.

In between we had an exhibition celebrating the life of Igor Sikorski, one of the great inventors of our time, who had made helicopters a reality, as well as many art shows on aviation and spaceflight, smaller exhibits on environmental matters and planetary flyby missions, an exhibition on the Hubble space telescope, and lots of other programs that were just plain fun.

8

The Impatient Veteran

A LETTER OF INQUIRY

In the spring of 1988, Robert C. Mikesh, senior curator of aeronautics at the museum, received a long and complicated inquiry from a World War II veteran, Ben Nicks, of Shawnee, Kansas, who said he was inquiring on behalf of the 9th Bomb Group Association, consisting of members of a 313th Wing, B-29 combat unit that had been based on Tinian. In the late stages of World War II, Nicks had been flying regular bombing and mine-laying B-29 missions to Japan. While he had not been associated with the 509th, he assured Mikesh that

> our Quonsets were within a half mile of the 509th's and we flew off adjacent runways.... Thus we have a rather deep and personal interest....
>
> Since our most recent convention at Tucson, members of the 9th BG have voiced growing concern about the plans of the Smithsonian with regard to display of the *Enola Gay* as an important American artifact. We understand that the project is, and has been for many years, "on hold." The 9th BG will meet again in October at Wright-Patterson. Can you advise me in enough time that I can make a formal report to the 9th BG as follows:

He then posed a list of seven questions, one of which consisted of six separate queries. Three of these questions were particularly significant.

Question 4 began to reveal an effort, like so many others, to have the *Enola Gay* transferred to another museum:

> 4. If the Smithsonian does not have funds to restore and display this bit of Americana, would it consider giving it to an organization or museum that would do so?
>
> (Example: Some funding has already been provided by the Jackson county, Missouri, legislature to explore the possibility of building an "atomic" museum next to the Truman Library and grave of the man who ordered the bomb to be dropped. An offer by the Smithsonian to turn over the *Enola Gay* might be just the impetus needed to put the project over the top and release the Smithsonian from a burdensome financial and production responsibility. A well-placed source confidentially advises that a museum in the State of Washington ... has indicated it might also find room and funding to accept it.) What is the Smithsonian's position on releasing the *Enola Gay* to a responsible entity?

As part of question 5, Nicks offered to have members of the 20th Air Force Association help with lobbying Congress to provide "authority to proceed with restoration and display." Could the Smithsonian advise?

Then came the most specific and detailed of his questions:

> 7. Just what is the specific financial and budget status of the Smithsonian with regard to the *Enola Gay* restoration? For instance, can you provide me with (or direct me where to obtain) a report covering:
>   - A. The number of man hours devoted to the *Enola Gay*, year by year, since the aircraft came into your possession?
>   - B. The labor dollars, year by year, for the same period?
>   - C. The material dollars, year by year, for the same period?
>   - D. Administrative and overhead dollars allocated to the *Enola Gay* project, year by year, for the same period?
>   - E. The *total* Smithsonian Air & Space expenses, year by year, for the same period?
>   - F. Budgeted total Air & Space expenses for the coming fiscal year, and amount budgeted for the *Enola Gay*?
>
> I could come to Washington and discuss this with you personally, if that would assist us in obtaining the information we seek. . . .[1]

Bob Mikesh responded to this, but evidently not to Nicks's satisfaction. Nicks wrote again on June 9. This time Mikesh, obviously irritated by Nicks's persistence, did answer more fully with a letter dated June 24.

> I will do my best to elaborate upon the questions that you do not feel were answered sufficiently in my first letter. As I recall, there were approximately 10 extensive and detailed questions, many of which were so involved in cost and time accounting that we do not have sufficient administrative staff to satisfy all your requests for this information.

> In hopes of informing your audience of former 9th Bomb Group members in October about this restoration, as of now, the nose section, extending through the bomb bay, should be completed by August. This includes a complete reworking of the cockpit with replacement of many of the special components that were removed following the historic bomb drop. . . .
>
> Exhibit space is a problem, and we do have a plan for the exhibit of this aircraft. That will be our Museum Extension, which is so necessary for not only the *Enola Gay*, but for many aircraft in our collection. Other persons are equally interested in seeing their favorite aircraft properly exhibited. This is particularly true with aircraft of World War II, where the wait to exhibit these has been so long. Rather than disperse aircraft like the *Enola Gay* to locations that either are not prepared to handle an aircraft of this size, or have expressed an unwillingness to do so, the best solution is to have the Museum Extension, which is in the planning stage, at a nearby airport, in order to handle future acquisitions. That is where the effort should be placed; seeing to it that a facility in which to exhibit this collection is provided.
>
> . . . Your question number 7 contains six sub paragraphs pertaining to labor costs, material dollars spent, year by year, etc., none of which I am able to answer. Since you persist in these time-consuming details, I have elected to forward your questions to higher channels within the Smithsonian for their decision and possible action. . . .
>
> Another member of our staff is preparing details whereby an outside organization called the "*Enola Gay* Restoration Committee" may be able to assist with questions and points such as you have raised. This may be in such forms as volunteer labor assistance, public relations support, and other constructive efforts. Once this is fully organized and functioning, you and your organization may find it appropriate to add your support with theirs.[2]

This apparently did not satisfy Nicks. On September 26, he wrote to Jay Spenser at the Museum of Flight in Seattle, Washington, and sent me a courtesy copy. Spenser had been a curator at the National Air and Space Museum until 1986 and had then joined the staff of the Museum of Flight. This evidently was Nicks's second letter to him. It indicated that Nicks by that time had joined Frank Stewart's *Enola Gay* Restoration Committee, which was trying to help the museum with accelerating the restoration of the *Enola Gay*, and also that he had been turned down on his first inquiry to Spenser:

> Thanks very much for your reply to my inquiry about the interest of your museum in the *Enola Gay* B-29. I and other members of the Indianapolis *Enola Gay* Restoration Committee are not nearly as optimistic as you are about the sincerity of the Smithsonian's intentions and interest in a quick restoration of the *Enola Gay*. It is true that the Smithsonian has said that it fully intends to restore and display the *Enola Gay* and has so said constantly and continuously—since 1949.

> Frankly, our committee is suspicious that the Smithsonian is more interested in sending a political message about strategic defense and nuclear weapons. Our position is that—like it or not—the *Enola Gay* did make history. Like it or not, the atom bomb was dropped. American citizens have as much a right to see the artifact involved as they have to see those other great artifacts of history, the Kitty Hawk Flyer (also controversial, as the Smithsonian well knows) and the Spirit of St. Louis. And let the morality of strategic nuclear defense be determined in the pulpit and the Halls of Congress. In the past 40 years, if the spirit were willing, the Smithsonian could have made financial provision for display of the *Enola Gay*. It spent plenty of other money in restoring and displaying other aircraft far less worthy. The conclusion: the Smithsonian is more interested in re-writing history and politics; has no serious intentions of displaying the *Enola Gay* in our lifetime; and will promise anything to stall critics demanding quicker action. Right now it tells the Indianapolis committee 1992—maybe. By then many of us World War II veterans who flew the B-29 in combat will be well into our 70s, dead and (hopefully if the Smithsonian has its way) quiet.
>
> I enclose copies of editorials which recently appeared in the *Smithsonian* Magazine and in *Air & Space* Magazine published by the Smithsonian. The anti-nuclear, anti-strategic defense bias of Mr. Adams and Mr. Harwit is painfully apparent.... Their concluding remarks deal not with history, the eminent domain of any museum—but instead, pure politics.
>
> I say this in way of explaining our feelings about the sad treatment the *Enola Gay* has received at the hands of its Congressionally designated caretaker. Should, through some political action initiated by our committee or by some other group, the *Enola Gay* be removed from the Smithsonian, perhaps your museum would reconsider its position and seek it for display. Shall we both keep our options open? Should you have occasion to see John Swihart at Boeing, give him my regards and tell him I expect to see him at our meeting in Dayton.[3]

On October 6, Lin Ezell responded to this letter on my behalf, and also gave answers to some of the questions Nicks had earlier asked—questions that Bob Mikesh had not been able to answer. She explained how the museum's accounting of cost and man-hours is done and invited Nicks to visit the Garber facility any time he could. For good measure, she sent a copy of her letter to Frank Stewart, since Nicks's letter indicated that he had by now joined the *Enola Gay* Restoration Association (EGRA).[4]

WORKING WITH DIFFERENT VETERANS GROUPS

Since Nicks's letter had attacked both Bob Adams's magazine article and mine, I conferred with Adams through memoranda we exchanged on October 5 and 6,[5] and then responded on behalf of both of us on October 14.[6] I told Nicks of the museum's contacts with the *Enola Gay* Restoration Committee and the visit by Frank Stewart, Rolland Nail, and General Paul Tibbets just four months earlier:

> Our visitors from the Committee expressed their approval of our overall restoration plan and our progress on the forward fuselage. We reviewed the condition of all the major components and discussed in detail the work remaining. . . .
>
> The *Enola Gay* is scheduled as one of the key artifacts for display in our first public building at an Extension. . . . Because of the great importance of the *Enola Gay*, we are anxious that it be appropriately exhibited as part of the *national* collection of historically significant aircraft in the nation's capital.

Like Ezell, I invited Nicks to visit us, saying that we'd be glad to show him around and discuss the matter further at that time.

This letter reached Nicks just after he had presented his report to the 9th Bomb Group. I present that report in its entirety because it appears to have influenced many B-29 veterans

THE *ENOLA GAY*
B-29 AIRCRAFT THAT DROPPED WORLD'S FIRST ATOMIC BOMB
HIROSHIMA, JAPAN
AUGUST 6, 1945
REPORT TO THE NINTH BOMB GROUP ASSOCIATION
AND RESOLUTION CALLING FOR ITS SPEEDY RESTORATION AND DISPLAY
ADOPTED AT THE ANNUAL ASSOCIATION REUNION, DAYTON, OHIO,
OCTOBER 13–16, 1988.
PRESENTED BY BENJAMIN A NICKS

> World War II came to a dramatic close in August of 1945 with the dropping of the Atomic bomb on Hiroshima and Nagasaki. The B-29 *Enola Gay* was the instrument that delivered the first of these weapons that brought to a sudden halt the continuing slaughter of millions of soldiers and civilians, giving the Japanese emperor the authority and excuse to move his government to end a war already lost.
>
> Today some view the *Enola Gay* as a monstrous symbol, embodiment of nuclear holocaust, visible evidence of brutal, unwarranted and immoral attacks by ruthless predators on a helpless populace. But in the hearts of Twentieth Air Force crews who carried the burden of aerial war, the *Enola Gay* is enshrined as a symbol of peace—not war. It is a reminder that many would not have returned to families and loved ones, were it not for the *Enola Gay*. It reminds our countrymen that, through strength, worldwide peace between the great powers has been maintained despite sad local lapses throughout the years.
>
> The *Enola Gay* should be displayed to citizens of America and the world, not as a symbol but as an artifact of history as it was—not as some would have it. Unfortunately its caretaker, the Smithsonian Institution, our national museum in Washington D.C., has not found it convenient to do so.

The Smithsonian has promised that it would restore and display the *Enola Gay* as soon as manpower and space permit, and has so maintained consistently and continuously—since 1949, the year it obtained the aircraft.

Since that time it has built a magnificent display hall on the Mall in the heart of Washington. It has acquired a staff of 6000 employees. [This represents the entire Smithsonian's staffing; the National Air and Space Museum's staff numbers only three hundred.] It has actively pursued and displayed hundreds of other aircraft of no similar historic significance, one of the latest in the summer of 1988 being a German Ju-52—"the pride of the Lufthansa fleet" and a workhorse of the Luftwaffe, according to the Smithsonian's own *Air & Space* Magazine. Ironically, *Air & Space* has never had a feature article on the *Enola Gay*, historically the world's third most important aircraft after the Kitty Hawk *Flyer* and the *Spirit of St. Louis*.

Documents attached to this report show that the Smithsonian has been devoting less than two equivalent headcount to the restoration project. This restoration is taking place at the Smithsonian's Paul E. Garber facility in Silver Hill, Maryland, some miles outside Washington. The public is welcome to view the *Enola Gay*—if they can find the facility, and if they care to travel away from the well-known tourist attractions in the heart of Washington. And, oh yes, if they make an appointment two weeks in advance! The Garber's public information brochure mentions various attractions that may be seen there—the *Spad*, the *Curtiss Jenny*, the Hawker *Hurricane*, the Able-Baker Missile Nose Cone and "much, much more!" Not a word about the *Enola Gay*! . . .

Many Twentieth Air Force veterans grow increasingly concerned that they will never see the *Enola Gay* properly displayed in their lifetime, that without this visible evidence of their valorous wartime service, America will find it easier to forget those days long ago when these gallant few bore the heat of battle.

An increasing chorus of complaint against the dilatory tactics of the Smithsonian has resulted in the coalescence of formal protest groups. Behind the scene museums across the country are cautiously exploring the possibility of obtaining the aircraft should the Smithsonian be divested of the *Enola Gay* by Congressional action.

This tide of protests has brought forth the usual regimented response from the Smithsonian. Again it sounds the pledge that the *Enola Gay* will be displayed "soon." Vigilant observance and action by our association in cooperation with other interested veterans' groups in time may achieve the appropriate public display of the *Enola Gay*. Therefore, I respectfully recommend adoption of the resolution presented together with this report.

Benjamin A. Nicks
9th Bomb Group

The attached resolution reads:

Whereas history acknowledges that the *Enola Gay*, the B-29 aircraft that dropped the first atomic bomb on Japan in 1945, speedily bringing to con-

> clusion the most costly and devastating war in human history, thereby saving the lives of millions of American and Japanese soldiers and civilians, and
>
> Whereas this historic airplane has been virtually secluded from view since 1949 .
>
> Therefore be it resolved: That the 9th Bomb Group Association supports in principle that the *Enola Gay* be speedily restored to insure that this historic aircraft will not fade into obscurity, but shall be publicly and proudly displayed as a symbol and a reminder for present and future generations that eternal vigilance is the price of freedom, and that securing of peace in this United States ofttimes demands the dedicated commitment of its citizen soldiers and the unfortunate awesome sacrifice of human lives, and
>
> Be it further resolved: That directors of the 9th Bomb Group Association be empowered to appoint a committee to monitor progress of restoration and display of the *Enola Gay* by the Smithsonian Institution, with authority to cooperate with other parties similarly interested in promoting the public display of the *Enola Gay*, and to make regular and periodic reports to the Association board of directors on the status of the restoration program, together with recommendations for further Association action.[7]

This resolution had been "respectfully submitted" by Benjamin A. Nicks, 9th Bomb Group Association.

Nicks sent me a copy of this resolution on November 3, noting that it had been adopted by "more than 400 members attending and voting at the recent annual reunion of the 9th Bomb Group Association (a B-29 unit in the 313th Wing, Tinian) held this year at Wright Patterson, in Dayton." He added,

> Following [the resolution's] adoption I was named chairman of the committee by the Board of Directors. . . . In addition to the activity cited, we are in communication with and soliciting support from the 20th Air Force Association and with other B-29 veterans' associations. The 504th Bomb Group Association has already agreed to consider adoption of a supporting resolution and, hopefully, establishment of a committee to interest itself in the restoration project. . . .
>
> I have been advised by the *Enola Gay* Restoration Committee, Indianapolis, that NASM and EGRC have reached agreement in principle that restoration shall proceed at an accelerated pace, and that after restoration if NASM does not have satisfactory display space, it might consider lending it to another facility for display until it does obtain suitable space.
>
> Obviously, therefore, our interests are mutual, Dr. Harwit. We B-29 veterans and you as director of our national museum both want the *Enola Gay* restored and on public display somewhere as rapidly as possible. Perhaps you can use this resolution of support to bolster your efforts to obtain necessary backing in your restoration efforts.[8]

Referring to the Lockheed Constellation, "Connie"—a popular postwar aircraft for long-distance passenger service—Nicks wrote that the "Save a Connie" association (Kansas City) had restored the same kind of engines found on the *Enola Gay*—Wright R3350 engines—and that their members, "active or retired TWA A&E mechanics" might be able to help us with engine restoration if we contacted them.

Two days later, on November 5, Ben Nicks issued an "*Enola Gay* Restoration Committee Report" on behalf of the 9th Bomb Group. He listed the members of the committee, gave their addresses and phone numbers, and assigned them their duties: "These committeemen are asked to monitor local media for references to the *Enola Gay*. . . . Also, should occasion present itself, they are asked to make known the views of the 9th BG concerning display of the *Enola Gay*, such as letters to the editor in the local press or stats in public meetings."

With evident satisfaction, Nicks also noted that General Paul Tibbets had addressed the group's reunion banquet on this occasion. Two of the members of his committee are identified as Bud Carey and Bill Feldman, and Nicks writes,

> Both Bud Carey and Bill Feldman mentioned "a potential museum" in Utah which would be interested in displaying the *Enola Gay*, should the Smithsonian ever voluntarily decide, or through Congressional action be impelled, to release ownership of the aircraft. . . . Incidentally, a museum in Seattle, when approached with an inquiry about its interest in obtaining the *Enola Gay*, understandably replied that it did not seek to obtain an artifact owned by another museum. However, since agitation is rising to seek release of the *Enola Gay* from the Smithsonian should its display program continue to lag, and should the *Enola Gay* "come on the market," it is possible that this Seattle museum might review its decision.[9]

Finally the report mentions that "the Indianapolis *Enola Gay* Restoration Committee ... is pursuing parallel paths in its campaign to get the *Enola Gay* on display."

A month later, Ben Nicks again wrote to offer help in the engine restoration for the B-29 by the "Save a Connie Association" and asked why I had not answered his letter to that effect, adding his advice on why we should not spend so much effort on the engines.[10] I asked Lin Ezell to answer this December 3, 1988 letter, which she did on December 9, telling Nicks that the San Diego Aerospace Museum would restore one of the engines for us at no charge.[11] Eventually the San Diego museum actually restored two of the four engines for us, demonstrating extraordinary workmanship.

Ben Nicks's approach to us was to continue along these lines for the next few years. Sometimes wishing to be helpful, sometimes implying threats, he tried to push the restoration schedule forward. He and most of the veterans' groups that sometimes independently, and sometimes in concert with others, were pursuing similar aims, seemed to labor under four misconceptions.

MISCONCEPTIONS

The veterans believed that the Smithsonian had been delinquent in caring for the *Enola Gay*. Protestations that the aircraft had been damaged while at Andrews Air Force Base, where enforcement of security was in Air Force hands, fell on deaf ears. The veterans never bothered to explain to us how the Smithsonian could have cared better for the aircraft without disassembling it, as the Institution had in 1960 in its final, desperate move to gain control over the continuing deterioration. Many seemed to assume that the Institution was wealthy beyond dreams and could have solved the problem had it been willing to throw money at it.

The veterans believed that restoration, decades after an aircraft had been decommissioned, was similar to maintenance applied to new aircraft in the field. The men on Tinian had been used to work wonders, fixing up damaged aircraft overnight. But their job often was simply to replace damaged parts with new, factory-fresh components. They did not have to disassemble them, arduously remove corroded components, and sometimes fabricate replacements, preserve deteriorating wiring, and do all that often without some of the special tools and jigs that had been developed to work on just this one type of aircraft. Forty years later, such equipment could only be located, if at all, through the intricate word-of-mouth grapevine that links professional and lay restorers all over the world, or by means of advertisements placed in specialized magazines. Many of the jigs no longer existed or could not be located and had to be fabricated in our shops before disassembly and reassembly could continue.

The veterans believed that we should not waste too much time on perfection. "Just fix her up to look good and display her so we can see her once more before we all pass away" seemed to be the repeated refrain. For us, as for other museums, the point of restoring a valuable object is to preserve it for future generations. In contrast to many World War II enthusiasts and airplane owners, the museum would not "fake" parts. We could not put a brand new engine into an old aircraft, just to make it fly, because one of the purposes of the museum is to show future historians the state of twentieth-century technology at the time any given aircraft was built. Such a historian of the twenty-fifth century might wish to see in exactly which year a given aluminum-titanium alloy came into use. If we replaced a 1945 part on the *Enola Gay* with a 1985 part fabricated from a different, far more advanced alloy, this type of historical investigation would become confused. The museum does, when absolutely necessary, replace irreparable parts, but always attaches a tag to the new component, identifying it as a replacement. The ordinary airplane enthusiast doesn't bother himself with such matters, and will throw into his restoration whatever is handy and looks good. That makes the work go much faster. These same enthusiasts then ask how the Air and Space Museum can be so slow in its restorations. Surely, they claim, these must be signs of foot-dragging or incompetence.

The veterans believed that once restoration was completed, the *Enola Gay* could be carted around the country and assembled or disassembled at will. In fact, the mating of the forward and rear fuselage and the attachment of the enormously long and heavy wings was a terribly difficult task without the specialized jigs that the World War II aircraft factories had developed to line up these massive parts. It had been done on one existing B-29 before, and the museum also expected to assemble the *Enola Gay* in this way, at the Dulles extension site; but the process was difficult and likely to involve enormous strains on the aircraft structure, even under careful assembly conditions without those now-obsolete jigs. This was not a process to be undertaken lightly. Moreover, the complete assembly would require the dedicated labors of a good fraction of the museum's restoration staff for several months. The museum could not afford to go through this costly process more than once.

Given all these misconceptions, we spent years soliciting the support of veterans' groups that were going to make it their business to seek speedy restoration of the *Enola Gay*, while also fighting their attempts to impose their own ideas on how the process should go forward and where the aircraft should be displayed. Our main difficulty throughout was that few of the veterans were willing to listen, and those few who were, tended to be shouted down by those who would not. We would see more of that in later years, as differences arose within and between different veterans' groups:

On such occasions, the more militant factions invariably prevailed.

9

An Enthusiastic Advocate

I can't pinpoint the day in the summer of 1988 on which Luc Smith first strolled into my office with Bob Faron. For the next two years we would be dealing extensively with both men.

Lucius Smith III, silver haired and seventy, was the quintessential fast-talking supersalesman/lobbyist. He spoke eloquently of his connections to Senator Pete Domenici of New Mexico. Domenici's leadership had helped the Senate pass legislation to benefit veterans who had been exposed to nuclear radiation in the armed services and later developed cancer.

As adjutant general of the National Association of Atomic Veterans (NAAV), Smith had also worked with Congressman G.V. "Sonny" Montgomery, Chairman of the House Committee on Veterans Affairs. By late May the legislation had passed both houses, had been signed by the president, and had become public law 100-321.

On June 28, Sen. Frank H. Murkowski of Alaska wrote Smith, "This legislation and its enactment would not have been possible without your support, perseverance, and faith. You deserve the heartfelt thanks of all atomic veterans, their families and survivors."[1]

Luc Smith was quite evidently elated by this success. The July 1988 issue of *World War II Times* carried a picture of him presenting a certificate of appreciation from the NAAV to Sonny Montgomery. With them in that picture was Robert S. Faron, secretary-treasurer and counsel to the Atomic

Library and Technology Foundation of Independence, Missouri, whose letterhead also listed Smith as cochairman of their steering committee on development.[1]

Of particular pleasure to Smith was that the first recipient of increased benefits would be Eleanor Shumard. Mrs. Shumard was the widow of former Master Sergeant Robert Shumard, assistant aircraft flight engineer on the *Enola Gay*'s mission to Hiroshima. He had died twenty-one years earlier in 1967, at age forty-six.[2]

Smith talked of an obvious tie between the National Air and Space Museum and the Atomic Library and Technology Foundation. He wanted the museum to restore the *Enola Gay* and then transfer her to the Foundation. To him this was an obvious move, since it would rightfully bring the *Enola Gay* to the home of former President Harry S. Truman, the man who had sent the *Enola Gay* on her mission, thus ending World War II. It was obvious too, because the *Enola Gay* marked the opening of the atomic age, and that was what the foundation was all about. A letter to Steven Soter made his intentions clear and offered the foundation's help to the museum along mutually agreeable lines.

To dispel any ideas Smith might have on transferring the *Enola Gay* to the foundation, but to also encourage him to work with us, I asked Soter to respond forthrightly. On August 19 he wrote to Smith,

> In order to prevent any misunderstandings, allow me to restate our position. As Dr. Harwit said during your visit, the Museum has every intention of displaying the *Enola Gay* in the Washington area as soon as it is fully restored. Only if this proves impractical would we consider a long-term loan to another organization. We are not in a position to offer any organization a guarantee of the right of first refusal to display or acquire the *Enola Gay*. . . . The Museum would of course welcome any help that you could provide towards accelerating the pace of restoration. The most important single need is for funding to hire additional technicians to conduct the arduous . . . work. . . . We also welcome your idea of forming a liaison group between the Atomic Library and Technology Foundation and the *Enola Gay* Restoration Association (and perhaps others who share an interest in expediting the restoration). . . .[3]

Soter added that he had enjoyed meeting with Smith and Faron, and sent a copy of his letter to Frank Stewart at the *Enola Gay* Restoration Association.

The following week, Bob Faron wrote to Elbert Watson at the *World War II Round Table* in Indianapolis on behalf of the Atomic Library and Technology Foundation and its board of directors. I had never heard of the Round Table, but Watson seemed to be ubiquitous. He was publisher of the *World War II Times*, a member of the *Enola Gay* Restoration Association, and a frequent correspondent with the Smithsonian. Faron's letter read,

> We welcome the action of the Board of the World War II Round Table taken August 15 and its adoption of a resolution supporting "The Placement of the *Enola Gay* at a site near the Truman Library in Independence, Missouri, under the auspices of the Atomic Library and Technology Foundation."
>
> We look forward to participating in a working relationship with the World War II Round Table and the Smithsonian Institution to expedite the restoration of the *Enola Gay* and her assignment to Independence at the earliest possible date. . . .[4]

To make sure we got the point, Faron employed a fairly common Washington tactic, and sent a copy to Senator Murkowski.

To answer Steven Soter, Smith wrote on August 30,

> Bob Faron and I fully understand Dr. Harwit's position at this time respecting the exhibition of the restored aircraft. We will, however, have more to say on the subject, and we look forward to meeting with you again.

His letter further stated that their plans for the Atomic Library and Technology Museum, which the foundation was trying to build, were

> progressing very well. We have received overt expressions of cooperation from the National Archives and Record Administration and the American Nuclear Society. We plan to develop our exhibits on reactor technology with the help of the Missouri University Research Reactor Facility at Columbia, Missouri. . . .[5]

It was now becoming abundantly clear that several different organizations, individually or in concert, were trying to remove the *Enola Gay* from the museum's control and transfer her elsewhere. To fully understand what pressures they might realistically apply and what defenses we would have, I wrote to the Smithsonian legal counsel to seek advice. A copy went to Margaret Gaynor, the Smithsonian's director of government relations, who often represented the Institution on Capitol Hill.[6]

Toward the end of September, Luc Smith wrote again, this time to state that the foundation had entered "a close working relationship with the *Enola Gay* Restoration Association. It is our desire to develop a joint effort that can facilitate the restoration program you began for that plane."[7] He attached letters of support from Sen. John C. Danforth of Missouri and Margaret N. Maxey, director of the College of Engineering, University of Texas at Austin.

Then, on October 1, 1988, he circulated a memorandum of intent to work with the *Enola Gay* Restoration Association (EGRA) to help raise $350,625 to support the restoration of the *Enola Gay* and produce a short documentary film on the process involved. This figure included $95,625 in overhead. The purpose of the foundation and EGRA would be "to fund the cost of an accelerated program to restore . . . the *Enola Gay* and to counsel with NASM and the Smithsonian [regents] in respect to the aircraft's public display, security, and perpetual care."[8]

On November 21, Luc Smith wrote me that Frank Stewart, two members of his board of the *Enola Gay* Restoration Association, and Ben Nicks, of Shawnee, Kansas, had met in Independence on November 18 and 19 with Bob Faron, general counsel of the Atomic Library and Technology Foundation, two other colleagues, and himself, and named "an ad hoc steering committee empowered by resolutions of the National Association of Atomic Veterans [NAAV] and EGRA to mount a fund-raising campaign to provide supplementary funds to be used by NASM to restore the *Enola Gay*." The NAAV and EGRA would each have two representatives. Ben Nicks would be acting secretary, and Smith himself would be chairman pro tem. He wrote, "I expect to step aside when a national chairman can be recruited." The campaign, he wrote, had been named the "United *Enola Gay* / War and Peace Restoration Fund."[9]

Ben Nicks, whom none of us at the museum had even met, had wasted no time. His first letter to Bob Mikesh had only arrived on April 8, and here, by November 21, he had not only managed to get himself appointed spokesperson for the 9th Bomb Group, he had also joined the Indianapolis-based *Enola Gay* Restoration Association and now, hardly seven months later, had been elected an officer of the joint ALTF-EGRA effort to help restore the *Enola Gay*.

At this stage Smith was firing letters at us every couple of weeks. I have not cited them all, but I was troubled by the tone of enthusiasm and militant stridency that permeated many of his letters, and wrote to him on November 22 to express my concerns.[10]

Smith responded on December 13, in a tone evidently meant to reassure:

> We associate ourselves entirely with your concern to display the *Enola Gay* in an evenhanded way—as an instrument of an imperishable moment in history to which the viewer will assign his or her own judgement. . . . On the other hand, an appeal for funds to restore this particular craft must fly through clouds of emotion from which we cannot hope to escape totally unscathed. Nor should we. Veterans of V-J Day may today, with hindsight, divide several ways over the ambiguities, but we are counting on sufficient numbers who feel good about it. They, after all, are the ones who will send checks, not letters to the Editor. To be effective, appeal "copy" will have to be targeted to the "pro-bomber" market. . . . But we hasten to add, not blatantly. Here, the "right touch" will be as important as was, to the 509th, the "right stuff." I'm not certain any of us or your people will wholly succeed. Still, we must move forward, and the principal market is WWII veterans.[11]

After five months of negotiations and the year 1988 coming to a close, we still were quite far apart in our approaches.

Early in January 1989, I received a further letter from Smith, who wrote,

> The Atomic Library and Technology Foundation has obtained zoning and use permits to develop approximately 8½ acres for its proposed museum next to the Truman Presidential Library in Independence. . . . ALTF has the support of the National Archives and Records Administration, the City of Independence, Jackson County, the American Nuclear Society, and members of Congress. Its advisers include a former general counsel, the director of the University of Missouri Research Reactor, and the chairman of the school's Nuclear Engineering Program. ALTF has received indications of private funding for its initial $9 million capital requirements; ground-breaking could begin this spring.
>
> You are invited now to consider locating an extension of NASM on land adjacent to the atomic technology center. (If more land is required, it can be obtained at this time.) We submit that a conjunction of an atomic technology education and demonstration center and an air and space center—at a midpoint of the continental United States—is worth study as an alternative to a remote annex, such as Dulles or Baltimore International Airports.[12]

I felt this might be a not-so-subtle suggestion that the help with fund-raising might be tied to relocating the extension the museum was trying to build at Washington's Dulles International Airport, and establishing it instead in Independence, Missouri. Smith went on to give all kinds of reasons for locating the extension at Independence. A long introductory paragraph said that the Mayor of Kansas City, Missouri, was thinking of making their Union Station into a $200 million tourist attraction, and Independence is only twenty minutes away.

I answered this on February 6, attaching a fact sheet on the *Enola Gay* that he had previously requested, but I wrote Smith that his main proposal was unacceptable:

> Regarding your suggestion for locating the NASM Extension in Kansas City, I'm afraid this cannot be done for various reasons, both practical and legal. Among these is that the Regents of the Smithsonian have given their approval to an Extension located in the vicinity of Washington D.C.[13]

I added that we looked forward to receiving the final version of a memorandum of understanding from their attorney, Bob Faron.

Finally, on February 22, 1989, I signed an agreement with the Atomic Library and Technology Foundation, which became fully executed within a week. It called for raising a net amount of $253,000, of which $15,000 would go into making a film about the restoration and its history. The ALTF would be permitted to keep no more than thirty percent of funds raised for administrative costs and spend no more than seven percent on marketing.[14]

Two months later, the first checks began to arrive. On April 21 I wrote to Smith, "On receiving the first check from the restoration fund last week, Lin Ezell authorized the start of work on the documentary short film on the *Enola Gay* project."[15]

On June 19, 1989, Smith sent Lin Ezell an account of the first ninety-nine donations, including two special donations of $2,500 and $1,000. The total was $5,727.55. He attached a two-page analysis addressed both to Lin Ezell and to me on how he had proceeded, what the main difficulties were, and what we must do to be successful. He wanted to approach larger donors to accelerate the process.[16]

This couldn't be easily done. Requests for large donations needed to be coordinated by the central Smithsonian development office, largely to prevent the different museums in the Institution's family from competing against each other or against the secretary's efforts on behalf of the Institution as a whole. Any approach Smith might make to a major donor would therefore have to be coordinated through the museum and the Institution.

Lin Ezell wrote to Smith in early July to explain this difficulty. She also wrote that she was sending an article to the Veterans of Foreign Wars *VFW* magazine that week.[17]

The enterprising Smith then turned to other innovative ideas. Nobody at the Smithsonian could object to his approaching different chapters of veterans' organizations, and one particularly promising one was an American Legion post on Maryland's Eastern Shore. Writing for the post on July 15, Maj. Gen. Andrew H. Anderson (USA Ret.) notified Smith,

> This is a follow-up to our recent conversation concerning the American Legion Post #91's donation to the restoration of the *Enola Gay*. Our members agree this is a most worthy cause. . . .
>
> Our Legion Post located in Cambridge, Maryland, on the Choptank River and off the Chesapeake Bay has more than 1,000 members. We will be holding our general membership banquet on Veterans Day, the 11th of November. The hall can accommodate 500 for a sit down dinner. . . . We would be most honored to have BGen Paul Tibbets, USAF (Ret) as our guest speaker. He certainly can give us some insight into one of the most dramatic events of modern times.
>
> As it turns out this is also the weekend prior to the opening of the waterfowl hunting season. One of the many advantages of living on the eastern shore of Maryland is the great duck and goose hunting available in this area. If General Tibbets agrees we will host him for a four or five day period that will feature a very receptive audience at the banquet and several days of some of the best duck/goose hunting on the east coast.[18]

Smith enclosed this note in a letter to me on July 21, noting that the post was contributing another $1,000 to our fund.[19] He wrote again on September 16 that General Tibbets would be the post's guest on the evening of November 11, 1989, and asked Lin Ezell to send Gen. Tibbets "a suitable report of the restoration stage as of mid-October," and forward it to the general at his Columbus home.[20] How he had persuaded Tibbets to go, I didn't know. Perhaps Frank Stewart had helped set up the contacts. Anyhow, all of us were pleased with the veterans' support.

Given this success, Smith had already seen another way to go forward if the Smithsonian would not let him approach major donors who could contribute $10,000 or more. On August 25, 1989, he wrote to Lin Ezell, with a copy to me, saying,

> With major corporations ruled out as a matter of fund-raising policy coordination with the Smithsonian's development programs, I am weighing the alternative, ... formation of an underwriting group willing to pledge sufficient funds to ensure completion of the restoration ... an alliance of a select few American Legion and VFW posts ... able to commit $1,000 or more annually for three years, together with a sprinkling of private firms or individuals having a special interest in the project.
>
> We share a concern that the purpose of restoring the plane not become twisted by super-patriots wired on war stories and the exploits of the Boeing superfortress....
>
> It is now clear that the decision to drop the bomb arose through a "mix of motives," to quote Yale historian Paul Kennedy: "the wish to save Allied casualties, the desire to send a warning to Stalin, the need to justify the vast expenses of the atomic bomb project." But, says Kennedy, "the point ... is that it was the United States alone which at this time had the productive and technological resources ... to invest the scientists, raw materials and money (about $2 billion) in the development of a new weapon which might or might not work. The devastation [it] inflicted ... not only symbolized the end of another war, it also marked the beginning of a new order of world affairs.

Smith concluded, "I feel comfortable trying to put together such a 'preservation' group in that context. Do you agree? It seems the most feasible extension of the fund-raising effort at this time."

Smith also pointed out successes along other lines:

> In an unrelated matter, the VFW, at its 90th convention, adopted a resolution endorsing the Atomic Library and Technology Museum at Independence. Adjutant General Howard Vander Clute will return to Kansas City Monday, bringing further details. The organization of course supports restoration of the *Enola Gay* and is circulating your article in the September issue [of *Air & Space* magazine].[21]

Others were beginning to verify progress on the establishment of the Atomic Library and Technology Museum. On September 26, Lin Ezell, Steven Soter, and I met with Luc Smith and Terrence R. Ward, President of the H&R Block Foundation, Kansas City, Missouri, who wrote the next day to thank me for the thought and time I had devoted to the Atomic Museum.[22]

At the beginning of November, Smith wrote again, saying his plans for an elite board to guide the Atomic Library and Technology Foundation were coming to fruition:

Shortly after we talked Tuesday, I was informed that Hans Bethe had accepted an invitation to join the Advisory Council ("as long as I don't have to attend any meetings").

Earlier we had heard from Glenn Seaborg of Berkeley; Richard Wilson, Energy and Environmental Policy Center, Harvard; John W. Landis, Vice Chairman, National Environmental Studies Project; Bertram Wolfe, General Manager, General Electric Nuclear Energy Corp.; Jay Kunze, Chairman Nuclear Engineering Program, University of Missouri; and Joseph Hendrie, former Chairman, Nuclear Regulatory Commission.

. . . I will explain matters more thoroughly following the organization of the "blue-ribbon" panel in San Francisco.[23]

Thus ends my correspondence file with Smith. A routine year-end financial audit for 1989 was submitted to the Smithsonian. Contributions kept coming in, but eventually at a diminishing pace.

AFTERMATH

A year later, the fall 1990 issue of *Atomic Veteran's Newsletter* ran two articles under the headlines "Management Change" and "The Atomic Foundation."

In an economy move, NAAV ended its five year relationship with Adjutant General Lucius Smith III. Management of NAAV has been taken over by the Board of Directors....

Adjutant General Lucius Smith III devised several ways to raise money from the NAAV membership for the Foundation headed by himself as its self-appointed president.... These included an honor roll, the COMrad fund, member contributions of atomic memorabilia for the Atomic Library and creation of the *Enola Gay* Restoration Fund. . . .

Following his dismissal, Mr. Smith left Independence without leaving a forwarding address....[24]

In mid-March 1991, Bill Dalton, staff writer for the *Kansas City Star*, wrote an article under the headline "Atomic museum 'up in smoke'?—County's $30,000, promoter are both gone":

Three years ago, Lucius Smith III lobbied Jackson County legislators to help build a museum honoring "atomic veterans"—soldiers exposed to radiation in the line of duty. Some legislators were wary, "We might just blow off $30,000," said legislator Ed Growney. But they spent the money anyway. Three years later, the county's $30,000—along with thousands more in private contributions by veterans and others—is gone. So is Smith, the tireless promoter of the proposed $9 million Atomic Library and Technology Museum. He is living in Indiana. And the site for his would-be project—"a high-tech way to explain the history and science of the atomic age"—remains a bare patch of ground next to the Truman Library. All that is left are architect's drawings and questions about what happened to the money. "I guess like detonating an atomic bomb, it all went up in smoke," Growney said Friday.

In a telephone interview, Smith said the library-museum was not dead, just delayed. "We slowed it down, not mothballed it, but put it on the back burner," Smith said, blaming a slow economy and the Persian Gulf war. . . .

"He's not even an atomic veteran," Rosen said.

At one point Smith solicited money to restore the *Enola Gay*—the B-29 that dropped an atomic bomb on Hiroshima—and perhaps bring the airplane to Independence. Smith estimates he raised $20,000 to $25,000 for the project, working with the Smithsonian and an Indianapolis group called the *Enola Gay* Restoration Association. They turned over the money to the Smithsonian, in charge of the renovation. . . .[25]

While these two reports, so full of insinuations of impropriety, later turned out to be unsupported, a rueful Ben Nicks wrote at once to Dr. Oscar Rosen of the National Association of Atomic Veterans enclosing a copy of the article from the *Kansas City Star*: "I am embarrassed to have to admit that I seem to have been a catspaw for ripping off my buddies. If I can be of help in any audit process that comes up, let me know."[26]

Nicks also sent a copy of this letter to Ezell, noting,

By copy of this letter to her I am asking what action the Smithsonian proposes to secure an audit of funds solicited, received and transmitted to the Smithsonian in the *Enola Gay* restoration fund-raising effort.

Your answer to this inquiry in time for me to make a report to my association annual meeting in Colorado Springs in May would be appreciated.

Ezell answered this letter right away:

In reply to your questions of 17 March, I have no evidence that supports the notion that Mr. Smith was involved in anything but a straightforward attempt at fundraising for the restoration of the . . . *Enola Gay*. At the end of the first year of activity, the Smithsonian received an audit. . . . To date, we have not received an audit of 1990 activity. If none is received by the end of the month, our General Counsel's office will investigate the matter. . . .

As Mr. Dalton's article says, the *Enola Gay* fundraising effort produced in excess of $20,000, most of which was used to fund a video on the restoration effort, as per the memorandum of understanding. Unfortunately, insufficient funds were generated to go on to our second and primary goal, which was to hire additional term-appointment employees for the *Enola Gay* restoration. However, we appreciate the efforts made in our behalf. There was another positive by-product of the fundraising effort which we should not overlook. Many people and organizations who mistakenly believed that the Smithsonian was not taking positive steps to restore and exhibit the *Enola Gay* were given more accurate information on the status of that project through the fundraising process.[27]

Nicks's report to the 9th Bomb Group Association, however, is strangely quiet on the entire affair, including his part in it acting as secretary of the ad hoc steering committee set up to collect funds on behalf of the museum. His only comment is,

> Money to speed the restoration process is really not a major problem. Oh, the Smithsonian would thankfully accept contributions for any of its many worthy projects. In fact over the past few years some $20,000 has been donated by Twentieth Air Force veterans for *Enola Gay* restoration. And you can send contributions directly to the Smithsonian, if inclined to assist the restoration process. Their budget does contemplate, however, enough manpower and material to finish the *Enola Gay* by 1995.

The rest of his report is his usual mix—a short but fairly detailed report on progress on the restoration, accompanied by harsh criticism.

> The nose section is completed now and actually is on limited display at the Paul E. Garber restoration facility in Silver Hill, Maryland, a few miles [from] Washington. Work is proceeding—at a more or less leisurely pace—on the rest of the aircraft. Two engines are already overhauled and a third in progress. Smithsonian officials are sticking to their guns in predicting that the *Enola Gay* will be completely restored by 1995.
>
> It might be noted—for the record—that had our national museum met its responsibilities when it first acquired the aircraft in 1949, it would not now be in need of a heroic restoration process. In fact it easily could have been kept flying all these years around the country—as a sort of traveling national museum honoring the men of the Twentieth Air Force and their exploits in World War II. For political reasons that are useless to dwell on now, the Smithsonian allowed the *Enola Gay* to sit outside in a swamp at Andrews Air Force base in Washington, completely unattended, a home to rats, rodents, birds and weeds.[28]

Nicks certainly wanted to see the aircraft rapidly restored and displayed, but his diatribes on the museum's alleged negligence were also rousing a large number of veterans against us. In the long run their antagonism would be decisive.

On April 1, 1991, Lin Ezell wrote to legal counsel,

> When we last talked about the state of our fundraising activities, I decided to wait through the first quarter of the calendar year for an audit of 1990 activities from Luc Smith. We did not receive such an audit, nor have I been able to contact him by telephone to discuss the matter; mail to the old address in Missouri goes to Oscar Rosen of the National Association of Atomic Veterans.
>
> We do know for a fact that the NAAV no longer wishes to be associated with the fundraising effort. . . . And we have not received any donations for several months now; Rosen forwarded the last he received some time ago.
>
> I recommend that we terminate the memorandum of understanding as per the terms of that document, based on the missing audit, inactivity,

> and NAAV's decision to not support this activity through the Atomic Foundation. . . .[29]

We never did receive a 1990 report, but on June 12, F. Mark Miller, secretary-treasurer of the Atomic Library and Technology Foundation, wrote to me to terminate the memorandum of understanding between the museum and the Atomic Library and Technology Foundation.[30]

On August 5, 1991, I replied, thanking Miller and other key members of the foundation for their help on our behalf:

> As you know, the first $15,000 raised during the *Enola Gay* restoration campaign funded a video exhibit. This exhibit, outside the entrance to the Museum's World War II gallery, highlights the restoration of this very important artifact. The remaining funds will be used to further support the actual work on the B-29 at the Garber Facility.
>
> Responding to your letter of 12 June and to advice given me by the Smithsonian Institution's General Counsel, I am officially bringing to a close the agreement between the National Air and Space Museum and the Atomic Library and Technology Foundation of February 1989. I acknowledge the hard work of Lucius Smith and the other officers of the foundation, which led to hundreds of contributions from World War II veterans interested in the preservation of the *Enola Gay*. . . .[31]

I cannot comment on Luc Smith's relationship to the Atomic Library and Technology Foundation or to the National Association of Atomic Veterans. But through his efforts, the museum had received 650 checks from veterans for a total of $22,500. These were greatly appreciated and helped us to restore the *Enola Gay* and keep the public informed.

The breakdown of the fund-raising effort, however, did have an important consequence. Statements like Ben Nicks's May 1991 report to the 9th Bomb Group that "Money to speed the restoration process is really not a problem," worked against us. Veterans who had been aiding us or at least had halted their opposition now either withdrew, like Frank Stewart, or worse, like Don Rehl and Bill Rooney, soon resumed active criticism.

I will return to that later.

EPILOGUE

Five years after I had last talked with him, and a few days after I had resigned from the National Air and Space Museum, I received a friendly three-line note, signed "Lucius Smith III, President of the Atomic Library & Technology Center, Independence, Missouri."[32]

I called Smith months later, and found him still as dynamic as ever. Asked about what had happened with the Atomic Veterans at the time, he said he had done as much for them as he could, setting up legislation to compensate their needy members. In the wake of that, he volunteered, their interests and his "were not totally compatible."

When I called Dr. Oscar Rosen, national commander of the NAAV, to check into the story, he stated only that the NAAV board of directors had felt that the fund-raising efforts Smith was conducting on their behalf were costing them too much, and that they felt they could do better on their own.[33] Contrary to the insinuations of the earlier news releases and Ben Nicks's letter, Rosen denied any impropriety on Smith's part.

At the time of writing, Smith is still actively pursuing the construction of the Atomic Library & Technology Center, with the strong backing of Independence, Missouri, Mayor Rondell F. Stewart.[34] Land for the project has been donated in memory of an Independence community leader, Roy S. Gamble. His eldest son, James W. Gamble, an architect, is building consultant for the project, which is included in the city's master plan.

10

Restoration and Authenticity

In June 1988, three years after restoration of the *Enola Gay* had begun, Richard D. Horigan, Jr., the restoration shop foreman, provided me with a history on the work that had been accomplished to date.

> The goal of the restoration is for the finished product to be faithful to the *Enola Gay* of August 1945. . . .
>
> The personnel initially assigned to the project noted that the aircraft was remarkably complete as regards its equipment and that the overall condition was excellent, with very little corrosion evident. With this . . . in mind, and considering the magnitude of the project, the decision was made not to remove every rivet and skin panel, but instead to concentrate on those areas and assemblies that required a thorough restoration (e.g. instruments, electrical equipment, fittings), as opposed to those areas that only needed thorough cleaning and preservation (e.g. main bulkhead, stringers, skin).
>
> Because of its size, the aircraft has been broken down into manageable assemblies for restoration: forward fuselage, rear fuselage, empennage, engines, center section and wings, and landing gear. The past three years, or roughly 8,000 man-hours to date, have been spent on the forward fuselage cockpit section and bomb bay. The two-man staff began their work in the aft cockpit area and worked forward. Once this was com-

plete, they turned their attention to the bomb bay. . . . There have been no unexpected problems to date, although . . . the bomb rack and associated pieces were made from scratch with vague blueprints as our only clue. The bomb bay has been thoroughly cleaned and preserved; what little corrosion was found was removed.

The corrosion we did find was discovered upon removal of inspection panels and interior blankets throughout the aircraft. The aluminum skin underneath appears to be virtually factory fresh, complete with manufacturing stamps. Apparently where dirt cannot gather and attract moisture, thereby starting the corrosion process, the aluminum will remain spotless. This, of course, speaks well for the long-term survival of the aircraft. Additionally with the use of such preservatives as "soft seal," "R2000" and Aeroloyd, the aircraft is doubly ensured long-term protection. Magnesium components, on the other hand, present a different problem because of this material's inherent ability to self-destruct over the long term. The restoration staff removed all magnesium components, notably the nose framework, cable pulley blocks and other assorted pieces, and suitably treated and repainted them to give them an extended life. Overall, the airplane, and specifically the sections already completed, have an excellent prognosis for surviving many years in a museum environment.

The cockpit area was in excellent condition, and we initiated what may be termed a "heavy clean-up." Whenever possible, the original paint was spared. If painting was required, exact matches for the mostly zinc chromate and interior green surfaces were used. All original wear and scuff marks were retained. The only items needing special attention were the pilot's control wheels, which had degenerated due to ultraviolet exposure during the years of outside storage. These were built up with plastic filler and smoothed, sanded, and shaped accordingly. Only a few small items were missing from the cockpit area, but all have been replaced from the NASM collection. . . . Also noteworthy is . . . that the staff made from scratch the bombardier's "Silverplate" control panel. Photographs obtained from an outside source assisted us in solving that particular problem.

. . . . All that remains to be done to the [forward] fuselage and bomb bay section is an external polish. Experiments were made to determine the best procedure to follow. The one chosen yielded a brilliant shine as per the original. Although looking rather dingy to the uninitiated, the thin oxide coat on the exterior at present is no problem and will be dealt with overall when the aircraft is complete.

The remaining sections of the aircraft currently in indoor storage at Garber appear to be in a condition similar to the forward fuselage. . . . The only problems that could severely impact the schedule would be hidden major corrosion underneath the panels or missing pieces. . . .

The landing gear is in very good condition. More importantly, the tires have not suffered from much ultraviolet and ozone deterioration, although the standard surface checks and cracks exist. Aircraft tires top the list as the most hard-to-find artifacts. . . .[1]

A DEBATE ON RESTORATION STRATEGY

Rich Horigan's optimistic tone might give the impression that everything about the restoration was routine. In fact, impassioned debate often preceded action, as different strategies were forcefully advocated. In the spring, I had asked that we consider the possibility of publicizing the restoration by carrying out part of it where visitors would be best able to observe it, in the museum on the Mall. Some years earlier, the staff had carried out such a restoration on the original 1903 Wright *Flyer* in public view. But restoring the Wright brothers' relatively tiny airplane, constructed primarily of wood and fabric, was one thing; restoring parts of a huge metal aircraft was quite another.

On April 29 and May 10, 1988, Lin Ezell held meetings primarily with the restoration staff at Garber to discuss the problems of restoring the *Enola Gay* at the Mall museum. A week after this second meeting, Lin Ezell and Steven Soter sent me a report citing concerns about health and safety hazards. These stemmed primarily from chemical fumes and particles loosened in the course of routine work, and the high noise levels that would accompany the process. They recommended instead that we produce the short video film already discussed in previous chapters:

> The film would include actual footage of the *Enola Gay* at Tinian, information on its role in the atomic bombing of Hiroshima, its postwar history, the progress and current status of its restoration . . . plans for its display in a major exhibition on strategic bombing . . . information on what there is to see of the airplane at the Garber Facility and on how to get there. . . . Suggestions for its location in NASM include the South Lobby or the WW II gallery.[2]

After some consideration, all of us agreed that while most museum visitors would be fascinated by seeing the restoration in progress, the associated problems were too great. We would continue to restore the aircraft entirely at the Garber restoration facility. We would also seek to produce the video film and display it just outside the World War II gallery on the museum's second floor, where interested visitors would be most likely to see it.

A more difficult point of contention in the summer of 1988 was whether the forward fuselage of the *Enola Gay*, now completely restored, should be polished so it would look better to visitors. The restoration staff did not think so, while Robert Mikesh, the curator in charge, wanted it buffed. On June 15, Ezell wrote Mikesh and aeronautics department chairman Von Hardesty, saying she did not think polishing the forward fuselage now was a good idea.[3] We risked having to do it all over again six or seven years later, when the rest of the aircraft was ready for assembly. That would mean double work for her people.

Two weeks later, Bob Mikesh responded with a long memorandum. Contrary to Horigan's assessment that work on the forward fuselage could now be considered completed, he proposed "a general list of major items that I see necessary to be accomplished before the nose section can be considered complete for this phase of restoration." He listed ten additional structural requirements and nine details on the markings painted on the fuselage. Before these markings were repainted, the aircraft would have to be buffed:

> I do not agree that there would be a problem of mismatch of surfaces when polishing the rear section is completed later. Dissimilarities already exist that cannot be corrected, i.e., note the wing surface compared to that of the fuselage. . . .
>
> It would be appropriate for visitors and media to view both the nose and aft sections side by side, showing what was accomplished with three years of work on the nose section, compared with that of the unrestored aft section. This would not be evident, however, if the exterior of the nose be left in its current condition. To accomplish this detailed marking phase, a certain degree of polishing will be necessary in these areas and areas now corroded. Therefore the polishing phase may as well be completed as intended for this phase of restoration.[4]

Richard Horigan was annoyed. On July 8 he responded with a memorandum to Lin Ezell, his immediate supervisor:

> The opening paragraph in the memo dated June 29 from Bob Mikesh clearly suggests that there is a lack of "completeness" to the *Enola Gay* forward fuselage. Also suggested, or at least intimated, is a lack of responsibility, a kind of shirking of duties on behalf of the restoration crew. Before I address these issues, one must keep in mind that any artifact or aircraft must be considered in total. It is complete when the aircraft is ready for roll-out from the restoration shop, and not when it sits on the shop floor with literally years to go before completion.

He then responded item for item to Mikesh's listing, agreeing with some, giving the reasons for disagreeing with others, and ended with,

> The closing comments to the memo in question are very revealing. Apparently, the best long term interests of the *Enola Gay* fade when compared to what is "appropriate for visitors and the media." . . .
>
> If the forward fuselage shines, then the "visitors" and media will be assured of concerted efforts. This is eye candy, an illusion of sorts. What about the 7,500 hours of work that has gone into the aircraft's interior. If it is not apparent to "visitors" then one can suppose nothing has been done. . . .
>
> Detail marking phase? Since when has that been done on any aircraft years before the completion date of said aircraft? It is done at the end of the project so as not to damage this work. . . .

> Yes, there is a very good chance of a problem with mismatched surfaces when polishing at different times (especially years). We all agree, techniques, different tools, different people, different polish, different chemicals all add up to a potentially different patina. Also, take the protective oxide coat off now, polish it again after pitting reappears, and wear away still more of the alclad surface of the skin. . . .[5]

In this dispute I sided with the restorers. I felt it more logical to proceed step by step—that they were right in delaying their polishing and detail marking until the entire aircraft had been restored. However, Bob Mikesh may well have had the right public-relations view. Completing the restoration on the forward fuselage in 1988 might have shown visitors and the media how well the final restoration would look. Veterans might perhaps have come to trust us to complete the job to the same high standards. Over the years, we might have been spared their constant stream of criticism. Might have . . . Who knows?

At any rate, we did produce the video film on the restoration process and placed it where far more visitors would see it—in the museum on the Mall. The Garber facility may have thirty thousand visitors in a given a year; the museum has more than two hundred times that number.

PRESERVING RATHER THAN RESTORING

As the variety of modern materials used in airplanes and spacecraft has multiplied, preserving components has become increasingly complex. Often, only a few industry specialists are familiar with the peculiarities of a given metal alloy or organic material. To help our restoration staff, we had established a small committee of volunteer industrialists who specialize in different materials and were willing to share their wisdom with the museum.

Preservation is particularly important for spacecraft, which are designed to operate in the vacuum beyond Earth's atmosphere, far from any moisture, oxygen, or atmospheric pollutants. But place a space vehicle in an ordinary terrestrial environment and you have to worry about oxidation and corrosion. That's where the conservator's art comes in. He has to decide how best to store or display the artifact to assure longevity against deterioration.

Walk around the restoration shop and you see chemical baths for the treatment of metal surfaces. And there are ultrasonic baths, in which strong sound vibrations loosen and shake off the dirt accumulated on an intricate part. As the restorer takes apart an artifact for cleaning and possibly repair, he takes detailed notes and video-records the appearance at each stage of disassembly to assure reliable reassembly. All of this takes a great deal of skill, patience, documentation, and education.

STEADY PROGRESS

In August 1989, roughly a year after his first report, Rich Horigan sent me a restoration update:

> Upon inspection of the rear fuselage, the team found a deteriorated cardboard box in the aft lower gun turret well. The box contained some of the missing parts needed for the restoration, including seat cushions, one oxygen hose, one recharger oxygen hose, one radar scope hood, one astrocompass, emergency engine hand cranking gear, and various other small parts.
>
> Due to the construction of the aft fuselage, we determined that it was more efficient to start restoration in the tail gunner's compartment and progress forward. We removed all equipment from the tail gunner's compartment, along with the fabric insulation blankets. This compartment is badly corroded because the escape hatch and pressure door were left open while it was in outdoor storage. This allowed the weather, birds, and mice to wreak havoc in this area. The fabric insulation blankets were badly deteriorated and will have to be replaced. A sample of the original fabric will be saved and put into storage. New fabric is now on order. At this time about 15% of the rear fuselage has been treated.
>
> We have removed all of the wood flooring from the rear fuselage. This material is being repaired and preserved and will be reinstalled. . . .[6]

By this time the restoration of two of the aircraft's four engines also was in full swing. At the Garber facility, sixty-seven-year-old veteran-restorer George Genotti, helped by volunteer Mike Nelson, had started on one engine. The other, Horigan wrote, was being "restored at the San Diego Aerospace Museum by a largely volunteer staff, with close oversight by NASM staff. . . . We are still on track for a completion date (1994–95) that will hopefully coincide with plans for the Extension."

FULL ASSEMBLY

Two years later, we were beginning to worry about the full assembly of the aircraft that still lay some years in the future but needed to be planned. On July 23, 1991, I met with Lin Ezell and her staff and called David Knowlen at the Boeing Company the same day. Knowlen had been a consistent friend and supporter who had persuaded Boeing to restore for the museum the "Dash 80"—the prototype of the Boeing 707, the most successful of the early passenger jets. In exchange for this costly restoration, we had told Boeing, they could keep the aircraft on display at their plant in Seattle until our extension was built and we could properly safeguard the aircraft. Boeing liked that idea, because the display of some of their early airplanes gave their younger work force a better idea of the corporation's history and instilled company pride.

The next day Ezell also wrote to Knowlen to explain more fully the problems we expected to face and ask whether Boeing still had the expertise on the B-29s to help us.

> As a first step toward preparing a minimum requirements document for the movement-assembly-disassembly process, we need the technical expertise of the Boeing Company. Archival documentation that details the wing assembly is of special interest; the technical manuals we have access to are written for field assembly of larger components. Tapping the memories and expertise of Boeing employees or former employees associated with the B-29 project would also be helpful. These are short-term needs. Ideally, we would like to invite a small team of B-29 experts to Washington for two days in August or September to discuss with the crews who are working on the *Enola Gay* the various problems we foresee—and to learn about those difficulties we have not thought about to date. . . .
>
> Longer-term needs include assistance with locating special fixtures, jacks and assembly equipment. Borrowing existing equipment would be a valuable savings for us if such equipment still exists. On-site consultation during the actual assembly in 1994–95 may also be requested. Prior to the exhibition, we will also need to recruit a team and the necessary equipment to assist us with polishing the *Enola Gay*'s aluminum skin—a huge task for a work force as small as ours! We had hoped that a modest fundraising effort initiated by a private group would produce the funds needed to hire additional term-position employees. . . .
>
> In return for your advice and assistance, I can offer to send a team of restoration specialists with experience on the *Enola Gay* to Seattle to give an illustrated briefing on the project during the [Boeing Company's B-29] planned 50th anniversary celebrations in August 1992.[7]

Coincidentally, Knowlen was in Washington the next day, so there was an opportunity to talk. I wrote again a day later to follow up on those conversations, and enclosed a list of needs Lin Ezell had assembled on the *Enola Gay* restoration for the summer of 1995:

> During your visit you mentioned the possibility of NASM making an approach to Boeing for financial assistance for the exhibition planned for 1995. I believe you suggested that it might be possible for us to request assistance from a fund you will be using in 1992 to support the 50th anniversary of the B-29, even though our project will not be completed until 1995. You have done so much for us already I am hesitant to ask, but the need is compelling. Is such a request appropriate and if so how do you recommend we proceed? . . . [8]

We soon were to decide that we would not approach outside corporations for funding for the exhibition, because we wanted to avoid the appearance of external influence on the exhibit. But we continued to seek Boeing's technical advice and expertise with this aircraft they had designed.

THE AUTUMN OF 1991

Old-time restorers were not entirely happy with the amount of time being spent on the *Enola Gay* to prepare her for exhibition in 1995. Walter Roderick, a former leading craftsman at Garber, who had left a few months before I had come to the Smithsonian in 1987, wrote me in August 1991 to express his concerns that the restoration of the *Enola Gay* would take too much time away from other worthwhile restorations, and that perhaps only part of the aircraft should be exhibited. I answered him in some detail, to tell him I felt the historical importance of the aircraft justified the admittedly enormous effort. In replying, I thought I might also be answering some of our own staff, too shy to raise the issue directly and instead, raising it through a former colleague.[9]

In October 1991, Rich Horigan sent me a further update on the *Enola Gay* restoration. By now two of the four engines had been restored, as had the rear half of the fuselage, except for final polishing of the skin. His staff was testing the best polishes available on today's market in patches on the skin. Three members of the staff and three interns had worked on the rear fuselage the past year. In the process they had removed, restored, and reinstalled all of the wood flooring, restored the tail gun turret system, the ground power unit, and the radar radios. They would continue to ask around for minor missing pieces (e.g., lights, oxygen bottles), which would be installed as they were found. George Genotti was working on his second *Enola Gay* engine, and the San Diego Aerospace Museum also was working on a second engine, the fourth and final engine to be restored. Very little corrosion had been found to date, but any corrosion discovered had been neutralized rather than removed. The two inboard wings had now been brought into the restoration shop and were being worked on by three further staff members with the assistance of volunteers and interns.[10]

By the close of 1991, restoration of the four engines had required between fifteen hundred and three thousand man-hours apiece. Both at the Garber facility and at the San Diego Aerospace Museum, the second engine had been restored about twice as fast as the first. The forward fuselage had consumed close to eight thousand man-hours, almost four man-years; the rear fuselage had required seven thousand man-hours; tail surfaces, two thousand; inboard wings, twenty-five hundred; and general support about four thousand.[11]

By the end of 1991, all in all more than thirty-two thousand man-hours—sixteen man-years—had been invested and great effort still lay ahead. Work progressed smoothly, although consuming a great deal of time, as was expected for an aircraft of such enormous proportions.

THE FINAL YEARS

In September 1993, William C. Reese, who had replaced Rich Horigan as restoration shop foreman, reported that work had begun on the control surfaces—the rudder, ailerons, and elevators. On a B-29 these surfaces were covered with fabric, to make them lighter weight. This does not show up on photographs, since the surfaces are painted silver to blend with the metal surfaces elsewhere. Much of this fabric had to be replaced.

The nacelles, the enclosures into which the engines fit on the wings, were also being repaired, cleaned, and preserved, as were a host of individual components like cabin pressurization units, oil coolers, and heat exchangers. The inboard wing sections into which the nacelles were built, the landing gear and propellers, all were being refurbished. Fuel lines, hydraulics, de-icing systems, and wiring were being reconditioned. This profusion of work enlisted the efforts of three members of the staff, six volunteers, and four young interns.[12]

Work continued at this accelerated pace in 1994, with the engine cowlings, the outboard wing sections, and the right main gear receiving particular attention, and the inboard wings being completed. Every part of the aircraft skin was now also being buffed, with particular attention to the forward fuselage, whose markings also needed to be repainted before display. The remaining sections of the aircraft were also being prepared for assembly as planned. The ten-year project was coming to a close. It had taken around twenty-five man-years at a cost of roughly a million dollars.

The museum's most difficult and most extensive restoration of all time was drawing to a close. All of us by now agreed that the museum should never again either accept an aircraft that required such massive restoration or accept one that would have to remain outdoors for long periods and then require similarly massive treatment. It was too draining an effort on the entire staff, diverting energies that should be devoted to a broader class of pressing needs. For the *Enola Gay*, such a large effort made sense, because she had played such a crucial role in world history. None of us could imagine another craft that would garner the same priority in the museum's schedules.

LITTLE BOY KIDNAPPED

Restoration of the *Enola Gay*, however, was not our only problem with artifacts from our collections required for the planned exhibition. From the earliest days, we had expected to exhibit the aircraft together with the museum's *Little Boy* casing. Unless we could display the bomb that the *Enola Gay* had dropped, the exhibition would be incomplete. The bomb casing would help to emphasize the aircraft's decisive historical mission.

Unfortunately, *Little Boy* had been kidnapped, and might not be returned.

The United States Air Force had given the museum a *Little Boy* casing in the 1960s. In the course of the years, it had occasionally been exhibited at the Smithsonian. Everything seemed quite routine.

Then, in the spring of 1986, a knowledgeable visitor to Silver Hill who had apparently worked on atomic bombs in the early days pulled aside members of the Garber facility staff and pointed out that the *Little Boy* casing in our possession still had classified elements. He urged the museum to get in touch with the appropriate authorities to discuss the matter.

So it was that twenty-four years after the Air Force had initially offered the museum the bomb casing, a letter had to go out to ask for more information. On May 20, 1986, E.T. (Tim) Wooldridge, then chairman of the museum's Aeronautics Department, wrote to John P. Cornett, Office of Security Police, USAF, mentioning that the museum had "two of these weapons in custody, either in unclassified storage or on display.... I have contacted ... [the] Office of Classifications, ... who will have the bombs examined in the near future to determine their classification."[13]

Three weeks later, much to the surprise of the collections management staff, who were ordered aside, a sizable, well-armed group from the Department of Energy (DOE) arrived at the Garber facility, picked up *Little Boy*, and left. Two weeks thereafter, on June 26, 1986, Don Ofte, principal deputy assistant secretary for defense programs at the DOE, wrote to museum director Walter J. Boyne, "Several weeks ago I spoke with you concerning our need to take custody of the *Little Boy* model which the Institution had on display at your Garber facility in Suitland, Maryland. That effort was completed on June 12, 1986, with the excellent cooperation of your staff. We plan to return the model to you as quickly as possible...."[14]

A FAST SWITCH

Nearly two years passed before the arrival of a letter from Richard E. Malenfant, dated March 7, 1988. It was addressed to Senior Curator of Aeronautics Robert Mikesh, and crisply informed him that

> The replacement for Little Boy, the replica of the Hiroshima weapon, will be shipped to the Smithsonian from Los Alamos as soon as shipping can be arranged. . . .
>
> Four photos of *Little Boy* are enclosed. The first color photo shows the "Smithsonian Unit" as received at Los Alamos in 1986. This should document the features of your unit that were on display at Silver Hill for several years. The second color photo shows the unit which had been on display at the Los Alamos Museum (Bradbury Science Hall). The Los

> Alamos display unit had been modified by removal of some tamper material, replacement of the gun barrel with thin wall tubing, and securing the external case by welding the unit shut. In addition, it had been painted white. We switched the two units; that is, the unit formerly at Silver Hill is now on display at Los Alamos, and the unit formerly on display at Los Alamos has been returned to you. . . .
>
> When the unit was received at Los Alamos from Silver Hill, we dismantled it to verify the internal condition. Please be assured that at no time was there concern that active components (uranium, explosive charge, fuzes, etc.) were in place, only the inner tamper and gun barrel. All units that had been released for display were "Type A," that is, a war reserve, or WR unit employing a Type B Mod 0, or fully functional gun barrel. Without modifications that were made, it was theoretically possible to use the components of the inert units to jury rig a functional weapon. . . .
>
> An unmodified Type B, Mod 0, unit is in secure storage at Oak Ridge National Laboratory, Tennessee. In addition, I have placed a set of disassembled components in secure storage at Los Alamos. The latter includes exact details of the active components and propellant charge. Hopefully this will provide correct and detailed archival information for the historical record. . . .
>
> During our visit to Silver Hill, there was considerable interest in the bomb release mechanism that was being reinstalled in the *Enola Gay*. I have enclosed a video tape prepared from a short film clip shot at Wendover Air Force Base about early 1945 that shows the bomb release mechanism, sway brace and loading procedure. . . .[15]

That last-mentioned video would be of use to the restoration crew at the Garber facility, since the museum was seeking ways to replicate these parts, which had been missing from the *Enola Gay* on her delivery to us in 1949.

Malenfant's second paragraph, however, was most disturbing. The agreement signed with the Department of Energy when *Little Boy* had been released to them was a loan agreement, which would see the unit returned as soon as the department had made modifications essential for security. Malenfant's unilaterally dictated swap violated that agreement. His letter provided no justification and calmly treated the switch as an accomplished fact.

When the substituted bomb casing arrived, we were deeply angered. Los Alamos had confiscated our *Little Boy* and sent us an artifact that we considered historically flawed.

A *Little Boy* bomb casing is not a solitary unit, but an assembly of several parts. To keep track of these parts at the time of fabrication during the war, each had been stamped with a serial number. The original casing owned by the museum had matching serial numbers on all its parts. The substitute from Los Alamos had a motley selection of serial numbers. Some parts could have come from an early batch, reflecting one model bomb, while others could have come from a batch manufactured later for a somewhat modified bomb type. A historian would learn little about any actual bomb that had ever been constructed by looking at this mix of parts.

Our unhappiness about this reached Malenfant, both directly and through a newspaper reporter who had stopped by the Garber facility. Malenfant decided to write to Mikesh again:

> Thank you very much for your "heads up" on the reporter and the *Little Boy* problem. Indeed we had been alerted to the matter through an inquiry to the Public Affairs Officer in the Defense Programs Office of the Department of Energy. Although it created a bit of a flurry at Los Alamos, . . . the matter was referred to DOE Security in Washington. . . .
>
> I expect that both the Smithsonian and the Department of Energy are sensitive to any publicity that could place the actions taken in a poor light. . . . Whether the matter was handled in the best of ways or not, the action has been accomplished to minimize the possibility of the serious consequences of no action. However, we are still receiving some indications that at least one person at the Smithsonian is very unhappy with the resolution of the problem, and still feels that the display you now have is in some way inferior. I can only assure you again that the unit now in your possession is as authentic as the one that you had for some years, and neither is an exact replica of the unit that was destroyed in its deployment in 1945. I must admit that I was quite surprised that I was identified with the switch. . . .Was this information included in the offhand remark by the docent?[16]

For a moment I dwelled on the matter-of-fact phrase "the unit that was destroyed in its deployment in 1945." Perhaps this way of describing the bombing of Hiroshima made good sense to one responsible for accounting for each and every warhead that had ever been produced—in its own way, a frighteningly important task.

SEEKING REDRESS

By the summer of 1989, it was evident that we were not making any headway. I met with legal counsel and Dominic Pisano, at the time acting chair of the Aeronautics Department, to discuss the bomb casing's return from the Department of Energy.[17] Four weeks later, on August 8, Pisano sent me a draft of a letter we might send to the director of the Los Alamos National Laboratory to seek his help in the matter.

I decided that I should first call Dr. Siegfried Hecker, director of the Los Alamos National Laboratory, to apprise him of the situation and then, having given him the background, write him a more detailed request. I called him at once. He had not heard of the affair, but promised to respond. I followed up in writing on August 16, 1989, attaching much of the previous correspondence as background:

> Since June 1986, the bomb has been on loan to Los Alamos National Laboratory, and for reasons we do not understand, has apparently been located in the Bradbury Science Museum. In March 1988, a replacement bomb was sent to NASM in place of the original. This was done against our wishes at NASM, and contrary to the stated agreement between the two agencies that the original would be returned to NASM after declassification. . . . The replacement weapon is in poor physical condition and unfit for exhibit purposes.
>
> I am relatively new at the Museum and would like to resolve the problem as soon as possible. I believe that the material I am appending, as well as additional documentation in our files, shows the bomb originally given to the Museum in 1965 to be the Smithsonian Institution's property. I would like to see the bomb declassified and returned to NASM as soon as possible. This action would honor the original agreement of 1965 and would return the bomb to its proper status as an important part of the National Collections of the Smithsonian Institution.[18]

Dr. Hecker answered on September 19:

> . . . Unfortunately, the situation is much more complex than you indicated in your phone call. Because of classification concerns, we are actually only custodians for the replica for the Department of Energy (DOE) and are holding it according to their direction. Any disposition of the unit would be at the direction of the DOE.
>
> I have transmitted your request for declassification action and return of the unit to the Smithsonian to the DOE in Washington, D.C., for their consideration.
>
> However, I must advise you that I believe the national security and the provisions of the Atomic Energy Act of 1954, as amended, make a return of the bomb claimed by the NASM impossible so long as it remains in its present form.
>
> Unfortunately, the issues deal not only with the act of declassification, but rather with the details of material and construction in the interior of the bomb. . . .
>
> I believe that the action of the Department of Energy in recovery of the unit from your facility at Silver Hill actually corrected the error of premature and illegal release in 1965.
>
> . . . I fear that return of the unit, while possible, could only be accomplished by modifying it to the point where it will be virtually identical to the unit you have now.[19]

THE DEPARTMENT OF ENERGY

This letter at least told us that we had been talking with the wrong people and should instead turn to DOE officials in Washington to seek resolution.

Dom Pisano paid a first visit and three days later sent me his notes on the meeting he had held with Malenfant and Glen Taylor of DOE.[20]

Over the next few months we held a number of internal meetings, and further discussions with the Department of Energy also took place. But as 1991 was dawning and we needed to know whether we would be able to display *Little Boy* as part of an *Enola Gay* exhibition, we decided that some form of resolution was urgently needed. On February 7, Tom Crouch and Dom Pisano, having discussed the matter with Gregg Herken, who had considerable previous experience with the Department of Energy, wrote me,

> We have to make a serious effort to find out what criteria DOE uses to determine that a bomb is "safe" and then convince DOE that we are able to meet those criteria. Any or all of the approaches Gregg suggests to alter the bomb might just work to persuade DOE of our good intentions.[21]

Having worked out a compromise strategy, Crouch, Herken, Pisano, and I met with John Tuck at the Department of Energy in his offices in the Forrestal Building to talk about a set of steps that might ultimately permit them to return our casing.[22] This meeting was cordial. When we left, we felt we had perhaps made some progress. Several discussions still had to follow, but by late summer that year, we had achieved success. By August 11, 1991, I was able to write Smithsonian under secretary Carmen Turner, "DOE now has assured us of the casing's return for the exhibition."[23]

After more than five long years, *Little Boy* was finally returning and could be included in our designs for an exhibition on the mission of the *Enola Gay*.

11

A Smithsonian Debate

STEPS TOWARD AN EXHIBITION

As 1991 approached, the museum had not yet found a place to display the *Enola Gay*. But time was pressing to decide *how* to exhibit her. The process ahead was lengthy. A whole army of curators, restorers, designers, carpenters, electricians, security guards, and other professionals and volunteers would have to be assembled and given clear instructions. The first step was the preparation of a *proposal* that all could understand and accept.

Acceptance means that the proposal falls within the museum's mission, that the public will find it interesting, that members of the staff have the professional skills to handle the exhibit or can attract experts who will, and that the required financial resources are at hand or can be raised in time for the opening date. Acceptance does not come overnight. At each stage, those responsible for producing the exhibition sign what in effect is an agreement that they understand the evolving plans and foresee no overriding difficulties.

These steps involve extensive debate, both within the museum and with knowledgeable colleagues elsewhere. As agreement emerges, curators and exhibition designers join forces to craft a *planning document* that outlines the exhibition in considerable depth. Where a proposal might be presented in two typewritten pages, the planning document will require another twenty. It must define the exhibition in sufficient detail to permit a thorough examination of the implied workload, which is then summarized in an *impact statement.* The impact statement shows whether and how the new exhibit might clash with other projects the museum is already pursuing. It defines the resources—money and work skills—needed, and specifies the times at which they will be required.

Exhibiting the *Enola Gay* was going to be difficult, with major national as well as international consequences. This meant that in contrast to more routine exhibitions, either tacit or formal approval of the secretary and the Smithsonian regents would be necessary. Only then would we be in a position to design the gallery layout; list all artifacts, images, and film footage to be shown; and write a *script* for all the labels that ultimately would introduce different sections of the exhibit and identify and provide information on all of the displayed objects and images.

At the start of 1991, four years before the exhibition was to open, it was high time to begin the process.

Curator of World War II aviation Michael Neufeld had been put in charge of a series of commemorative exhibits on significant World War II aircraft. With the support of the P-47 Pilots Association, Neufeld first curated a display of the museum's Republic P-47 Thunderbolt, affectionately known as the "Jug." He followed this popular exhibit with one dedicated to the sleek, German Arado reconnaissance bomber, the first fully operational jet aircraft of World War II, beautifully restored by the museum's Garber facility staff. The third in this series of airplanes was to be the *Enola Gay.*

On February 4, 1991, Neufeld sent me "A Proposal to Exhibit the *Enola Gay* on the Mall." It briefly described an exhibition along lines we had long discussed, and explained,

> The creation of nuclear weapons, and their employment against Hiroshima and Nagasaki in August 1945, are perhaps the central events of recent human history. As we approach the fiftieth anniversary of the atomic bombing of Japan, the National Air and Space Museum has an obligation and an opportunity to help visitors better understand the meaning and implications of the decisions and events that have shaped the subsequent history of the twentieth century.
>
> The *Enola Gay,* the airplane that dropped the Hiroshima bomb, should be the centerpiece of any thoughtful commemoration of this important anniversary. One of the most significant aircraft in the NASM collection, the *Enola Gay* has become an icon and symbol of a critical turning point in our history. . . .
>
> . . . Important artifacts and documents may be displayed, if available, such as the 1939 letter from Einstein to Roosevelt discussing the possibility of a bomb. There may also be a theater to show a movie to visitors about some aspect of the story. . . . Through the use of proper lighting, sound insulation, black-and-white photographs, artifacts from the mission

> and Hiroshima, and a restrained text, a quiet and contemplative mood will be established in the exhibit. . . .[1]

TALKS WITH THE PARK SERVICE

At the time, I still had great hopes that we might find a way to exhibit the entire aircraft, and the remainder of Neufeld's proposal depicted such a display.

These hopes were given a boost on April 24, 1991, when the museum's associate director Wendy Stephens, the Institution's congressional liaison officer Brad Mims, and I met with Robert Stanton of the National Park Service at his Haynes Point offices in Washington, D.C.

The Park Service administered a small plot of land separated from the National Air and Space Museum only by 4th street. We asked Stanton whether the museum might obtain use of the plot for as long as several months to a year to exhibit the *Enola Gay* in a temporary building. Stanton thought this might be difficult; but he noted that the land was to be turned over to the Smithsonian Institution sometime around 1996 or 1997 for the construction of the National Museum of the American Indian (NMAI). If the Institution were to petition the Park Service to make the land available a year or two earlier so we could mount our exhibition there in 1995, the service might be willing to relinquish control. He suggested we pursue the matter through Smithsonian channels and submit a proposal.

I began to seek the Institution's support. One question was whether the American Indian community would view the exhibition as a desecration of their grounds. I had previously written to W. Richard West, Jr., director of the National Museum of the American Indian, to see whether he would have any objections, and now, after the more recent consultations, I wrote him again to tell him about the Park Service's suggestion. I explained, "Before we go any further, however, I would like to know how you and your board feel. If you should have any concerns, I would very much like to discuss them with you, to see whether they could be alleviated." I enclosed a copy of Neufeld's proposal, which by now had been revised and had a new title, "A Time for Reflection: Fifty Years Past Hiroshima—A Proposal to Exhibit the *Enola Gay* on the Mall."[2]

On May 23, 1991, I also wrote to Secretary Adams to inform him about this new possibility and included a further revised proposal to which two new paragraphs had been added:

> The Museum has been restoring this historic aircraft during the course of the past seven years. The restoration has had the support of some 650 veterans, many of whom flew B-29s in World War II and by now have sent contributions amounting to more than $22,000 to help restore the airplane. The Museum has also had periodic guidance from General Paul Tib-

> bets and his crew, who flew the *Enola Gay* on its fateful mission on August 6, 1945.
>
> Veterans from all over the United States have written the Smithsonian requesting that the National Air and Space Museum restore and exhibit the airplane in time for their aging group to still see this historic aircraft once more. Serious scholars and citizens concerned about the proliferation of nuclear weapons also have asked us to restore the airplane which symbolizes the introduction of atomic weapons to modern warfare.[3]

Copies of this letter went to all the senior members of the museum's and Institution's staffs who would need to be informed, including Smithsonian public affairs officer Madeleine Jacobs. Within a few days, on June 3, Kim Masters, of the *Washington Post,* wrote an article reporting,

> The Smithsonian hoped to have an Air and Space annex up at Dulles in time to mark the 50th anniversary of the bombing of Hiroshima and Nagasaki in 1995. But with the annex caught in political funding wrangles, that won't happen. Accordingly, museum director Martin Harwit has come up with an alternate proposal to display the *Enola Gay*, the plane that dropped the first A-bomb, in a temporary facility on the Mall. [4]
>
> Harwit suggests an exhibit tentatively titled 'A Time For Reflection: 50 Years Past Hiroshima,' according to Smithsonian spokeswoman Madeleine Jacobs. Visitors would see the *Enola Gay,* which is now being restored and would be kept under cover, after viewing "a whole exhibit about the origin of the bomb, the Manhattan Project and the decision to drop the bomb," Jacobs says. She emphasizes that this plan has not been reviewed by senior management and may never materialize.
>
> Jacobs recognizes that the public may have mixed feelings about such an exhibit. But she adds that the Smithsonian has received hundreds of letters from veterans who want the plane displayed. "They feel we've been dragging our feet," she says.[4]

On June 13, Richard West called me and followed up with a memorandum: The NMAI Board of Trustees position regarding the NASM's *Enola Gay* Exhibition:

> . . . At its meeting held on June 7, 1991, the Board of Trustees chose to take a position of neutrality regarding location of the *Enola Gay* Exhibition on the future site of the NMAI. The Board's rationale was that it is not appropriate for the NMAI to become involved in this kind of institutional decision prior to formal transfer of the land to the NMAI.[5]

CARMEN TURNER

On July 23, 1991, I sent a detailed memorandum to Under Secretary Carmen Turner, who had been at the Smithsonian for only a few months and needed to know the history of the project. She might be asked about our plans in Congress, and had to be prepared to give an informed response.[6]

As a result of my writing, Turner scheduled a meeting for 4:30 P.M. on August 12. In preparation for it the museum assembled a sizable folder to provide her and other Smithsonian participants at the meeting a full overview. It contained letters, memoranda, advisory committee reports, information on related projects and their funding, restoration updates, news releases, and newspaper and magazine clippings. We also included the proposal, as again reedited on July 26, 1991.[7]

The meeting raised useful questions about the effect the planned exhibition might have on the bills for the museum extension pending in Congress. In hearings that Congressman Bill Clay had held on July 30, the Institution had just suffered a setback in its plans for the extension. With Colorado congressman David Skaggs challenging the Smithsonian's choice of a Dulles site and insisting that Denver could provide a much more attractive offer, it seemed that a final decision on the exhibition should be postponed pending a settlement on the extension in Congress.

The question was taken up again a few weeks later at a September 5 meeting with Adams, Turner, and Smithsonian head of government affairs Margaret Gaynor, where we confirmed the decision.

Unfortunately, within days, Mrs. Turner was found to be terminally ill with cancer, was confined to her home, and passed away early in 1992. I was to see her only one more time, when she came to her office for a few hours in late October 1991. Until her successor as under secretary, Constance Berry Newman, could assume the position in the summer of 1992 and gain sufficient familiarity with the situation, a final Smithsonian decision on the exhibition had to be put on hold.

The museum nevertheless had to continue work if we ever hoped to open the exhibition by 1995. On October 11, 1991, I drafted a long memorandum for Adams, placing the exhibition in perspective. I showed how other exhibitions and symposia on warfare had prepared us to take on this challenging display, how we had been able to attract some of the best scholars to help prepare a thoroughly documented exhibit, and how questions of security were being handled. By this time, the idea of displaying the *Enola Gay* across 4th Street from the museum had been abandoned. Erecting a separate, freestanding structure and providing an independent set of security guards had turned out to be excessively expensive. We were now contemplating a temporary structure abutting the museum on the west side. On October 22, I wrote to Adams:

> As you know, the National Air and Space Museum has attempted, during the past few years, to examine and gain a deeper understanding of the interrelationship between aerospace technology and modern warfare. This is a complex topic, still inadequately understood. Aerial warfare has clearly shaped world history in this century, but a properly documented picture of successes and failures, and the distinction between doctrine and reality is only just beginning to emerge.

As the nation's primary aerospace museum, . . . the National Air and Space Museum has an exceptional opportunity to explain some of the pivotal aerospace issues of our times to a public which does not otherwise have ready access to open and balanced presentations. That kind of explanation on historically decisive issues is particularly important in a nation with a democratically constituted government whose very structure requires an informed citizenry.

Starting in late 1989, the Museum organized a successful fifteen-month-long series that dealt with one of the more complex issues of World War II. *The Legacy of Strategic Bombing* included scholarly symposia, films, and lectures. It brought before the public distinguished participants with opinions as diverse as those of Kurt Vonnegut and Curtis LeMay. We videotaped all of these presentations, and are expecting to make use of particularly informative fragments to provide balance to a series of planned exhibitions. The first of these is *Legend, Memory and the Great War in the Air,* which is about to open and includes sections on the origins of air power and strategic bombing in World War I. The next is a sequence devoted to *Air Power in World War II.* It will include the display of some of the more significant aircraft from the war, with accompanying exhibitry that will place them in the context of that conflict. This display will be placed in the west end of the Museum building, and different aircraft will be rotated in at a rate of about one per year, during the fiftieth-commemoration of the war between 1992 and 1995. The series' lead curator is Michael Neufeld, an expert on World War II matters, and in particular on the V-2 development—Germany's attempt to perfect the ballistic missile as a wonder-weapon substitute for air power, in the desperate last months of the war in Europe.

As the final installment of *Air Power in World War II,* we plan to exhibit the Boeing B-29 *Enola Gay.* Though the aircraft might just barely fit into the largest gallery in the building, its weight and that of the cranes required to assemble it in place would far exceed the floor loading capacity of the building, threatening the most costly consequences if installation inside the building were undertaken. However, it should be entirely possible to display the airplane in a structure abutting the west end of the Museum building, and immediately adjacent to the area where earlier and smaller World War II planes will have been displayed up to that time. . . .

We have begun considering security measures that will need to accompany the exhibit, and are coordinating those with Richard Siegle [Smithsonian Director of Facilities Services]. While letters concerning the display of the *Enola Gay* seemed divided between proponents and opponents, mailings in opposition seem to have almost entirely subsided in the past few years. Articles on the restoration of the airplane are continually appearing in papers all over the country, without eliciting any comment. And two features, respectively in the *Washington Post* and in the *New Yorker,* on the planned display of the aircraft in 1995, resulted in not a single letter for or against, no phone calls at all to the Institution, and only one phone call to the Museum from a lady who mistakenly thought that the exhibit would be celebratory and wanted to advise against it. Nevertheless, security surrounding the exhibit will be a strong concern, and we will make sure that nothing is left to chance regarding the safety of visitors or artifacts.

Visitors will be able to approach the planned exhibit only from within the Museum proper. As they approach the west end of the building, they will first enter the area where earlier displays in the *Air Power in World War II* series will have been mounted. Here we will introduce the visitor to the history of aerial bombardment, in an understated, sober and, if anything, quietly somber display. Although strategic bombing had been introduced in the final stages of World War I, we will emphasize the period starting with the Guernica bombing in 1937, which started a steady escalation, as the contesting nations successively sought to vanquish their opponents through raids of ever increasing ferocity. We will document the public's response to these raids, both among the Allies and the Axis nations.

We plan to faithfully recreate the mood in the United States toward the end of the War, by exhibiting a representative display of newspaper and magazine clippings, recordings of speeches and broadcasts, and newsreel clips of the day. This will lead to a display presenting the choices President Truman faced in the waning days of the War, as he had to decide for or against the use of the atomic bomb. You may recall that Gregg Herken, chairman of our Space History Department, some years before coming to the Museum authored *The Winning Weapon*, a book that dealt with Truman's dilemma about sharing the atomic bomb with the United Nations after the war. He is fully familiar with the Truman archives, and his expertise on the subject will undoubtedly make itself felt at this stage.

. . . We intend to provide the thoughtful visitor sufficiently detailed information so that the context of the bombing can be understood. The exhibit will deliberately avoid judgment or the imposition of any particular point of view, but will give visitors enough information to form their own impressions.

We will end this portion of the exhibit with a brief account of the Nagasaki raid, the end of the war, and the beginning of the nuclear arms race.

Only upon emerging from this exhibit will visitors be confronted with the *Enola Gay* itself. They will now have the information they need to make their own interpretation of the gleaming aircraft, which, on its own, might merely be an impressive piece of aviation technology. Labeling may be limited to a description of the aircraft. The tone will again be somber. When visitors leave the building they will have not only seen the *Enola Gay* but they also will be better informed about the decision to drop the bomb and the origins of the nuclear arms race. However far armaments reductions go by 1995, it is clear that the nuclear dilemma will still be with us and that this legacy of World War II will persist.

As you had requested, I am accompanying this memorandum with a plan of the Museum's west end showing the *Enola Gay* in its intended location. The exhibit is expected to be on display for four or five months.[8]

ADAMS WORRIES

Secretary Adams must have been mulling this over for some weeks. He called me early on December 11, 1991, while the staff and I were meeting with retired Vice Admiral Tom Kilcline, president of the Retired Officers Association, and other Navy officers to discuss a film we hoped to make on aircraft carriers and naval aviation.

On returning to the office I called him back. He was clearly concerned that the museum needed to forge strong relations with backers from the military, and began telling me I no longer was at a university and should really begin establishing close ties.

This was an easy one. I told him I had just returned from a meeting with half a dozen retired admirals who had come over to see the beginnings of a film we would like to produce, and that the Air Force chief of staff would be over for lunch.

Adams harrumphed and said he guessed I had things under control.

Gen. Merrill A. "Tony" McPeak did in fact come over that noon, for a meeting of the museum's congressionally mandated advisory board. The board's ten members include the chiefs of the Army, Navy, Air Force, Marines, Coast Guard, NASA, and the Federal Aviation Administration, or their designees. Three other members are presidential appointees. Normally, the Air Force chief of staff attends in person, while the other military services are represented by the chiefs of their air arms. NASA and the FAA are usually represented by the deputy administrator. On this occasion Vice Admiral Richard F. Dunleavy was representing the Navy, while Rear Admiral Richard Herr represented the Coast Guard. Samuel W. Keller was there for NASA, James "Buddy" Thompson, from Louisville, Kentucky, Peter F. Schabarum, from Los Angeles, and Theodore C. Barreaux, from Washington D.C., were the presidential appointees. Three members had been unable to attend—a fairly normal matter with such a high-level group that often had to be at the administration's beck and call.

From the museum, Wendy Stephens, Linda Ezell, Tom Crouch, and Gregg Herken joined me. Two topics occupied the agenda, the extension and the exhibition of the *Enola Gay*. On the exhibition, the minutes of the meeting record,

> Lin Ezell gave an account of the work that had been going on for several years to completely restore the airplane by 1995. Tom Crouch and Gregg Herken added a description of the kind of exhibition in which the Museum hopes to be able to display the aircraft. The main thrust would be for a thoughtful retrospective which places the role of the aircraft and the bombing of Hiroshima in the context of its times, focussing on factors affecting Truman's decision to drop the bomb, and presenting the factual information on World War II that has emerged from archival documents over the years. The Museum would refrain from editorializing or seeking to influence conclusions individual visitors might reach.

In responding, Committee members urged care to avoid revisionism, and had concerns about how Truman's decision might be viewed by visitors in the light of developments of the past fifty years, rather than in terms of information the President had available to him at the time. There also was discussion on whether the Museum would not be acting more like an armed forces museum, rather than as a museum of technology in its planned exhibit.

> Martin Harwit explained that the Museum's Congressional mandate is to 'memorialize aviation and space flight,' rather than to highlight technology. The historical emphasis that has traditionally marked the Museum's exhibits since its opening in 1976, therefore, is in line with the intentions Congress had in establishing the Museum.
>
> The Committee members were unanimous in agreeing that the *Enola Gay* is an artifact of pivotal historical significance and that it should be exhibited. It was clear, however, that the manner in which the exhibition ultimately would be mounted deserved a great deal of thought in the years remaining until the restoration is completed.[9]

Attached to the minutes was a lengthy appendix on the status of the extension in Congress.

On receiving these minutes, which were distributed on December 31, Ted Barreaux wrote a note to advisory board members and me in which he wanted to be sure his advice would be heeded:

> To repeat what I said at the meeting, I believe that in order to place the *Enola Gay* in its proper historical context, the museum should include as part of any exhibit, the historical aspects of the decision to use the atomic bomb.[10]

Adams apparently still had concerns. He asked his senior staff to meet at the museum on December 23, 1991, to thoroughly examine the new World War I gallery we had just opened in November, and then to meet to discuss controversial exhibitions in general. He thought the World War I gallery might provoke visitors, and would be a useful starting point for a discussion on other exhibitions.

Adams had already come over by himself to see the exhibition alone, but we showed the other senior Smithsonian staff around at 9:00 A.M., before the museum opened to the public, and then they, the secretary, and I met for a private discussion in the museum's main conference room. I was surprised that Adams didn't want any other museum staff members around.

For the assembled group, I listed the various precautionary steps we had taken to assess visitor sentiment as we prepared the exhibition on the *Enola Gay*. Many of these programs that dealt with related military subjects I had listed in the third paragraph of my October 22, 1991, memorandum for Adams. I added that we were proceeding cautiously. Adams voiced a variety of concerns, and we discussed the issue for perhaps an hour before the meeting broke up.

CONCERNS ABOUT THE MUSEUM EXTENSION

With no under secretary for much of 1992 and the fate of the museum extension still under debate in Congress, the internal debate on the exhibition of the *Enola Gay* went no further. Adams did attend the Museum Advisory Board's meeting on June 10, 1992, to listen to the board's opinions on the exhibition; but not much more happened until November 1992, as the museum and the Institution waited to see how Congress would act. Our plans for building the museum extension at the Dulles International Airport—a project of overriding importance for the future of the museum and the safety of the airplanes, spacecraft, archives, and aerospace artifacts in its collections—had to be accorded preeminence.

12

A Battle for the Museum Extension

A COMPETITION BETWEEN STATES

The National Air and Space Museum desperately needed an extension to properly house its collections. Many of the buildings at the Garber Restoration and Storage Facility were unheated, with leaky roofs and nesting birds and insects. The pursuit of the extension, however, was difficult. A fierce competition for ownership ensued and seriously delayed our plans for mounting the exhibition on the *Enola Gay.*

On January 4, 1988, Congressman Sidney Yates had told Secretary Adams and me that the museum was now free to go ahead with master planning for an extension at Washington's Dulles International Airport, located in nearby Virginia, or at the Baltimore Washington International Airport (BWI), in nearby Maryland. This was formalized by the regents at their meeting at the end of that month. They instructed Adams to initiate a comparison of Dulles and BWI, and encouraged the Institution to seek the best supporting offer from the two states, both of which had expressed an interest in attracting the facility.

I met with Richard Siegle, who was in charge of all new construction at the Smithsonian, and we and our two staffs worked closely together to obtain the best deals we could. We undertook two studies. The first was a \$120,000 preliminary investigation "Planning for the National Air and

Space Museum Extension," conducted in the summer of 1988. This determined the requirements of an extension that would meet the museum's most urgent needs. It envisaged a complex that would provide adequate storage for all the museum's collections, and feature an additional four galleries. One of these would be "a thought-provoking and possibly controversial exhibition on the history of strategic bombing, from the pre-WWI era to modern strategic arsenals, and feature the fully restored B-29 *Enola Gay* as the centerpiece."

This study was followed the next year by a more extensive, $396,000 effort, which further defined these requirements and then compared the Dulles and BWI sites to determine how well they would serve the extension. This study looked at every aspect of site selection, including suitability of the site for construction, wetlands preservation, archeological features that might require special attention, noise pollution from the airport, visitor access by public transportation, numbers of visitors expected, as well as the ethnic and cultural makeup of the expected visitorship, and many other factors.

Museum and Institution staff worked closely with teams representing Maryland and Virginia. We tried to distinguish for them factors that were essential rather than merely desirable; we prioritized our needs; and we encouraged each side to independently come up with the strongest bid to help us. That meant the most suitable parcel of land and the best financial terms.

The bids for land were analyzed and submitted to Secretary Adams in the closing days of 1989. The final packages of financial support offered by the State of Maryland and the Commonwealth of Virginia went to Adams a month later, in time for the regents' meeting of January 1990.

These proposals were an example of the generous help the museum could expect from friends—whether they were individuals, communities, corporations, or entire states. Many were ready to provide aid, though they were also expecting to benefit from the museum's popularity, some of which might rub off on them.

There were others who were more intent on helping themselves, rather than the museum.

POPULARITY HAS A COST

Everybody loved the National Air and Space Museum. Some loved it so much that they wanted to wrest away pieces of it. That might be an atomic bomb casing, like *Little Boy*; it might be an entire aircraft, like the *Enola Gay*; it might be a complete facility, like the extension; or, more subtly, it might be a dominant influence on what the museum displayed and how we displayed it. So intense were some of these ambitions that the museum had to spend inordinate effort to fend off the many aspirants.

Just before the Virginia and Maryland bids went to the secretary, the Institution received an invitation to meet with two members of the staff of the Senate Appropriations Committee's Subcommittee on Interior and Related Agencies, Rusty Mathews and Charles D. Estes. The subcommittee, chaired by Senator Robert Byrd of West Virginia, reviewed the Institution's annual budget, giving its chairman considerable influence over the Institution.

Senator Byrd wanted to know what the Smithsonian could do for West Virginia. On his behalf, Mathews and Estes suggested that Sheperd Field at Martinsburg, West Virginia, might be an excellent site for the museum extension. Fortunately, I had been forewarned, and had come with a letter the museum had written to Senators Byrd and Randolph Jennings in November 1983, explaining that the requirements for the extension called for a minimum runway length of 8,500 feet, in order to land the NASA Space Shuttle mounted atop its Boeing 747 mother-craft. The Martinsburg runway was only 7,000 feet long and would not qualify.[1]

This did not fully satisfy the Senator's staff, and negotiations with the Institution continued for some months before the museum was totally off the hook. Fortunately, Senator Byrd did not press much further to have the extension located in West Virginia. But other localities and their congressmen did, and proved far more persistent.

Just prior to the regents' meeting in January 1990, Denver, Colorado, submitted a last-minute proposal that the extension be based at the city's Stapleton Airport. Denver was about to build a newer, larger airport, and Stapleton would become available for the intended purpose. The regents had no time to study this proposal, which in any case was far less complete than the other two. At their meeting they reaffirmed their preference for Dulles. They saw nothing to challenge the merits of the originally selected site and preferred the federal status of the land at Dulles in contrast to the state or city ownership of land at other airports.

The Stapleton bid was pushed by Congressman David Skaggs, Democrat of Boulder, Colorado, who approached the chairman of the Appropriations Subcommittee on Interior and Related Agencies, Sidney Yates, with a Denver offer. He portrayed this bid as a far better deal for the government than either Virginia's or Maryland's. In the autumn of 1990, Yates, therefore, asked the Institution to stop planning for Dulles, to investigate more deeply Denver's Stapleton Airport as a site for the extension. To that end he authorized expenditures up to $50,000.

The Smithsonian Institution now commissioned a comparison of Stapleton to Dulles, and for that purpose a team of us visited Denver on two occasions. On our first visit, we arrived in Denver in the early evening on December 11, 1990. To our amazement, we were met by two gleaming white, stretch limousines and whisked off to one of the city's premier hotels. That evening we were guests at a generous reception hosted by Denver civic leaders.

The next morning a hectic round of meetings began at breakfast with then Mayor Federico Peña, who spoke very effectively on behalf of his city. The leading person behind the entire effort appeared to be Walter A. Koelbel, a local real estate developer, who felt that a National Air and Space Museum extension would be an enormous tourist attraction for Denver.

During the day, we met with airport personnel to see what airport buildings, if any, might be suitable for museum use. In most cases, ceilings were too low, floor loading capacities inadequate, or, in the case of available hangars, there were inadequate temperature and humidity controls to store the airplanes and spacecraft safely for anything like the centuries we had in mind. Late in the afternoon, we were joined by Congressman David Skaggs, who had flown in from Washington for the occasion.

In the evening, three of us were invited to the governor's mansion to have dinner with Governor Romer and Mayor Peña. We had expected a closely guarded building, but when we arrived, the front door was unlocked and we let ourselves in. The governor was there by himself and came to greet us and fetch us into the living room. It was all refreshingly informal, and we had a pleasant evening with this governor known for his innovative ideas.

After further discussions the next day, we returned to Washington with our work cut out for us. We made a variety of estimates on the cost of operating an extension at a distance of two thousand miles from Washington. The added duplication of staff that would be needed seemed prohibitive, and because the available buildings at Stapleton were largely inadequate to our needs, we would require extensive and expensive additional construction.

The final summary of our findings concentrated on the operating expenses anticipated at Denver, though it also reflected more broadly on operating expenses at any remote site. This report was provided not only to the secretary, but at Congressman Yates's request also to the United States General Accounting Office (GAO). We concluded that whatever support for construction Denver promised, the increased operating costs of having to maintain an entirely separate staff at such a great distance would nullify those advantages in a matter of a few years. Most importantly, however, the Denver site would not provide badly needed improved care and housing for the museum's Washington-based collections, nor could it house the archives and exhibition production services needed for the museum on the Mall.

There was another striking difference. In contrast to the Virginia and Maryland proposals, the Stapleton submission to the regents contained guarantees neither from the governor of Colorado, nor from the state legislature, nor even from the mayor of Denver, but was instead signed by Walter A. Koelbel, Sr., head of a specially formed consortium of developers named "Air and Space West," whose origin, stability, and financial resources were not made available to us. I couldn't believe that Congress would hold us up for another two years to consider this kind of offer. Yet that is precisely what ensued. Mr. Yates held hearings on February 5, 1991. Mr. Skaggs asked to be heard and spoke enthusiastically on behalf of the Denver site. I was surprised by his statements in favor, which ran directly counter to the body of factual information we had carefully compiled. Then the GAO report was presented. We had not had an opportunity to see it before the hearings, and it also came out against us.

Fortunately, Under Secretary Carmen Turner asked Mr. Yates whether the Institution might not meet with the GAO to clarify a number of discrepancies. Yates agreed to that, and under Turner's diplomatic and resourceful leadership we met a few times with the GAO representatives and convinced them that we had a case. They had asked us to produce a thirty-year projection, to show the difference in cost of operating an extension at Dulles and at Denver. Our analysis showed a long-term deficit that ranged up to $200 million. On the basis of such figures, the GAO team reversed itself and issued a second report to Congress, dated March 21, 1991. In this letter, the GAO confirmed that "the choice of Dulles International Airport as the preferred site can be objectively defended by the Smithsonian."

Throughout these discussions, the museum had maintained that without a nearby extension, we would be drastically curtailed in our ability to serve our army of visitors. A nearby extension was also needed to permit the museum to house and display airplanes too large for transport along roads into town. Visitors to the Washington area would be able to see not only the small spacecraft and airplanes that can be displayed on the Mall; at the extension they would also be able to view the large commercial and military aircraft and satellites that in the course of this century have totally changed the way we live and work. No book and no other educational institution can bring to life this perspective on America's history as vividly as the Air and Space Museum.

None of this seemed to deter Congressman Skaggs. Together with Maryland congressman Ben Cardin, whose district included the Baltimore Washington International Airport, he introduced a bill in the House to reopen the competition for an extension site nationally.

And again—incredibly, I felt, in view of the GAO's confirmation on Dulles—it was the Skaggs-Cardin bill on which the House Administration Committee's Subcommittee on Libraries and Memorials chose to hold hearings on July 30, 1991. Again Skaggs spoke, as did Walter Koelbel and others. Again I was disturbed by the figures they presented, none of which matched ours.

With the Institution inactivated by Carmen Turner's tragic death, many of Skaggs's colleagues in Congress saw a possibility of attracting the extension to their own districts in Texas, Pennsylvania, and Florida, to name just a few. By this time, the Institution had spent about half a million dollars just to investigate the merits of three different sites, Dulles, BWI, and Stapleton airports. With as much interest as was being shown, I figured we might need to look at another two dozen sites, and run up a bill of $5 million in selecting the best one. Where that money was to come from, nobody had yet indicated.

The late Speaker of the House Tip O'Neill used to say "all politics is local." With so many localities interested in capturing the museum extension, and so many years of political skirmishing spent on gaining advantage in the contest, it was not clear who, if anyone, was looking after the nation's broader interests in a national institution like the Smithsonian.

SETTLING ON A SITE

In the waning days of the 102nd Congress, before the elections of 1992, Congressman Skaggs, anxious to get his bill on the floor of the House for a vote, asked Speaker of the House Tom Foley to put the Skaggs-Cardin bill on a fast track, where passage would require a two-thirds majority, in place of the usual simple majority. The Speaker complied.

On the afternoon of the vote, we all were quite anxious. But as the votes were tallied, it became increasingly evident that the bill did not have the support of even a simple majority. The bill was heading for certain defeat. When it was all over, the Skaggs-Cardin bill, instead of gaining a two-to-one victory, had suffered a three-to-one defeat.

The following year, in the 103rd Congress, the Institution's bill for the extension finally prevailed. By midsummer 1993, both the Senate and the House passed authorizing legislation, which President Clinton, in early August 1993, signed into Public Law 103-75, authorizing the Smithsonian to plan for a National Air and Space Museum Extension at the Washington Dulles International Airport.

THE COST OF DELAY

Denver's aspirations for the extension had caused the museum a four-year delay in our efforts to confirm a site and begin plans for construction of a facility where, among other large aircraft, the fully assembled *Enola Gay* would be displayed.

Perhaps the most serious collateral cost was the animosity toward the museum that sprang up among veterans who wanted to see the entire aircraft fully assembled and quickly displayed before they passed away. They were frustrated by the delay in congressional authorization for the extension, resumed their attempts to wrest the *Enola Gay* away from the museum, and persuaded themselves that we were deliberately dragging our heels. Their animosity would cost us dearly.

The delay had also set us back for more than one year, from the fall of 1991 to the end of 1992, in our plans to shape the exhibition on the *Enola Gay* and define its contents. Fearing that the very mention of the exhibition might muddy the waters, the Smithsonian and museum had all but shelved activity on the exhibit pending resolution of the extension issue in Congress.

When we finally could go ahead with plans for the exhibition, we were less than two and a half years from a planned opening, a relatively short time for planning and mounting such a complex exhibition. It was high time to start.

GOING AHEAD WITH THE EXHIBITION

The decision to go ahead with the exhibition came about unexpectedly.

On November 18, 1992, the Smithsonian's director of government relations, Margaret Gaynor, wrote to Secretary Adams and the Institution's new under secretary, Constance Berry Newman,

> Last Thursday several Congressional staff and Regents' liaison people visited the Garber facility and the Museum Support Center. Because of commitments here I did not join them, but Brad did. He reported, and it was later confirmed, that during lunch Rusty Mathews had a lengthy conversation with Dr. Harwit about the *Enola Gay* and NASM's plan for its exhibition in conjunction with the 50th anniversary in 1995 of the bombing of Japan. Apparently Rusty urged caution with respect to those plans, which could include bringing the plane to the Mall.
>
> Rusty called me yesterday to make certain I was aware of his concern, which acquired considerable depth over the weekend. We had a lengthy exchange, in the course of which I recalled that the issue of exhibiting the *Enola Gay* on the Mall had arisen earlier and that you had ruled against it.
>
> Before moving into another authorization and appropriations cycle it would be useful to bring together those shown below to clarify the issue of the plane, the strategic bombing exhibit, and the anniversary, particularly in light of the extension proposal, the 1994 exhibition in Japan, and the Institution's anniversary in 1996. Unless we are clear about these interrelationships, and can defend them cogently, the bomb and bombing are likely to color and to dominate the discussion about the extension, our budget requirements, and our future plans, perhaps to the detriment of all.[2]

Rusty Mathews was a senior congressional staffer on the Senate Appropriations Committee's Subcommittee on Interior and Related Agencies, where he reported to Senator Robert C. Byrd, the subcommittee chairman. The Smithsonian liked to maintain good relations with him. This memorandum, which went to nine other members of Adams's staff, including Connie Newman, therefore raised considerable concern. Both Bob Adams and Newman immediately called me. Not having been informed, I could not respond until the next day. Referring to Jim Hobbins, staff director for the secretary, and Brad Mims of the Congressional Liaison Office, I wrote,

> I am writing because I was taken aback by your respective phone calls last night. . . . I do not know what you had been told when you called me, but fortunately both Jim Hobbins and Brad Mims were present at Garber, and I would welcome your checking with them on their version of my account:
>
> I joined the group of visitors, which included Rusty Mathews, Kathy Johnson and three others, just before lunch, having taken a two-hour leave from an all-day session at the National Academy of Sciences. As we sat down to eat our sandwiches, Rusty launched into a probing conversation on the *Enola Gay,* whose ongoing restoration he had evidently just seen. He wanted to know all about when, and how it could be exhibited, what the public perceptions were, what logistical problems needed to be faced, who all had been consulted, what scripts were in existence, what the potential costs might be, and so on. This part of the conversation may have taken twenty minutes, before the discussions switched to other matters.
>
> I answered Rusty as truthfully as I could. The Museum has, in fact, considered and analyzed most of the possibilities and alternatives that can be brought up in a conversation of that length, and it seemed to me that Rusty seemed satisfied that we knew what we were doing and had done our homework.
>
> One of the last questions Rusty asked dealt with the deadline by which a decision to exhibit or not to exhibit would need to be made. I told him it was highly unlikely that we would be able to mount an exhibit that included the entire aircraft in 1995, unless a decision was reached by the end of the calendar year, and even then it would be difficult to see how we could ever raise the necessary funds in time—most likely in the range of $5 million—since this was not the kind of noncontroversial project that industrial donors favor. Rusty then mused that Bob might want, in any case, to check with various people on the Hill, particularly with Senators Stevens and Hatfield, to see how they felt about an exhibition.
>
> Perhaps I should have passed that message on. But since Bob had told me he was reluctant to broach the topic of the *Enola Gay* before the Extension is authorized, I did not feel you'd be inclined to follow up.
>
> As we were ending our lunch, Rusty suggested to me that we might get together again. As a result we've arranged to have him and Kathy come down to the Museum for lunch on separate days. . . .[3]

Kathy Johnson was an aide to House Appropriations Subcommittee chairman Sidney Yates, in charge of the Smithsonian budget.

This exchange, though causing a flurry at the time, brought sufficient attention to the *Enola Gay* to get the Institution moving again. I had a meeting with Newman the next day. She recommended that the issue of an

Enola Gay exhibit should go to the regents. We needed to obtain a clear decision for or against an exhibition.

Three days later, on November 23, 1992, the two of us held a further meeting, this time with Adams and Assistant Secretary for Museums Tom Freudenheim, in the secretary's office. By this time, all of us at the museum were convinced that we could not mount an exhibition of the entire aircraft at any place affordable in 1995. I told Adams that we might only be able to display the forward fuselage, which would readily fit into the building, and would not require an additional structure. I mentioned our plans for the "Air Power in World War II" series of displays in the west end of the museum, which would include the P-47 Thunderbolt and the German Arado, and be mounted during the World War II anniversary years. Displaying the *Enola Gay* as part of such a series and in the same part of the museum would be relatively straightforward.

Adams jumped at this possibility with evident relief, saying that if we exhibited the *Enola Gay* as part of an ongoing series of World War II aircraft, and only displayed its forward fuselage, it would appear to be less of a symbol. He had been afraid that because it was such an American icon, the *Enola Gay* would be especially controversial. I understood that well, but regretted the turn of events for just this reason. Nevertheless, it was the best we could do. I agreed that the museum should provide Adams with a revised proposal reflecting current intentions.

By December 1, 1992, Michael Neufeld had again reworked the proposal, now titled, "A Proposal for a Fiftieth-Anniversary Hiroshima Exhibit in the West End (Gallery 104)." It differed from earlier versions in several respects:

> Unfortunately, assembling the whole airplane for an August 1995 exhibit no longer seems feasible. Therefore I would like to propose that we display the forward fuselage of the aircraft in the west end of the Museum. This impressive artifact allows the visitor to see the *Enola Gay*'s cockpit and bombsight, the evocative name painted on the port side of the aircraft, and the forward bomb bay with our reconstruction of the special atomic-bomb sway braces and latch.
>
> For maximum visual impact, the bomb bay needs to be adequately lighted and the internally lit glass nose of the B-29 needs to be visible along the corridor as the visitor approaches the west end of the Museum. . . . The primary exhibit space would be to the right of the fuselage, facing the side with the name *Enola Gay.* Additional exhibitry could be put along the gallery's left wall, if it was fairly flat, and toward the rear of the starboard side of the fuselage, if needed.
>
> The exhibit itself—which could be part of the 'Air Power in World War II' series, if that label were convenient—should take as its theme Hiroshima and Nagasaki as watershed events in the origins of the Cold War and the nuclear dilemma. ("The Dawn of the Nuclear Age" is a possible title, but it is a bit clichéd and perhaps too positive sounding.) Placing

> it within that context helps to minimize the controversy that could be engendered by an exhibit that explicitly made a discussion of the decision to drop the bomb its central focus. Controversy is nonetheless unavoidable, because the exhibit cannot avoid an examination of the process by which the United States government decided to use atomic bombs on Japan, nor can we, or should we, avoid photographs of the physical effects of the bombs on the inhabitants of the two cities. The point of those parts of the exhibit should be to convey to the general public some of the moral and political dilemmas of the decision to drop the bomb, while giving the historical context of 1945. This material could be supplemented by a movie or by videos of participants and later commentators giving their views. This material could be integrated into the larger theme by indicating that Hiroshima and Nagasaki opened the age of the nuclear dilemma: can nuclear weapons be used in war again, and is there any way to 'ban the bomb' when the genie is out of the bottle? The political effects of the decision to conduct the Manhattan Project without Soviet participation and the diplomatic dimensions of the atomic-bomb decision also must be part of the exhibit, because one of the most crucial effects of the bombing was to strengthen the rapid move towards a Cold War immediately after 1945. . . .[4]

Two days later I met with Nadya Makovenyi, head of the museum's Exhibits Department, Tom Crouch, Gregg Herken, and Mike Neufeld, to discuss the proposal, before forwarding it to Adams with a cover letter.[5]

To allay some of Adams's fears, I mentioned the support we had from veterans for the restoration, and—as I will mention below—preliminary contacts with the Japanese. I also referred to safety measures we had discussed with the Smithsonian security office, as well as the apparently relaxed mood of the public.

The following week Jim Hobbins replied on behalf of Bob Adams and Connie Newman:

> In talking with Bob and Connie Newman I get the sense that neither has any major disagreement with any aspect of your proposal, but they remain especially wary of the potential political implications of this show. Accordingly Connie has endorsed the Secretary's position that he will sanction the further development of the exhibition *provided* that you agree to show both of them plans and layouts as well as all labels and other descriptive materials, before any commitments are made.
>
> Please let me know if you agree to this, because the Secretary wants to inform the Regents' Executive Committee in early January that the show will be proceeding with this understanding.[6]

I replied on December 21, 1992,

> We've discussed it here at the Museum, and all of us feel pleased at the prospect now offered to go ahead with this exhibition, subject to your request that we show both Connie Newman and Bob Adams 'plans and layouts, as well as all wall labels and other descriptive materials, before any commitments are made.'

> We realize that this exhibition is extraordinary, with implications that go well beyond the interests of the museum and the Institution alone. This makes the additional scrutiny you wish to direct at this exhibition both understandable and reasonable.
>
> Accordingly, we will go ahead on this basis.[7]

I was relieved that the hiatus was over and that we could now openly go forward. Three key factors appeared to have contributed to the turn of events. After more than a year, the Institution once again had a fully effective under secretary. The new proposal to display only part of the *Enola Gay* seemed to rob the exhibit of its emphasis on a national icon and might reduce tensions. And, most importantly, the waning days of the 102nd Congress had seen the defeat of Congressman Skaggs's bill to move the extension to Denver, and gave assurances that the new Congress would finally give us authorization to plan a structure at the Dulles International Airport.

The year 1992 was closing on an optimistic note.

EARLY 1993

One of our first actions of 1993 was to draft a request on January 4 to Assistant Secretary Tom Freudenheim to ask for Special Exhibition Funds, SEF. The actual proposal was submitted to him on the due date, February 19, and reiterated earlier thoughts. Some of the requested information, however, was new:

> Target Audiences
>
> In the case of an exhibit like this, it is clear that the target audience is very broad: all thinking citizens (and children) of this country and others. While there will undoubtedly be other commemorations and demonstrations in connection with the fiftieth anniversary, this exhibit can provide a crucial public service by reexamining the atomic bombings in the light of the latest scholarship. Almost everyone has an opinion on this matter, but these opinions are often shaped by limited knowledge and personal prejudices. Hopefully the exhibit might even contribute to a deeper examination of these questions among the general public of the United States, Japan and elsewhere. As a purely secondary objective, this exhibit will also attract to the National Air and Space Museum underserved audiences who have little interest in aerospace history or who may believe that the Museum only offers an uncritical examination of the difficult aspects of aerospace technology.
>
> Through a team approach, bringing together the principal curators, the exhibit designer, and audience advocates from the NASM Exhibits and Education Departments, extensive efforts will be made to ensure that a wide range of individuals will be engaged by the exhibit at all intellectual and interest levels. The educational aspects of the exhibit can also be supplemented with a brochure, a catalog and public programs, plus the

> school-oriented activities of the Education Department. The scholarly content of the exhibit will be ensured through the participation of . . . Michael Neufeld, . . . Tom Crouch, . . . Gregg Herken, . . . the many historians interested in this topic at the National Museum of American History, and other scholars inside and outside the Institution. . . .[8]

My cover letter to Tom Freudenheim noted,

> . . . Because of sensitivities about this exhibition, the Museum was asked, more than a year ago, to refrain from any further planning. That moratorium has now been lifted, but we find ourselves with little time left to complete the exhibition. Raising money for the exhibition would have been difficult in any case, but is practically hopeless on this short notice.
>
> In order to succeed we will need to obtain $25 thousand in planning money now. We will also want to return next year with a one-time request for production funds. We hope to keep that request modest, knowing that there may be little exhibits production money available to the Institution for such purposes next year.
>
> At any rate, we would appreciate any help you could provide for this exhibition, which is going to be challenging and—if we succeed—should also prove important.[9]

Later that year we would be asking for a considerably larger commitment, but starting this late, we needed this seed money right away to meet the 1995 deadline.

THE 50TH ANNIVERSARY COMMEMORATION COMMITTEE

Claude M. "Mick" Kicklighter, a burly, brown-haired lieutenant general retired from the Army, was executive director of the congressionally mandated "50th Anniversary of World War II Commemoration Committee." The committee had been established to sponsor and coordinate commemorative activities throughout a five-year period that spanned the fiftieth anniversary of Pearl Harbor and the fiftieth anniversary of the war's end. Though Kicklighter now was retired, he operated out of offices in the Pentagon.

Since the exhibition we were mounting on the *Enola Gay* was a commemoration of the aircraft's pivotal mission in the war, I thought our two groups should get together to see whether we might be able to help each other. On March 29, 1993, we held a first meeting with Kicklighter and four members of his staff, including Army colonel Kevin Hanretta and Navy commander Luanne Smith. Linda Brown, a staff member from the National Archives, also attended. Our visitors arrived at the museum at 1:00 P.M., and stayed for two hours to hear about our plans and for us to hear theirs. Their motto was "A grateful nation remembers," and they honored war veterans, their families, people who had worked on the home front, as well as significant events and campaigns. To organizations working with them the committee distributed a commemorative flag bearing their motto. After some months, they gave us such a flag as well, hoping that we would mount it with our exhibition as a sign of their support. Later, they probably wished they had not.

Kicklighter was enthusiastic at that first meeting, however. Nobody else was touching the atomic bombings, and yet somebody should commemorate them. So here we were, willing to jump into the breach and take full responsibility. Kicklighter asked whether we might be able to help his effort, whose congressionally mandated mission called for both commemoration and education, and included the distribution of educational materials to schools. If we could produce some pamphlets on the atomic bombings, they could begin distributing those in late 1994. They could also use similar pamphlets on Air Power in World War II, and another on the Tuskegee Airmen—the first Black pilots permitted to fly in combat in the U.S. Army Air Forces. Our curators were expert on both subjects and could readily provide the material Kicklighter's team needed. In return, Kicklighter was willing to offer resources to print in large numbers, at no cost to the museum, a pamphlet we could distribute to visitors to guide them to all our World War II offerings. We also discussed our staff's helping the committee with producing brochures on aviation in World War II, which the committee could print and distribute to schools.[10]

I was particularly pleased to hear that the commemoration committee had money to disburse, and that Kicklighter thought he might be able to help fund the exhibition. Accordingly, I wrote to him two days later, thanking him for his visit, and adding,

> Meanwhile, if you have suggestions of possible underwriters for the *Enola Gay* exhibition, I would very much like to hear from you. As I mentioned, we plan to fund the exhibition out of the Museum's base; however, if there were someone who would be interested in sponsoring this exhibition, that would be great for all of us.[11]

On October 22, 1991, I had written to Adams,

> While letters concerning the display of the *Enola Gay* seemed divided between proponents and opponents, mailings in opposition seem to have almost entirely subsided in the past few years.[12]

I was reporting this only for the United States. Japan was quite another matter. We knew almost nothing about how the Japanese would react, and I felt we needed to address ourselves also to any antagonism on their part. Even as we were speaking with Kicklighter and his staff, Tom Crouch and I were about to go to Japan to attend to this issue.

In the eighteen months since October 1991, a small vocal group, which later referred to itself as "five old men," had also begun to write in again; but they appeared to be voicing largely their own opinions, rather than those of a wider community. They certainly did not sound at all like General Kicklighter, who was in touch with all the veterans' commemorative groups and who now also was meeting with officials from Japan. I was glad to hear that he too wanted to make sure that the commemorations he supported would not offend their country. We seemed to have a lot in common, and agreed to meet again in a few weeks.

In the meantime, the five old men were launching a letter-writing campaign.

13

Only Five Old Men

Over the years, a handful of men persistently wrote to anyone they thought would listen and might act. In one of his letters, a member of this group referred to himself and his comrades as "only five old men." But those five were to prove remarkably influential on the fate of the exhibition.

These men's correspondence defined many of the complaints that would repeatedly arise in veterans' accusations. Here I will raise those issues, more or less in turn, though they often appeared intertwined.

ISSUE 1: NEGLECT OF THE AIRCRAFT

As far as I can tell, William A. Rooney, of Wilmette, Illinois, who had been so outspoken in his early correspondence with Bob Adams, had not written in more than three years. But on May 11, 1990, he took to his keyboard again, this time addressing himself directly to the three members of the regents' executive committee, including Chief Justice William H. Rehnquist.

> I am addressing this letter to you because of certain facts that seem not to be taken into consideration by the management of the Smithsonian. . . .
>
> Messrs. Harwit and Adams seem to ignore that by their actions, a great number—into the millions—of Americans who fought in World War II will never see this aircraft on display. With most now aged 65, they will be

> dead before the Smithsonian acts. It is this constituency that the Smithsonian seems to ignore. I am one of these.[1]

On behalf of the executive committee, Secretary Adams replied on June 25, in a two-page letter in which he described the substantial progress that had been made on the restoration of the *Enola Gay*. To convince Rooney, he enclosed a number of photographs.[2]

That Rooney was not persuaded is evidenced by his next letter, addressed on September 10 to Congressman Sidney Yates, chairman of the House Appropriations Committee's subcommittee that was most influential in deciding on the Smithsonian's budget. Rooney again accused the Smithsonian of neglecting the *Enola Gay*, and then recalled the vow he had made to Adams, more than three years earlier.

> I have told the Smithsonian that it is my intention to see the *Enola Gay* proudly restored and displayed and that I intend, once again, to see Gen. Paul Tibbets sit in the pilot's cockpit and wave out of the pilot's window as he did on that historic day, August 6, 1945. Gen. Tibbets and I are now [respectively 75 and] 73. We do not have much time and we deserve better than to have the Smithsonian take this occasion from us.[3]

Rooney's two letters defined the first of the issues. Did the Smithsonian owe the veterans an exhibition that would satisfy their nostalgia, or even celebrate them, on this fiftieth anniversary of the mission of the *Enola Gay* and the virtually coincident end of World War II?

Rooney clearly felt we did.

All of us at the museum, certainly I as its director, felt that the veterans were owed their day of glory. Doubtlessly commemorations would and should abound across the country on the fiftieth anniversary of the end of this long and terribly costly war. But did this mean that every organization had to follow a set format without exception?

The National Air and Space Museum had a large number of historical artifacts that could be displayed on this fiftieth anniversary, including the *Enola Gay*, which we were spending great effort to restore to go on public view. We were at the same time also restoring a Japanese Okha piloted suicide bomb, to illustrate the costly resistance through kamikaze tactics Allied troops were facing toward the end of the war. We had both a *Little Boy* and a *Fat Man* bomb casing we planned to place on display. With these and a host of smaller artifacts, the museum was in a position to mount an extraordinary historical exhibition on conditions at the end of World War II and the mission of the *Enola Gay* in the closing days of the war.

No other organization in the world was as well equipped to do this as we were. Should we be denied the opportunity to provide the public a truly thoughtful display that included many recently declassified documents on World War II that now provided new insights? Or should we simply become one more site of commemoration and celebration?

Given our Congressional charter, which specifies that the museum should "memorialize the national development of aviation and space flight ... and provide educational material for [its] historical study," I felt that the museum's choice was clear. On this fiftieth anniversary we could serve the public best by mounting a retrospective exhibition that would give insight into the steps that had led to the development of the atomic bomb, the design and construction of the B-29 Superfortress that was needed to deliver it, the mission of the *Enola Gay*, the destruction of the two cities, and the accommodations the postwar world had to make to find a way to live with the enormously escalated explosive power now at our disposal. The human race now knew how to destroy itself.

We would certainly honor the memory of our servicemen—there could be no question about that; but our prime thrust would be to provide a history of the times.

ISSUE 2: ONE VIEW OR A BALANCE?

For two years, nothing more was heard. Then, in mid-1992, arguably the most prolific letter writer of them all, W. Burr Bennett, Jr., entered the fray. He addressed himself to Adams on July 23, and made sure he'd gain sufficient attention by sending copies to Senator Paul Simon and Congressman John Porter of Illinois. Bennett, a former master sergeant who had earned an Air Medal and Distinguished Flying Cross as an aerial photographer in World War II flying over enemy territory from bases in China and India, introduced himself as a disciple of Rooney and got straight to the point. He didn't like the direction he thought our exhibit would take.

> Way back on March 10, 1987, you responded to William Rooney concerning exhibiting the *Enola Gay*. Mr Rooney asked me to respond.
>
> From correspondence, messages in your magazines, press releases, and newspaper interviews, I have deduced that the *Enola Gay* will be the draw for an exhibit to expose visitors to the Smithsonian's belief that strategic bombing is ineffective in terms of cost and casualties.
>
> This is anathema to B-29 veterans. The war with Japan was brought to an abrupt end because a B-29 was available to deliver the bomb. The bomb became available because this country put together special teams that created the first controlled nuclear chain reaction, invented a manufacturing process for the fissionable materials, and found a way to make a nuclear bomb.... The result of their success shocked Japan into surrender, saving millions of lives and the probable total destruction of Japan.
>
> The first target was Hiroshima, a relatively small city. Warning leaflets were dropped. *It should be noted that Tokyo was not the target selected.*
>
> Let us put all this in perspective. For the first two years we took a terrible beating by Japan.... As the war progressed we were shocked at the death march and the dreadful treatment of American prisoners. We should

not have been surprised. Most of us were aware of the "rape of Nanking," and other atrocities.... When the mobilization of our entire nation was achieved, the tide began to turn, and the resistance of the Japanese stiffened....

My point to all of the above is that we should be proud of what the United States accomplished, in a race against time against two determined adversaries with impressive technological credentials. Had either produced a bomb before the United States, I seriously doubt if they would have selected Scranton over Washington D.C. as their initial target.

Americans should be proud of these accomplishments, including the human sensitivities woven into the decisions ending the war.

Am I wrong in assuming that the Smithsonian may be ashamed of these achievements? My assumption is based on the years you left the *Enola Gay* outside, unlocked and unprotected.... It is also based on your anticipated display of the plane in an exhibit that quotes from staff may be interpreted as one aimed at changing popular views on the success of strategic bombing.

Mr. Adams, this simply is not acceptable. The *Enola Gay* cries out to be proudly displayed.

... If you will reconsider, and display the *Enola Gay* as the vehicle that accelerated the surrender of Japan, great! If you are unwilling to do that, then perhaps you should consider donating or loaning the plane to another exhibitor who will. I would be more than willing to devote significant time to find such an organization.

I have a warm invitation from Linda Ezell to visit Garber the next time I visit Washington. I cannot bring myself to see the *Enola Gay* in its present pathetic condition. But I would welcome a chance to sit down with you to see if there isn't some way in which the past 43 years can be put aside. We need to focus all available energy and funds on a positive, upbeat plan to proudly display the *Enola Gay* to Americans as well as Japanese (after all it also saved untold thousands of their lives and the almost certain total destruction of their country).

This is the very least we owe to that wonderful aircraft and the men who flew her. I liken the B-29, as epitomized by the *Enola Gay*, to the Long Bow which allowed English infantry for the first time to defeat mounted armored knights at the Battle of Crécy; or the United States Frigate Constitution, which out-sailed and out-shot the British to quickly end the War of 1812; and the pinpoint bombing of our aircraft in Desert Storm, a really short war. It is hard for me to understand why the Smithsonian doesn't appear to share my pride of the *Enola Gay*, and its crew.

I would deeply appreciate a blunt, straightforward reply explaining the precise context in which you plan to display the *Enola Gay*.

In short, if the Smithsonian will not, or cannot display the *Enola Gay* proudly for its own accomplishments then there is little point in my visit and I will begin to broaden the discussion elsewhere.[4]

Bob Adams responded to this long letter on August 27, 1992. He spoke of being aware of

> . . . previous correspondence with the National Air and Space Museum. . . . But as you know, no firm plans have yet been made for any . . . exhibit. Therefore, it is impossible for me to state what the precise content would be. . . . When we do undertake this task, and you should be assured that we have every intent to show the *Enola Gay* as soon as it is feasible, the Smithsonian will naturally take into account and present the viewpoint of the veterans. It cannot, however, present this view only. Rather, it must present in a balanced and sensitive manner all points of view about bombings of Hiroshima and Nagasaki, including those of veterans.
>
> Regarding the storage and restoration of the *Enola Gay*, the Smithsonian Institution has never hidden or neglected the airplane because it was "ashamed" of it. . . . Since 1984 the National Air and Space Museum has undertaken the largest and most difficult restoration project in its history, with the aim of completely restoring the *Enola Gay* by 1995. Far from being in a "pathetic" condition, many of its sections are now finished or are nearing completion. . . .
>
> In conclusion, I would like to reassure you that the Smithsonian understands your concerns and will present, in any exhibit that it might undertake on this subject, a balanced and thoughtful exhibit about one of the most important events of the twentieth century.[5]

Bennett was not satisfied. He responded more aggressively on September 6.

> I . . . wish you could be a bit more specific about the thrust of the strategic bombing exhibit in which the *Enola Gay* will star. Mr. Crouch put it this way, *'I can assure you that we would not attack the decision to drop the bomb. Nor would we celebrate the event.'*
>
> The second sentence, 'Nor would we celebrate the event,' is incomprehensible to me. . . . Help me Mr. Adams, I am trying to understand the Smithsonian's apparent shame of the *Enola Gay* and the bombing that saved so many lives.
>
> Your more diplomatic explanation of the exhibit . . . anticipates reflecting all points of view . . . in a balanced and sensitive manner, including those of veterans. I think you are saying the same thing as Mr. Crouch in more polished, gentler language. . . .
>
> No war is fought in a sensitive and balanced manner. Most wars are fought to win. . . . The *Enola Gay* should be displayed with the sensibilities of the American veterans in mind. To do otherwise is to bruise our sensibilities, which should be paramount if the word *National,* in the *National* Air & Space Museum means the United States of America. . . . A lot of the world fervently celebrated the end of the war with Japan![6]

This second letter received an answer from Assistant Secretary Tom Freudenheim, who wrote on Adams's behalf the following month:

> Given the extensive previous correspondence between yourself and the Smithsonian, I will keep my reply brief. . . . The nub of the problem is that you think there is only one possible American view of the bombing of Hiroshima and Nagasaki. In fact there is a lively debate among scholars

> and historians in this country about whether it was necessary to drop the atomic bombs to end the war or prevent an invasion. Opinions are also divided among the general population as to the necessity and the morality of the bombings.... General Curtis LeMay, before his death, also questioned the necessity of using the atomic bombs when his fire raids were destroying Japanese cities, including Tokyo, since March 1945.... The Smithsonian as a national institution cannot therefore take only one point of view on this question, given the diversity of opinion in this country. Such an exhibit would also not be consistent with one of our central aims: presenting the best scholarship in a form that engages the general public. In conclusion, I can only repeat that we understand your concerns and that any exhibit of the *Enola Gay* will examine these questions in a balanced and thoughtful manner.[7]

Bennett apparently had not yet received Freudenheim's letter when he wrote to Adams again on October 12, 1992:

> In case you are at home celebrating the Columbus Day Holiday, enclosed is the editorial from today's Wall Street Journal ... [which states, "The once-respected Smithsonian Institution—in danger of becoming the Woodstock Nostalgia Society—produces an exhibition that is multiculturally correct down to its tiniest sensitivity."]
>
> The Journal's view of the Smithsonian, and your exhibit 'The West as America: etc.' is perhaps more caustic than mine. Still, I imagine that you place those who want the *Enola Gay* displayed proudly in the same category as the Journal.
>
> I hope you make the connection between the two and give credence to the possibility that you have placed the Smithsonian out of step with a major segment of the population. I also hope that you will take another look at your decision to join the "PC forces." As disgust of this rewriting of history builds, it could erode both your credibility and your financial support.[8]

By the following week, Freudenheim's letter had arrived, and on October 20, Bennett wrote to Adams again:

> Many thanks for Mr. Freudenheim's response.... Paraphrasing (his) points:
>
> 1. *Can't specify content of an Enola Gay exhibit having no date or location.* If that is a problem after owning the *Enola Gay* for 43 years, then at least send me the findings of your symposium on strategic bombing as requested in my September 7 letter.
> 2. On the *lively debate of today's scholars and historians, and today's population:* Is it honest to judge what happened in 1945 by the morality of today?
> 3. I was surprised at Freudenheim's LeMay quote. Fact is, up until August 6, fire bombing had not yet ended the war. Fact is, armed forces around the world were gearing up for an invasion. Fact is,

most wars require invasion. Fact is, *Enola Gay* ended war with Japan. Fact is, the *Enola Gay* averted invasion.

4. Is it moral to prolong war, even one day? Is it not the height of sophistry to look back now and say that bombing of Hiroshima and Nagasaki was unnecessary?

We always had diversity of opinion. That is why leadership is needed. We had it in 1945. You can provide it now for the Smithsonian. The Wall Street Journal dubbed the Smithsonian *The Woodstock Nostalgia Society*. Is being politically correct worth the damage that is being done to the reputation of your Institution?

The *Enola Gay* deserves to be exhibited proudly. The United States of America deserves that type of exhibit. You should feel no shame that the United States won the war with Japan.[9]

Two days later, on October 22, 1992, Bennett wrote to Congressman John E. Porter:

Thank you for your letter of October 16 concerning the *Enola Gay*, prepared by 'sec'.

Unfortunately I have little "enthusiasm anticipating the exhibition of the *Enola Gay* after restoration." . . . The exhibit will display the *Enola Gay* only to draw a crowd for their negative position on strategic bombing. . . .

There is this terrible movement afoot by the so-called "intellectuals" in this country to mock everything ever accomplished by the United States. The Smithsonian's cryptic explanation of their planned exhibit implies that they plan to do the same to the *Enola Gay*. If so, then I would prefer that the plane not be displayed by the Smithsonian.

I also trust, that should you find any merit in the above, you will act accordingly on funding proposals for the Dulles facility.[10]

It was my turn now to answer for Bob Adams. On November 23, I responded to two of Bennett's letters:

Regarding your specific questions, first, it is unfortunately not possible to send you the proceedings of the "Legacy of Strategic Bombing" symposium, because the editing of this volume is not yet complete. However, we will keep you in mind when the volume is published, hopefully within a couple of years.

Your other questions all speak to the same issue, namely, how to treat the history of the *Enola Gay* and the atomic bombings in any future exhibit. It is both unavoidable and useful that any explanation of historic events not only take into account the perspectives of participants at the time, but also the perspective of the time in which the historical work or exhibit is formulated. Often a much clearer view of events is available years later, when many secret actions and documents have come to light, and that is certainly true in this case.

You also assume that there was only one American viewpoint of the bombing in 1945, which is not true. Most notably, many scientists in the Chicago laboratory of the Manhattan Project objected to the dropping of the atomic bomb on civilian populations without warning. Any exhibit we might do would not take any particular view of the morality and necessity of the atomic bombings, but must take into account the diversity of views—then and now.

Finally, it is precisely the questions you ask, whether dropping the atomic bombs shortened the war significantly or not, and whether it is morally justified or not to bomb civilian populations to achieve that end, that are being debated by scholars. Because these questions are so perplexing, fascinating and important, an exhibit on the *Enola Gay* and Hiroshima is very much worth doing. All we can say, once again, is that we will definitely take your point of view into account, when the decision is made to put on an exhibit on this subject.[11]

ISSUE 3: SAVING MILLIONS OF LIVES

Now Donald C. Rehl, who had not written in a long time, took up the correspondence once more. On December 2, 1992, he wrote me,

> It has been a year or perhaps two since you and I have traded letters concerning the *Enola Gay* so permit me to reintroduce myself as one of the fifteen B-29 pilots in the 509th Group and one of the MANY veterans (among others) that have pressed for the restoration and display of the plane in the *correct* historical context. Of the many people you have heard from you should be aware by now that they are not in agreement with the manner in which you plan to display the plane. Also, most are not pleased with the location where you plan to display it—Dulles Field.
>
> I also am quite certain that you have used the famous name of the plane in your news releases aimed at getting support for the funding of the hoped for Dulles project....
>
> Need I remind you that the *Enola Gay* represents the fact that the war was ended "overnight" and put to rest LeMay's contention that the fire bombing of Japanese cities *might* end it? Yes, might do it. No one could say *when* and meanwhile the plans for the invasion proceeded.
>
> President Truman's brave decision to use the bombs was meant to remove the word "when" from the controversy and use the word "NOW." Perhaps just as important as the immediate end of the war, which saved millions of allied and Japanese lives that would have been lost in the invasion, was the fact that the Russians were thwarted in their plans to sit at the peace table and no "Berlin wall" was built in Japan. Please tell me why there is a problem in deciding in what historical context the *Enola Gay* should be displayed.[12]

Rehl's letter received two replies. I asked Lin Ezell to respond to issues he had raised regarding restoration of the aircraft, and I answered him on other matters.

Ezell's letter, dated December 9, included two annual reports on the restoration, for 1991 and 1992. She noted, "You may recall that I sent you similar updates in previous years." She also invited him to come to Garber and said, "We are confident that the restoration is on schedule."[13]

My letter to Rehl was dated December 18. It recounted some of the practical difficulties we had faced in trying to find the best location for the exhibition, and mentioned our plan to mount the exhibit in "1995 inside the museum, using the forward fuselage of the *Enola Gay*, which includes the bomb bay." I also explained the situation on Dulles, before turning to the display of the aircraft:

> The question . . . regarding the best historical context in which to display the *Enola Gay* is indeed a difficult one. Unfortunately there is no one correct view of the so-called "decision to drop the bomb" by President Truman. Although the fanatical resistance of Japanese forces on Okinawa and other islands, plus the sudden ending of the war in August 1945, have lent much credence to the view that only the atomic bomb prevented an invasion of Japan, historical studies since that time have shown the problem to be more complex. The Japanese government initiated a peace offer in July through the Soviet government which may or may not have been half-hearted, depending on which scholarly view one follows. Others, including General LeMay, have argued that the Japanese would have surrendered anyway if only conventional bombing and the blockade had continued. Army staff studies of the invasion plans for Japan done in 1945 also predicted casualties far below the "millions" often asserted, and many historians now view an invasion as unlikely in any case. Others support the correctness of President Truman's decision. In addition, there are many points of view regarding the morality of bombing civilians.
>
> All of this is merely to indicate that in any exhibit we might do, we will be careful to give a full discussion of the viewpoints in this debate, while giving due weight to the views of the veterans of the 509th Composite Group like yourself, and all veterans. . . .[14]

But Rehl had in the meantime also written to George Larson, Editor of *Air & Space* magazine, identifying himself in a cover letter:

> I was one of the B-29 plane commanders in the 509th Group that flew the missions on Hiroshima and Nagasaki. I was co-leader of an effort that began in 1986 to "rescue" the *Enola Gay* from 40 years of neglect and deterioration while in the hands of the Smithsonian. The effort was joined by many veterans' groups, bomb groups, and other organizations. In 1988 the American Legion and the VFW passed resolutions in favor of our effort to expedite the restoration of the *Enola Gay*. It was passed at their national conventions, not just state or local.

In his proposed "letter to the editor," Rehl took the museum to task for not taking care of the *Enola Gay* after receiving it in 1949:

> . . . I and MANY others believe that the Smithsonian is using the famous name of the plane in their efforts to obtain funding for the hoped for extension of NASM at Dulles Field. They have used the name, *Enola Gay*, in

every news release pertaining to plans for the extension. It appears they are holding the *Enola Gay* as hostage for their goal.

Rehl then went on to comment on a eulogy I had written for the magazine after Paul Garber's recent death. In recalling his life's work, I had mentioned the prodigious collection of aircraft Paul had assembled for the museum, including the *Spirit of St. Louis* and the *Enola Gay*. Rehl wrote:

> In doing so it seems clear that Harwit ranks the *Enola Gay* of equal importance with the *Spirit of St. Louis*. He is correct in doing so and I congratulate him for recognizing the fact. . . . [But] he fails to *fully* explain the historical importance that the *Enola Gay* represents. He merely states that it dropped the first atomic bomb on Hiroshima. He leaves the field wide open to the revisionists of history to continue their cries about how horrible and unnecessary President Truman's (brave) decision to use the bomb was. His statement about the *Enola Gay* should have been—"that dropped the first atomic bomb on Hiroshima which saved millions of lives that would have been lost in the planned invasion of Japan and also thwarted Russia's plans to sit at the peace table and build a 'Berlin Wall' in Tokyo." If Dr. Harwit felt that such a statement was too lengthy for his "eulogy" he should not have mentioned the *Enola Gay* at all.[15]

Air & Space editor Larson answered Rehl's letter on December 14, 1992:

> The suggestion that the occasion of Paul Garber's death presents an opportunity for debate about the role of the *Enola Gay* is unlike anything I have read or heard in the years I've worked here, but it helps to clarify, for me at least, that for many people, the status of the airplane eclipses many other human concerns. This was, after all, a man's funeral. Many of us who felt Garber's passing most acutely believe that his career was one of the Museum's singular treasures.[16]

On December 11, Burr Bennett wrote to Larson on the same topic:

> Dr. Harwit's editorial [eulogizing Paul Garber] leaves me puzzled. He calls August 6, when the *Enola Gay* dropped the first atomic bomb on Hiroshima, *that fateful day in August*.
>
> On August 6, I was on the high seas headed for Okinawa from India, bracing for the invasion of Japan. When the surrender occurred on August 15, we really celebrated! August 6 was one glorious day! I don't know if the Japanese celebrated, but *that fateful day in August* saved massive casualties for their country as well as ours. . . .
>
> Could it be that Harwit is anti-nuke? Hard to believe. Japan was working on the same type of bomb. They might have beaten us to it had it not been for the destruction of the Japanese laboratory through the B-29 strategic bombing that slowed them down. In any event someone had to be first. If not the United States, it would have been some other country. Perhaps the Smithsonian's Legacy of Strategic Bombing Symposium has turned Harwit against strategic bombing.

This, and so many other actions today about the use of the atomic bomb are incomprehensible to me. . . .[17]

Larson replied on December 15:

I am unable to respond to your speculation about what Martin Harwit thinks or why he chose the words he did. He is certainly not the first or last person to use the word "fateful" in connection with the bombing of Hiroshima.

In fact, there are enough ambivalent feelings loose in the land concerning the event that initiated the age of nuclear weapons to keep anyone busy for years trying to adjust popular attitudes so that everyone can learn to embrace the use of the first atomic weapon with your enthusiasm and affection. Like so much in life, it's all a matter of perspective. . . .[18]

I also wrote to Bennett that day, though on one of the other topics he had repeatedly emphasized—Japan's effort to construct an atomic bomb:

Thank you for your latest letter. I am sorry that your second letter did not receive a full reply. Regarding the so-called Japanese atomic bomb project of Yoshio Nishina, it is clear that he was very far from building a bomb. The Japanese were not even as far along as the Germans, and the German physicists no longer worked seriously on a bomb after early 1942. Nishina was not successful in very small experiments in isotope separation he had attempted before his laboratory was destroyed in April 1945. . . .

In any case, the state of the Japanese and German programs do not have much bearing on the questions we have been discussing. Few people have ever argued with the morality of the United States and the British Commonwealth beginning an atomic bomb program in World War II, when the Allies could have faced a nuclear-armed Nazi Germany—something that seemed a real possibility early in the war. The questions came only at the end of the war, when a debate arose, at least within the tiny circle of people who had full knowledge of the Manhattan Project, as to whether bombs should be dropped without warning on civilian populations. That debate obviously has continued since the war among a much expanded audience, accompanied by much more evidence about what the American and Japanese governments were thinking at the time. The debate about the so-called "decision to drop the bomb" has not ceased then, nor will it ever do so. As I have indicated in previous letters, our task in any exhibit will therefore be to convey the complexity of that debate to the general public, while giving due weight to the viewpoint of veterans.[19]

Bennett wrote to Adams again in March 1993:

As you know, your position on the *Enola Gay* has puzzled me. I get almost vehement support from friends around the world to exhibit the *Enola Gay*. They find it absolutely incomprehensible that the plane has never been exhibited proudly. People who deal with reality daily view this plane as an historic artifact that started the chain of events that ended the war with Japan. They don't glorify war or bombing, but they do know what this plane did and the millions of lives it saved.[20]

The Smithsonian had the tradition of answering every letter that came in, no matter what. But Adams also had other duties. This time he replied, "The reconstruction efforts that the Smithsonian has expended on the *Enola Gay* have been by all odds the most time consuming and expensive we have ever devoted to any item in our collection. I am pleased to report that they are, finally, approaching an end...." He spoke of plans for an exhibition in 1995, most probably of only the forward fuselage of the aircraft, since no site was available for exhibiting the aircraft in its entirety. He then spoke his mind on the usefulness of further correspondence.

> There is a degree of repetitiveness in our past correspondence that suggests these are issues on which further explication is not likely to produce agreement. I do hope, however, that the slight bits of elucidation offered above will be helpful, and I want to thank you once again for the friendly and constructive spirit of your letter of March 12th.[21]

ISSUE 4: COMPLETE OR PARTIAL DISPLAY?

Don Rehl now wrote to Adams the following month:

> There have been rumors that further restoration of the plane will not be made and that you plan to display the front section of the fuselage only. Would you be so kind as to confirm this "rumor" if it is true and tell me where the plane is to be displayed; the earliest date it will be done and in what historical context it will be displayed.... Please be as specific as you can in replying to this letter since I will be passing the information on to other veterans and civic groups that have inquired about the present and future status of the plane.[22]

I replied to this letter on May 17, 1993:

> I hope that the Post Office has not lost my letter to you of December 18, 1992, because it largely responds to the questions you have asked. But to answer your concerns briefly, we have not stopped the restoration of the *Enola Gay*, nor do we have any plans to do so....
>
> We are planning to meet the historic 50th anniversary of Hiroshima and Nagasaki by putting on a major exhibit within the Museum using the forward fuselage of the aircraft, an engine and other parts, plus atomic bomb casings and numerous other artifacts. This exhibit will give a balanced account of the decision to drop the bomb, the 509th Composite Group, the missions themselves and the aftermath. All points of view will be represented, including of course the viewpoint of veterans such as yourself.[23]

Rehl's taking up the correspondence where Adams's refusal to continue the exchange with Bennett had left off was unlikely to be coincidence. The two were in intimate touch as evidenced by a letter Rehl forwarded to me shortly thereafter. It was written by Alfred A. Yee, of Honolulu, and addressed to Bennett. At the top Rehl had typed, "This is a copy of a typical letter from a petition signer received by Burr Bennett, chairman of Restoration Committee."

ISSUE 5: BOMBING AS A BOON FOR JAPAN?

Yee's letter, dated May 10, 1993, described a long train ride in China, on which he had met a Japanese man, Mr. Fukushima, and had many discussions:

> I brought up the question of morality in the atomic bombings of Hiroshima and Nagasaki. Mr. Fukushima, who served in the military throughout the Sino-Japanese campaign and WW II, responded emphatically that the atom bomb brought the greatest single benefit to the Japanese people as it ended the war almost abruptly! He opined that had it not been for this superior weapon, which completely overwhelmed the Japanese psychologically, the Japanese people would never have surrendered and would continue the fight until the last survivor. He estimated that Japan would have lost untold millions of civilians and military personnel but that they would never give up and would have continued relentlessly on their mission to eliminate every single foreign soldier occupying their sacred land. America would have lost millions of lives.
>
> Mr. Fukushima completely convinced me that the atomic bomb incident not only was justified but looked upon by many Japanese to be the largest single factor in saving the nation of Japan by completely revising the Japanese psychology against war and towards industry. As a result, the Japanese were able to direct their entire emotional and physical energies toward the reconstruction of Japan, operating manufacturing facilities and participating in world trade. According to Mr. Fukushima, this new direction obviously became the basis of Japan's economic successes and the Japanese people are forever grateful that they had taken this direction.
>
> Please accept my best wishes for success as you endeavor to have the *Enola Gay* exhibited proudly at the Smithsonian Institution.[24]

How "typical" this letter was is not clear. On July 9, Burr Bennett sent a copy of the very same letter to Adams.[25] It represents an attitude often expressed by Bennett and Rehl. In their letters, the dropping of the atomic bomb is described as having not only saved millions of American lives, but also far more Japanese lives.

Bennett, Nicks, and Rehl never claimed that the bombing of Hiroshima and Nagasaki ought to be seen as a humanitarian gesture; but they did point to the enormous advantages they believed had accrued to Japan as a result of the bombing. Yee's letter could be construed as implying that Japan's postwar economic boom and welfare were a direct result of the bombings of Hiroshima and Nagasaki. Its tacit implication was that these benefits would not otherwise have come to Japan. Japan would now be worse off had we not dropped the bomb.

I personally never met any Japanese person who advocated these views. I doubt they were widely accepted in Japan.

ISSUE 6: LOCATION OF A FINAL HOME FOR THE *ENOLA GAY*

On August 10, 1993, Ben Nicks wrote to Don Rehl immediately after the Institution had finally received authorization to begin planning the museum extension with the Washington Dulles International Airport as its definite site:

> Great news from the Smithsonian Institution! On August 2 President Clinton signed a bill authorizing the Smithsonian to proceed to develop plans for constructing the new Air & Space Museum annex at Dulles airport right outside Washington D.C.! All the pork-barrel congressmen gave up trying to get it in their own trough.
>
> I know you were holding out for another location but I am positive that the Smithsonian would never turn loose of the *Enola Gay*—Martin Harwit told me as much last May. So our long battle to get the *Enola Gay* out on permanent public display is almost over, despite our misgivings about the purity of the good intentions of the Smithsonian! As you know, it goes on partial-temporary display in mid 1995 for the 50th anniversary celebration of the end of WW II. I say partial because space limitations will permit only the cockpit portion to be exhibited. The new annex will permit display of the completely renovated and assembled *Enola Gay*—in five more years, maybe, but we're rolling down the runway now!
>
> I know every 20th AF veteran will greet this news with satisfaction, and hoping not to appear immodest I take personal pride in the role I managed to play along with you in nagging the Smithsonian until they took action.[26]

William A. Rooney answered Nicks's letter three days later:

> Sure thank you for your letter of the 10th about the Smithsonian and the *Enola Gay*. I am enclosing some material of possible interest to you.
>
> It would appear that me and you ain't on the same wave length when it comes to the Smithsonian, Harwit, the A&S Museum and the *Enola Gay*.
>
> It sounds like you are happy about the fact that the plane will go on partial temporary display on the occasion of the 50th anniversary. Space limitations limit the ability of the A&S Museum to display the entire plane? Bullshit. . . .
>
> You talk about this appropriation for Dulles. It took them 17 years to get the A&S Museum built. If they start today, how old will you be by the time the Dulles facility opens? Further, the appropriation was only for preliminary engineering studies.

Our original goal in preparing the enclosed petition was 5,000 signatures. These came in so fast from all over the world that we have raised our sights to 10,000.

I plead with you not to wrap me up in any pride taking over what has been done with the Smithsonian vis-a-vis the *Enola Gay*. There is no joy in Mudville over what has happened.

With respect to the Smithsonian never giving up the *Enola Gay*, remember this: Anything hit with a big enough hammer will break.

All good wishes, Bill Rooney [27]

The attached petition reads,

We, the undersigned, petition the National Air & Space Museum of the Smithsonian Institution to complete restoration of the *Enola Gay* and display it in a patriotic manner that will instill pride in the viewer for the outstanding accomplishments of the United States and the *Enola Gay* in ending World War II without the need for an invasion of Japan and without the additional invasion related casualties.

If the Smithsonian cannot display the *Enola Gay* proudly and patriotically, we respectfully request that the plane be turned over to a museum that will.

Return to: The Committee for Restoration and Display of the *Enola Gay*, c/o W. B. Bennett, Jr., 1902 Techny Ct. Northbrook, IL 60062 (USA).

Attached to this form, which asked for name, state, and country if other than U.S., was a sheet on the *Enola Gay*. It quoted the Smithsonian as saying, "At this point it is impossible to say what form an exhibition of the airplane might take. However, I can assure you that we would not attack the decision to drop the bomb. Nor, on the other hand, would we celebrate *the event*." The sheet underlines the last two words and then goes on:

The event saved untold hundreds of thousands of lives. Behind *the event* was a technical tour de force by the scientific and engineering community of the United States and its Allies that produced a nuclear reaction leading to the development of the atom bomb. An equally significant event was the development of a bomber (the B-29) capable of delivering an atomic bomb over a long distance. Germany and Japan tried to reach these same goals without success.... The *Enola Gay* should be displayed proudly....[28]

ISSUE 7: COMPLETE RESTORATION

Rehl's response to Nicks was dated August 18, 1993:

I received your letter of August 10 with mixed emotions most of which was disappointment (and perhaps anger) over the fact that you are satisfied and, in fact, pleased with the news about the future of the *Enola Gay*. After reading your letter I reviewed the file of correspondence between us and it is clear that you never really saw the goal we have been trying to reach nor joined in our effort to reach it. You have always been dancing to a different fiddler and I have never figured out why.

Ben, from the very beginning the aim has been to *EXPEDITE AND COMPLETE* the restoration of the plane in its *entirety* and have it on display in the correct historical context before you and I and all the others (including Paul Tibbets) are dead and gone. If the Smithsonian could not or would not fulfill that obligation and duty then the *additional* goal was to have the plane released to another air museum or organization that would. And, I assure you there has been no shortage of locations or organizations that would have met that obligation if the Smithsonian had admitted their errors and financial problems and opened the door. It is just that simple, Ben. How can I make it any clearer to you? I have certainly tried in the past and it has been extremely disappointing that you have not used the 9th Bomb Group roster as a tool to help out.

You began your letter with the words "great news." What is great about the plans to exhibit just the nose section in 1995 "on a temporary basis"? What happens then? What's so great about Buhbu Clinton signing the legislation for the annex at Dulles? Doesn't that tell you that the restoration will now definitely be stalled until the annex is built? The news article states, "The *anticipated* completion date is *about* the year 2000"— Note the words "anticipated" and "about" in the same sentence. You also state that the new annex will permit display—"in five more years, maybe; but we're rolling down the runway now." Well, the "anticipated" date of 2000 is not five more years, it adds up to seven (plus ten?) to me and, I don't know about you but my runway is running short.

Having held a respect for you in the past I suppose I should apologize for the tone of this letter but your letter was anything but "great news" to me and I am only giving my honest reaction. I enclose a petition for you to make copies of and use. We have nearly 7,000 signatures to date and have a goal of 10,000, Why not join the parade?

Best regards—still;
Don Rehl [29]

To this Rehl also attached the same petition blank that Rooney had already sent to Nicks.

Nicks answered on September 1, but was not entirely successful and wrote again on October 23, 1993, to mollify Rehl:

Thanks for your letter on the *Enola Gay*. I think you're too hard on yourself at times, Don. I know it isn't coming out like you'd hoped, but, Don, I consider that you are the one most significant person who got the *Enola Gay* out of dead storage and on the road to restoration and display. Sure it's been a long and tough haul, but you never gave up for 10 or 15 years and your leadership in the campaign to display the *Enola Gay* may be the only reason it is coming out of the closet. If it had not been for you and your efforts to inform and engage us veterans of the Twentieth Air Force it might still be sitting at Andrews. When the *Enola Gay* is displayed in 1995,

> it would not be too much out of line to have your name painted on the nose along with those of the crew members there. When it is displayed the rest of us will know the time, effort, money, blood and sweat you've put in it. . . .[30]

In the meantime, Nicks sent Lin Ezell copies of Rehl's and Rooney's correspondence and a list of others significantly involved. I wrote back on September 3, 1993, to let Nicks know that

> I've begun writing letters to all these individuals to give them a better idea of where we stand. I understand why the B-29 veterans are angry. My letter may not help to change that, but at least the Museum will have done its best to provide information on the actual state of matters.[31]

That day I had already written identical letters to William Rooney and Donald Rehl. In these I recalled the entire history of the *Enola Gay*'s restoration for them, and recounted the difficulties we had faced in trying to find an adequate building to display her. I mentioned the clear understanding the museum had of the aircraft's historical importance and concluded by writing,

> I am saddened that veterans have seen it necessary to circulate a petition asking the Museum to display the *Enola Gay* "in a patriotic manner that will instill pride in the viewer. . . . Do veterans really suspect that the National Air and Space Museum is an unpatriotic institution or would opt for an apologetic exhibition? Eight to ten million people continue to enthusiastically throng into the Museum each year, and no such criticism has ever been leveled against us.
>
> I would be pleased to send you the document that details the plans for the exhibition. I sent Ben Nicks a copy some time ago and told him that he was free to share it; I believe he may have done that already. But if you are interested, I certainly will be glad to send you a personal copy, because I would very much like to clear the air on this matter.[32]

ISSUE 8: PORK-BARRELING AND THE EXTENSION

To this, Don Rehl responded on September 21, 1993:

> . . . I am pleased that Ben Nicks sent my letter to him to your attention because . . . It gave you a sense of my frustrations as well as that of hundreds of veterans who are disturbed over the past treatment of the *Enola Gay*. . . .
>
> The *Enola Gay* represents President Truman's painful but well thought out decision to use the only weapon that would bring the war to an end "overnight" and avoid the catastrophic loss of Allied (as well as Japanese) lives that would have resulted from the planned invasion of Japan. That in itself should be enough to display the plane with pride not to mention the fact that the Russians never had time to reach Japan and demand a place at the "peace table" and build a wall as they did in Berlin. Have

you read that chapter of WWII history or have you forgotten it? . . .That is the historical context in which the *Enola Gay* should be displayed. It is so evident and simple. Why do you constantly ignore it?

You state that restoration plans were delayed for three years while a congressionally mandated search was made for another location for the NASM extension. With all the military air base closings underway—bases with hangars, buildings and runways—do you mean to tell us that another location could not be found?—a location that is more centrally located where the average American family can more affordably and more easily reach?—a location where the facilities merely need updating and perhaps some expansion rather than build from scratch with the taxpayer's money? Or, could it be, Dr. Harwit, that "pork barrel" legislation was underway to benefit the state of Virginia and, in pursuit of that legislation, the well known name of the *Enola Gay* was used in every single news release concerning the planned extension of NASM at Dulles? We think that therein lies the reason for the foot dragging on the restoration of the EG and not that NASM or the Smithsonian did not have money or display space or location. . . .

. . . Six years ago you failed to instruct your staff to restore the plane *AS SOON AS POSSIBLE*. Instead you arbitrarily chose a date to coincide with the end of WWII with no feeling of guilt over the fact that the plane had already been ignored and abused for 45 years and that hundreds of us who wanted it restored were getting older and might never see the plane on display. You disregarded all of that while you restored other planes of *foreign* countries which used up time, labor and money that could have been spent on the *Enola Gay*, the plane which, by your own admission in the past, was the most historically significant aircraft in the inventory that was not yet restored. . . .

In closing I want to respond to your comment about having received "close to thirty thousand dollars from around seven hundred B-29 veterans" to help on the restoration. I believe that some $20,000 of that was the result of the efforts of the original ENOLA GAY RESTORATION ASSOCIATION which my navigator and I organized in 1988 and, if I recall correctly, none of that money was spent on restoration but on making a filmed record of the restoration work. Correct me on that assumption if you wish. . . .[33]

I answered Rehl on October 7, 1993, taking care to go point by point:

You mention the ease with which military bases could make hangars available to us. . . . Unfortunately, the Smithsonian was never able to get Andrews Air Force base to provide the protective cover of a hangar. Moreover, most military facilities are understandably reluctant, for reasons of security, to have a stream of visitors flocking through their gates. . . . You commented on the Commonwealth of Virginia, where the Museum extension is to be located: In offering its support . . . the State undertook to provide all the infrastructure for this large facility at a cost estimated by the Smithsonian's architectural consultants to be roughly $40 million. Additional commitments by Virginia of cash contributions toward the construc-

> tion amount to $12 million. Together this package represents roughly one third of the total cost of the planned facility. I do not doubt that Virginia hopes to benefit from the Museum extension, but I believe it significant that Governor Wilder placed responsibility for the State's contribution under his Secretary for Education, rather than assigning that task to an office coordinating, say, tourism.
>
> It is difficult to appreciate the size of the restoration we have undertaken on the *Enola Gay*, without seeing both its initial state and the enormous progress we have made. Admittedly we have also conducted restorations on other aircraft, some even foreign. One is a Japanese piloted suicide bomb, which will be part of the *Enola Gay*'s exhibition, making clear the ferocity of the desperate fighting our forces faced in the Pacific at the end of the War. Another was a P-47 fighter, which the P-47 Pilots' Association had long asked us to display as part of an ongoing World War II commemoration here at the Museum. I am sure that you will understand that we could not drop everything to restore the *Enola Gay*. I felt that having the *Enola Gay* available for display in 1995 was a reasonable decision in view of the size of the project, the limited resources of the Museum and the responsibilities we had to complete other restorations as well.[34]

I added that we were seeking contributions of memorabilia and invited him to come for a personal tour of the restoration facilities, saying, "you might be in a position to share with us experiences that would help us improve the exhibition."

ISSUE 9: FAVORABLE PORTRAYAL OF STRATEGIC BOMBING

On October 13, Rooney wrote me a four-page, single-spaced letter responding to my letter of September 3, 1993:

> Burr Bennett and Don Rehl have shared with me their replies to you. . . .
>
> All of us have one point of view with regard to you, Mr. Adams, and NASM. It is: Through wanton neglect, you allowed the *Enola Gay* to deteriorate and be vandalized. . . . It is an absolute disgrace to our nation, our history, and to the military that the very repository of our nation's history could allow this to happen. To the end that you are a party to this, you and NASM are responsible and guilty. It is an issue of betrayed trust. It is altogether too easy to disavow responsibility because the majority of the time of neglect did not occur on your watch. . . .
>
> It is evident that you intend to portray strategic bombing in a negative way and that you intend to use the *Enola Gay* as your tool to tell your story. . . . And, Dr. Harwit, you can use the NASM as an instrument to sell your story of the horrors and ineffectiveness of strategic bombing using the *Enola Gay* as your attention getter but remember one thing: The Strategic Air Command, for more than 20 years, was the single unit in this world that kept the peace. . . . You are saddened by our petition circulation effort. . . . There were 80,000 men in the XXth Air Force in WW II. Until its stand

> down, there were 240,000 men and women in SAC. We shall continue to collect signatures.[35]

On November 27, Rehl wrote me that he was resigned to the 1995 display of the forward fuselage, adding "Needless to say your time table is a great disappointment to me. . . ." He asked me to send him the same information I had sent to Ben Nicks, which I had offered him. He also wanted to know whether in 1995 there would be "some sort of ceremony (and invitation) featuring the surviving members of the 1945 509th Composite Group?"[36]

I responded on December 21, 1993, and sent him a copy of our planning document. I wrote, "I would appreciate your responding with any comments you may have. We want to make certain that this exhibition is both accurate and balanced, and tells the story in true human terms. Since July, we have circulated this document widely, and benefited from the responses we have received."

As a postscript, I added, "You asked about the 509th. One of the museum's curators has been in touch with most of the surviving members of both the *Enola Gay* and *Bockscar*. We are pleased that they have promised us important memorabilia for our display and invited us to the reunion in Chicago in 1994."[37]

But despite his avowal of being resigned, Rehl seemed set on starting all over again in a letter dated January 30, 1994. On February 10, I answered,

> We have now corresponded many times and it seems pointless to continue exchanging letters, since my explanations to you seem not to satisfy you. Instead, let me reissue my invitation:
>
> The Museum would be extremely pleased to give you an opportunity to inspect the restoration of the aircraft yourself, any day of your choosing, except Christmas day, when we are closed. You would be able to talk to the many members of our staff who have taken part in the restoration if you came on a working day, but we could also show you the *Enola Gay* any weekend.
>
> Please let me know your travel plans and we will be glad to pick you up anywhere in downtown Washington and drive you to our suburban Maryland restoration facility.[38]

ISSUE 10: THE MUSEUM'S CHARTER

As 1994 began, W. Burr Bennett, Jr., wrote to Chief Justice William H. Rehnquist and Barber B. Conable, both members of the executive committee of the board of regents.

> This is a plea for the "proper" display of the *Enola Gay* by the Smithsonian's National Air and Space Museum. . . .
>
> There are two mutually exclusive positions on the *Enola Gay* and its bombing mission. There are those assigned to wade ashore in the sched-

> uled invasion of Japan, and those who were unborn, too young or back home in their job. Paul Fussell in his essay, "Thank God for the Atomic Bomb," doesn't demand that the anti bomb folks "experience having their ass shot off, I merely note that (they) didn't." Those on their way to the invasion knew that the *Enola Gay, Bockscar,* the hasty entry of the Russians, and the atomic bombs saved their lives. . . .
>
> The Smithsonian has selected the philosophical side. Their document, "Proposal: Hiroshima and Nagasaki: a Fiftieth Anniversary Exhibit at the National Air and Space Museum" reflects that position, as did the Symposia underpinning the exhibit.
>
> I am one of a five member committee of veterans gathering petition signatures seeking the proper display of the *Enola Gay* by the Smithsonian. These petitions will be presented to Dr. Adams and Dr. Harwit in a few months. With only five old men on this project, we have attracted 8,000 signatures. This indicates that there is a significant world-wide constituency that is not ashamed of ending a war swiftly with the best weapon at hand. The signatures are by realistic people from foreign nations and across our country.
>
> It is an insult to every soldier, sailor, marine and airman who fought in the war against Japan, or who were on their way to the invasion, to defame this famous plane by using it as a centerpiece of a negative exhibit on strategic bombing. We have no quarrel with the Smithsonian having such a negative exhibit without the *Enola Gay.* They have a tradition of politically correct exhibits including Tom Crouch's "A More Perfect Union" and the exhibit on "Opening the West."
>
> The Proposal for this exhibit violates Statute 20 USC, paragraph 80a, under which military equipment is to be displayed by the Smithsonian. This statute states "*The valor and sacrificial service of the men and women of the Armed Forces shall be portrayed as an inspiration to the present and future generations of America.*" This exhibit, as currently planned, does not meet the Statute.
>
> Dr. Harwit wrote to me on Dec. 15, 1992, on the Statute that "how it (the Statute) is implemented is a matter of the Regents' discretion."
>
> Therefore, may I most respectfully request that the Board of Regents review this matter and direct the Smithsonian to fully meet the statute in their display of the *Enola Gay.* If they are unwilling, or unable to do so, then the *Enola Gay* should be released to a museum which will complete the restoration and present the plane in its correct historical context as the aircraft that set in motion a chain of events that led to the surrender of Japan in nine days and the canceling of all those invasion plans.[39]

Bennett's reference to my December 15, 1992, letter concerned an inquiry he had made at the time. In his letter, he had asked about the same statute he was now quoting to Chief Justice Rehnquist and Conable. He had wanted to know whether it applied to the National Air and Space Museum.

The statute from which Bennett's quote came is designated Title 20, United States Code §80a. It concerns a National Armed Forces Museum which was authorized by Congress in 1961, but was never funded and never built. Given this lack of activity for more than thirty years, it may not be surprising that I had never heard of this legislation when Bennett first wrote. I inquired with the Smithsonian's legal counsel and duly received its text. In my response to Bennett I had included the entire text, because it is quite clear that it refers to a nonexistent museum:

> §80a. Display of contributions of Armed Forces; study center; historical collections; National Air Museum provisions unaffected.
>
> (a) The Smithsonian Institution shall commemorate and display the contributions made by the military forces of the Nation toward creating, developing, and maintaining a free, peaceful and independent society and culture in the United States of America. The valor and sacrificial service of the men and women of the Armed Forces shall be portrayed as an inspiration to the present and future generations of America. The demands placed upon the full energies of our people, the hardships endured, and the sacrifice demanded in our constant search for world peace shall be clearly demonstrated. The extensive peacetime contributions the Armed Forces have made to the advance of human knowledge in science, nuclear energy, polar and space exploration, electronics, engineering, aeronautics, and medicine shall be graphically described. The Smithsonian Institution shall interpret through dramatic display significant current problems affecting the Nation's security. It shall be equipped with a study center for scholarly research into the meaning of war, its effects on civilization, and the role of the Armed Forces in maintaining a just and lasting peace by providing a powerful deterrent to war. In fulfilling its purposes, the Smithsonian Institution shall collect, preserve, and exhibit military objects of historical interest and significance.
>
> (b) The provisions of paragraphs 80 to 80d of this title in no way rescind §77 to 77d of this title, which established the National Air Museum of the Smithsonian Institution, or any other authority of the Smithsonian Institution.

Both the full title and the last paragraph made it clear that the legislation specifically excluded and was meant not to affect the National Air Museum or its successor the National Air and Space Museum in any way. Paragraph 77a, the governing mission statement for the museum reads,

> §77a. Functions of the museum
>
> The national air and space museum shall memorialize the national development of aviation and space flight; collect, preserve, and display aeronautical and space flight equipment of historical interest and significance; serve as a repository for scientific equipment and data pertaining to the development of aviation and space flight; and provide educational material for the historical study of aviation and space flight.

This legislation gave us the rationale for having collected the *Enola Gay* and preserved her with the help of our restoration staff, and authorized us to display her in an educational exhibition on her history and historical significance. It did not direct that we must suppress material that failed to portray "the valor and sacrificial service of the men and women of the Armed Forces ... as an inspiration to the present and future generations of America."

This issue was raised again in early 1995, shortly after Adams's successor, Secretary I. Michael Heyman, had taken up his office. In response to Heyman's queries, Smithsonian general counsel Peter G. Powers wrote,

> It might be noted that the specific purposes enumerated for individual Smithsonian museums have always been understood to be demonstrative, rather than restrictive, since they cannot diminish the basic Smithsonian trust responsibility "for the increase and diffusion of knowledge."[40]

Here, Powers was citing the original bequest from James Smithson that established the Smithsonian in 1846. It directed that the funds be used "to establish at Washington an institution for the increase and diffusion of knowledge among men." Given that overall directive, it was possible to establish museums for specific purposes, but those purposes could never restrict legitimate inquiry leading to the increase of knowledge, nor could they restrict the dissemination of that knowledge—its diffusion among men—through exhibitions or by any other means.

Bennett had asked whether the statute applied to the museum. Having quoted both sections (a) and (b) for Bennett's benefit, I concluded,

> The Institution's Associate General Counsel advises me that the legal implication is that this statute applies to the National Air and Space Museum. How it is implemented is a matter of the Regents' discretion. But, to answer your specific question, any exhibit the Museum decides to mount on strategic bombing will in any case be in full consonance with the law of the land.[41]

Perhaps my statement to him had not been sufficiently clear. In the many letters Bennett would write over the next two years, he would only quote the parts of §80a that he liked, and then accuse the museum of violating them. In this he was not alone. Veterans and veterans' organizations would repeatedly admonish the museum over the months ahead that in their view, the exhibition on the *Enola Gay*'s mission failed, in the words of the legislation, to portray "the valor ... of the Armed services ... as an inspiration ... to America."

These veterans were convinced that the mission of the *Enola Gay* had not only ended World War II, but had also saved millions of lives—Japanese as well as American. For them this was the central, inspirational theme to be highlighted in the museum's exhibition. But other, equally concerned citizens saw the *Enola Gay* as the aircraft whose mission had instantly killed eighty thousand people, mostly civilians. For them, the exhibition's central theme had to be that nuclear weapons should never again be used.

This divergence of views was to become the museum's major problem. We had set ourselves the task of mounting a historical exhibition on the mission of the *Enola Gay* that a majority of visitors would find honest, informative and balanced. But this effort was to be made increasingly difficult as first one side and then the other began to insist that we heed their wishes and adhere to their views.

14

Japan

While letters from the "five old men" were arriving more frequently in late 1992 and early 1993, I felt that they did not pose an extreme threat to the exhibition. We were beginning to work with General Kicklighter, who as executive director of the 50th Anniversary of World War II Commemoration Committee was in touch with a large number of veterans' groups. Kicklighter felt that the demand for our exhibition would be strong.

For me, a more worrying unknown was Japanese reaction. Earlier, I had mentioned the warning from Morihisa Takagi, in the summer of 1988:

> The Hiroshima and Nagasaki bombings remain firmly imprinted in the Japanese consciousness, much as the Holocaust does with the Jewish people.[1]

His message could not be more clear! I appreciated his candor: If we, as a national museum, exhibited the *Enola Gay*, we needed to consider Japanese sensitivities to avoid the risk of precipitating a potentially serious international incident between the United States and Japan. Such concerns never seemed to have occurred to the five old men and other veterans.

I wanted to make certain that an exhibition of the *Enola Gay*, the bomb casings of the *Little Boy* and *Fat Man* bombs, and many other artifacts of World War II would not cause an uproar. In 1994 and 1995, when commemorations of World War II were at their height, still-vivid memories and long-suppressed rage did once again come to the fore to revive hard feelings between the U.S. and Japan.[2] To avoid adding to these I sought to keep the Japanese informed, and decided to seek their participation in this exhibition linking our two nations' histories to one of the decisive moments of the twentieth century.

A FIRST APPROACH

In the summer of 1991 Gregg Herken was in Japan, where he visited the Hiroshima Peace Memorial Museum and met with its director, Yoshitaka Kawamoto, to whom he told our plans.

The exhibition's design, though still vague, called for a forthright display of the destruction and suffering on the ground. Over a hundred thousand people had been killed and others horribly maimed at Hiroshima and Nagasaki. This had to be acknowledged, however we displayed the *Enola Gay*. In a museum setting, this meant not just an enumeration of the casualties, but images of the cities and their people in the immediate aftermath of the atomic explosions and a display of severely damaged objects.

On learning of these plans, Mr. Kawamoto told Herken he would be willing to work with us.

My next step was to write to Jun Wada, director of the Center for Global Partnership of the Japan Foundation, an organization dedicated to strengthening international relations for Japan:

> From the outset, planning for this exhibition has been guided by our concern that the display of the *Enola Gay* be understood as a commemoration, rather than a celebration, of an event that is seen as pivotal in the history of both our countries. . . .
>
> Last month, the chairman of our Space History Department traveled to Japan to acquaint the Director of the Hiroshima Peace Memorial Museum, Mr. Yoshitaka Kawamoto, of our plans, and to request . . . a temporary loan of artifacts from his Museum. . . . Mr. Kawamoto, who is a survivor of the atomic bombing, . . . graciously acceded to our request. . . . I hope that the Japan Foundation . . . may be able to assist our two Museums in this project—which, by putting a painful past in perspective, may also serve as a compelling symbol of present cooperation and future hope.[3]

Attached to my letter was an updated version of Mike Neufeld's proposal, now with the title "A Time for Reflection: Fifty Years Past Hiroshima." I also wrote to Smithsonian under secretary Turner to keep her informed.[4] The foundation did not reply.

I still felt I knew too little about Japanese feelings, and though I had not met him, decided to turn for advice to former senator and ambassador to Japan Michael J. Mansfield. I outlined the background, and wrote,

> Everyone we have talked with agrees on the historic importance of the Hiroshima bombing. Many however feel that there is no way of avoiding the appearance of a U.S. celebration, if the *Enola Gay* is exhibited. That is a serious problem, and we have tried to allay fears by seeking to mount an exhibition sponsored both by the Japanese and the United States.
>
> We have had preliminary conversations with the director of the Hiroshima Memorial Museum, and on the director's advice, I have also written the Mayor of Hiroshima who, I understand, is also greatly interested in drawing the United States into a significant reconciliatory discussion.
>
> The exhibit we have in mind has been tentatively named "A Time for Reflection: Fifty Years Past Hiroshima." . . . Ultimately, its main title might well be changed to "A Time for Reflection and Reconciliation," given that August 6th has for all these years remained a day of protest and recrimination.
>
> A single exhibition will not change all that, of course. But it could serve to resolve, for Japan and the United States, some of the unarticulated issues that persist to this day. And in the process, it may also help to focus attention on broader questions of war and peace for the many visitors from all over the world who visit the museum each year.
>
> I understand that you serve as a Special Advisor to the Japan Foundation, which I have recently written in the hopes of obtaining both moral and financial support for the proposed exhibition. In view of your long service as Ambassador to Japan, and your knowledge of Japan and its people, I was hoping that I could call on you one of these days to seek your advice on how best to proceed both with the exhibition and in contacting the Foundation.[5]

The reference to August 6 came from my first two years at the museum. In 1988 a group had arrived on the August 6 anniversary of the Hiroshima bombing to throw a red liquid, we thought chicken blood, at the huge rockets on display. The museum's security force was not surprised. They told me such protests were annual events. The next year, 1989, the same group tricked us. They waited until August 8, when our security guards would not be expecting them, and created a disturbance then.

Mr. Mansfield's assistant called me on receiving my letter, and on August 22, 1991, I visited the senator in his Pennsylvania Avenue office.

This was a time when we were still searching for a site to display the aircraft in its entirety. When I told Mansfield we were thinking of exhibiting the *Enola Gay* elsewhere, because she would not fit in the museum on the Mall, he appeared greatly agitated. He said he would be alarmed if the aircraft was not exhibited in the museum, simply labeled, like the *Spirit of St. Louis,* or, alternatively, exhibited at a military site like Andrews Air Force Base. I could not elicit from him why he felt that way, but it seemed he thought it would give the *Enola Gay* the appearance of having a special, more prominent aspect, to which we were erecting a shrine.

Mansfield's fears reminded me of my reaction to Gary Mummert's proposal three years earlier, to display the *Enola Gay* in a separate building as a shrine to peace and a warning of "never again." I too wanted to avoid the semblance of erecting a shrine, but I saw no way of avoiding a separate building if we wanted to display the whole enormous aircraft. Nor could I see an exhibition on a military base that would not appear militaristic. I saw no clear way to satisfy Mansfield's concerns.

Two weeks later, as I was still pondering this problem, two unexpected events coincided. On September 5, Gregg notified me that he had just received a message from Japan: Mr. Kawamoto would be pleased to meet with me in Hiroshima.[6] That same day, however, Bob Adams and Carmen Turner and I had met and reluctantly agreed that we needed to avoid publicity and potential controversy for some time, in order to obtain congressional authorization for the museum extension at Washington's Dulles airport. Major activity on the exhibition, including a trip to Japan, would have to wait.

PREVAILING ATTITUDES

This did not mean that we dropped everything.

We were concerned about public attitudes, some of which were now surfacing through surveys in anticipation of the fiftieth anniversary of Pearl Harbor. A survey conducted by pollsters Martilla & Kiley, Inc., in the United States and Japan in December 1991, reported on this question and response:

> As you may know, this December will mark the 50th anniversary of the beginning of the war in the Pacific between the U.S. and Japan. . . . Do you feel that the U.S. atomic bomb attacks on Hiroshima and Nagasaki were *a justified means of ending the war* or *unjustified acts of mass destruction*?[7]

In Japan, twenty-nine percent of those asked thought the bomb a justified means of ending the war, compared to sixty-three percent in the United States. Sixty-four percent in Japan considered it an unjustified act of mass destruction, whereas only twenty-nine percent of Americans were of that opinion. Seven to eight percent on both sides were not sure.

A New York Times, CBS News, and Tokyo Broadcasting System survey asked, "Should Japan apologize for the attack on Pearl Harbor?" Forty percent of Americans said yes, and so did fifty-five percent of the Japanese. Asked whether the U.S. should apologize for dropping the atomic bombs, only sixteen percent of Americans thought so, while seventy-three percent of the Japanese thought we should. And eighty-three percent of the Japanese felt that dropping the bombs was morally wrong.[8]

A poll, "U.S.-Japan Relations 50 Years After Pearl Harbor," conducted by the ABC News Polling Unit and the Culture Research Institute of NHK Broadcasting, found these results:

> Despite the passage of time, 39 percent of Americans ... say they still hold it against Japan for mounting the Pearl Harbor attack. Among Americans who are 60 years old, this lingering ill will soars to 60 percent.
>
> Across the Pacific a greater level of bad feeling remains from the devastating atomic bombings of Hiroshima and Nagasaki. Fifty-seven percent of Japanese say they hold it against the United States for the bombings. That includes a majority in every age group except those under 31.... Among Americans and Japanese alike, this residual anger is roughly twice as prevalent in the oldest group as it is in the youngest.[9]

It seemed that the younger generation, at least, would be reasonably open to an exhibition of the kind we were planning, but the attitudes of older people would also have to be carefully considered.

Another indication came from Sadao Ishizu, manager of the cultural projects division at the Osaka department of the newspaper *Asahi Shimbun*, who wrote Gregg Herken on November 15,

> The year 1995, the 50th anniversary of the end of World War II, is already coming into sight, and we are thinking of organizing some event to mark the occasion....
>
> I have heard from a Hiroshima authority that you have a deep interest in this connection, and I wonder whether it might not be possible for us to cooperate in some way.
>
> ... May I ask you, then, to let us know your own Museum's/Institution's intentions concerning the 50th anniversary? ... One plan we have in mind, though it is still very tentative, would be to organize an exchange of documents and other materials on the War between the Hiroshima Peace Memorial Museum and your Museum, as a basis for examining World War II as a whole....
>
> I feel that a close tie-up between your Museum and the *Asahi Shimbun* could be most fruitful as a means of obtaining a broad perspective on the war and of making a meaningful appeal to the international community....[10]

Mr. Ishizu's offer seemed a good omen. But for the moment, our preoccupation with congressional authorization for the extension was keeping us stalled.

In March 1992, I had one more opportunity to assess Japanese attitudes. A close Japanese colleague and astrophysicist visited Washington for discussions on a joint astronomical project. He and I had to drive out to NASA's Goddard Space Flight Center in the Washington suburbs, and that gave me a chance to ask him privately how he thought Japanese people would react to an exhibition of the *Enola Gay* along the lines we were planning. Though I had known him for almost twenty-five years, he looked uncomfortable, but he also appeared too polite to express an adverse opinion. This worried me.

CONTACT RESUMED

Throughout 1992, the Smithsonian was planning a major exhibition to open two years later in Japan. Titled "An American Festival," it would display a wide variety of perspectives on the United States through artifacts in the Institution's collections. The National Air and Space Museum had promised the loan of an airplane, a spacecraft, and some other artifacts for this occasion. Perhaps to show Japanese appreciation, my wife Marianne and I were invited to a dinner at the Washington residence of Ambassador Kuriyama on August 5, 1992.

One of the two guests of honor was Shinichiro Asao, retiring from his post as Japan's ambassador to Italy and about to take on the position of director general of the Japan Foundation. Seeing an opportunity to involve the foundation, which had not responded to my earlier inquiry, I sought out Asao, told him of our plans for exhibiting the *Enola Gay*, and asked whether the foundation might be willing to support us. He suggested that I write to him, and a few days later I did:

> Last week we spoke at the dinner held at the Embassy of Japan in Washington.... I mentioned to you that the National Air and Space Museum is currently restoring the *Enola Gay*. We expect the restoration to be complete by 1995. The question of then exhibiting the aircraft, most probably in Washington, has been widely discussed. And it is of great concern to our Museum to make sure that this exhibition does not strain relations between our two countries....
>
> I believe that an exhibition would be considerably more significant if it could be produced by the National Air and Space Museum, with access to information that would only be available in Japan. We wish to assure an accurate presentation, and that would be facilitated through the availability of Japanese archival information which might be more accessible if we were to work with leading Japanese scholars....
>
> If you and the Japan Foundation were willing and able to put us in touch with Japanese scholars whom we could consult, particularly on Japanese archival data but possibly also on other matters, we would greatly appreciate your efforts.
>
> I would be prepared to come to Tokyo to meet with you and your staff, to see whether some such arrangement could be worked out.
>
> I believe we have an opportunity to produce a truly significant exhibition which, if properly planned, could help to define the events of the last few days of World War II and clarify the common history of our two nations in a way that few other opportunities can offer.[11]

Three months later, I received a remarkably helpful answer from Jun'etsu Komatsu, deputy director of the Center for Global Partnership at the Japan Foundation. Komatsu had followed up on my letter by writing Yoshitaka Kawamoto, the director of the Hiroshima Peace Memorial Museum. Kawamoto, in turn, had recommended that the Radiation Effects Research Foundation, a cooperative Japan–United States research organization, be asked to make its own recommendations to the museum.

In mid-October Komatsu had accordingly called Dr. Itsuzo Shigematsu, president of the Radiation Effects Research Foundation, who had sent him a list of several possible collaborators for our project. This list included two research sociologists, a person familiar with sources of reference materials, and Dr. Shigematsu himself. Unwilling to put the three others into an embarrassing position, Shigematsu had asked the Center for Global Partnership to first check with all three to see whether they would be willing to work with us before recommending their names to the National Air and Space Museum. This having been done and all the responses now in hand, Komatsu listed several Japanese colleagues we might contact, some of whom we later got to know.

Komatsu also attached a clipping in which the mayor of the city of Hiroshima was quoted as expressing reservations about lending materials if the exhibit were to in any manner glorify the destructive power of the atomic bomb. Komatsu pointed out that the article indicated serious concern among the *hibakusha* (victims and those who were exposed to the atomic bomb) about the museum's restoration of the *Enola Gay* and our intention to exhibit archival material. There were, he noted, not only worries about the restoration itself, but the possible repercussions an exhibition might have at a time when, with the termination of the Cold War, global forces were moving toward significant disarmament. In view of these sentiments, Komatsu felt that we might wish to proceed with even more sensitivity than we might have initially imagined.

In concluding, Komatsu quite needlessly excused himself for the length of time it had taken to get back to us and conveyed the Center's most sincere wishes for the success of our project.[12]

THE RADIATION EFFECTS RESEARCH FOUNDATION

By this time the takeover attempt by the city of Denver had failed, the extension was likely to be authorized for the Washington Dulles International Airport in the new session of Congress, and I was able to persuade the secretary and under secretary that the museum had to move forward on the exhibition to have any hope of opening in 1995. Their agreement came at the close of 1992, and in January 1993, I contacted Dr. Shigematsu.

After introducing myself, I wrote,

> . . . the National Air and Space Museum is considering an exhibition for the summer of 1995 that would commemorate the pivotal events surrounding August 6 to 9, 1945, which had catastrophic effects for the people of Hiroshima and Nagasaki. These bombings are widely credited with abruptly ending the war between our two nations, but also led to a forty-five-year-long legacy of escalating nuclear arsenals of unimaginable destructive potential.
>
> One of the main thrusts of the exhibition will be to portray the mood in the United States and Japan toward the end of the war, to help the visitor understand the public attitudes that shaped President Truman's decision to drop the bomb. We would also seek to provide a realistic picture of the human suffering that followed the bombings, and contrast that to the attitudes in the United States, where there was a sense of relief that the war had finally ended. To conclude, the exhibition would take a deeper look at the enduring legacy which the dropping of those two bombs produced. . . .[13]

I added that we wished to discuss the loan of artifacts from the Hiroshima Peace Memorial Museum, proposed that Mike Neufeld and I come to Hiroshima to discuss these and other issues, and appended an outline of the exhibition's thrust.

The Radiation Effects Research Foundation is an organization now jointly administered by the United States and Japan. In 1975, it became the successor to the Atomic Bomb Casualty Commission, established shortly after the war, and funded and operated solely by the United States. Both organizations have aimed to understand the physiological effects of atomic warfare. For many years, the citizens of Hiroshima and Nagasaki objected to the ABCC, which only studied the medical effects of the bombs without offering help to the affected. Some of these resentments have subsided since the establishment of the binationally administered RERF, and the foundation is now recognized as the world leader in its field. As Dr. Shigematsu put it,

> The mission of the Radiation Effects Research Foundation is to clarify the health effects of radiation on Atomic Bomb survivors. For this purpose, a broad research program in epidemiology, clinical medicine, genetics, immunology, etc. has been pursued with the joint participation of scientists from Japan and the United States during the past 46 years, in itself a historically unique event.[14]

When the giant nuclear power plant at Chernobyl erupted in April 1986, Soviet leaders turned to the RERF for advice on medical matters. The foundation had been tracing the effects of atomic radiation on seventy thousand survivors of Hiroshima and Nagasaki for forty years, by then the most massive medical record ever assembled.

On February 26, I telefaxed Dr. Shigematsu, proposing,

> The topics we would like to put on the agenda for the first day would be an exchange, where we . . . portray for you the kind of exhibition we would like to mount, and . . . also ask you for your help and suggestions on topics that you think need to be included in order to represent Japanese perceptions on the bombing of Hiroshima and Nagasaki in proper perspective. We will need to show the radiological aftermath and suffering due to the bombings, and to understand some of the historical and sociological implications that affected the two cities.
>
> We will also wish to establish, in our exhibition, the general mood in Japan in the last months of the war, so that a largely American audience will be able to understand why Japan was still willing to fight on, when many in your country already believed that the war no longer could be won.[15]

Having taken the plunge, it was time to plan the trip as carefully as we could. On February 20, Tom Crouch, Mike Neufeld, and I met for lunch with history professor Akira Iriye of Harvard, one of the truly knowledgeable experts on United States–Japanese relations in the twentieth century. A few days later, Iriye sent me a list of four scholars in Japan who might be able to help us, particularly in obtaining archival material.[16]

Early in March, I learned that the Smithsonian Institution maintains a liaison officer in Tokyo. I wrote at once and introduced myself to Mrs. Hanako Matano, whom I came to respect as a truly gifted lady. She seemed to know everyone and was trusted by all. A stream of daily telefaxes began to flow across the Pacific as we worked out a detailed itinerary. She had good ideas on people I needed to meet to make headway.

On March 5, I received a surprise telefax from Dr. J.W. Thiessen, who introduced himself as Vice Chairman of the RERF:

> I decided to write to you directly in the hope that you won't mind frank opinions expressed by an American who has been involved in A-bomb survivor research in Japan for nearly six years, and who is much in favor of a forceful U.S. exhibit to memorialize the 50th anniversary of the use of A-bombs.

He attached a translation of an article from *Asahi Shimbun* and commented,

> One can predict that local reactions will be negative about anything that glorifies, promotes, condones or even explains the use of atomic weapons. If you or your exhibit create this impression, then you can expect little or no cooperation. Such public attitudes were recently exacerbated when it became known that the Hiroshima and Nagasaki bombings were listed under "experimental explosions" in a Department of Energy publication. . . . The key to your success will be projecting the appropriate image. In this regard, you should realize that your plans about the *Enola Gay* are unlikely to be viewed favorably here.
>
> . . . Initially we were very much a part of the Occupation, and even today the more politically active A-bomb survivors view us that way, i.e., as an act of domination by the victors over the victims.

> . . . Given the situation here . . . you are better off with a direct approach uncomplicated by attitudes toward us. I recommend that you consider a meeting with the Mayor(s) as the "centerpiece" of your attempts to obtain cooperation of the local communities. . . . Other meetings—including the one with our foundation—should be considered as secondary.[17]

Thiessen's reference to "experimental explosions" was in line with questions I would later encounter. Many in Japan asked whether the bombings of Hiroshima and Nagasaki had not been tests designed to measure the effects of atomic bombs on the populations of the two cities? Why else, I would be asked, had Hiroshima and Nagasaki been kept off the lists of cities to be subjected to regular bombing—if not more easily to assess the power of the atomic bombs? Why had two bombs been dropped, one uranium and one plutonium—if not to have a clear-cut comparison? Why was the RERF set up—if not to find out the effects of radiation on humans?

I was glad to accept Dr. Thiessen's advice on writing the mayors of the two cities, but decided, as a matter of etiquette, to first call the American embassy in Tokyo. At a February 25 reception in Washington I had met Mrs. Michael Armacost, wife of the U.S. ambassador to Japan, who had mentioned cultural attaché Robin Barrington as the person to call. Barrington worked for Paul P. Blackburn, who had two portfolios, as director of the U.S. Information Service for Japan and also as minister-counselor of the embassy for public affairs. Barrington soon left Tokyo to return home, but during the next two years, I would frequently call on Paul Blackburn for advice; he was invariably helpful.

I now called Barrington to let him know that I intended to contact the two mayors and to ask for his comments. With that done, I wrote to Mayor Hiraoka of Hiroshima and Mayor Motoshima of Nagasaki. To both letters I attached a copy of the exhibition proposal:

> I am writing you on a matter which is both urgent and, as you will understand, difficult to describe.
>
> For the summer of 1995, the National Air and Space Museum is planning an exhibition whose central theme is the bombing of Hiroshima and Nagasaki in August 1945. We realize that this will be a controversial exhibition:
>
> For the population of Hiroshima and Nagasaki, the bombings were a terrible disaster.
>
> For the country of Japan it became a national tragedy.
>
> For many in the United States and all over the world, the bombings came to be seen as ending a long war and a final cessation of further death and destruction.
>
> For all of us it soon became clear that the awful shadow of atomic weaponry could not so easily be turned off. The threat of world-wide nuclear destruction remained with all of us for forty long years of Cold War between major powers, and the threat of nuclear terrorism still is with us today.

> These are the topics of our planned exhibition. . . . The events of August 1945 inextricably link the histories of our two countries with the events that befell Hiroshima and Nagasaki. For our exhibition to have a serious impact it must provide a faithful account seen not just through American eyes, but also from a Japanese perspective.
>
> . . . We would . . . wish to exhibit objects and materials that we might be able to obtain on loan from the Hiroshima and Nagasaki Memorial Museums, in order to vividly portray the immense amount of damage and suffering that the bombings produced.[18]

I gave some details about our planned trip and asked for the opportunity of a meeting. Similar letters went to the two museum directors in Hiroshima and Nagasaki.

Before the week was out, local newspapers in Hiroshima and Nagasaki were writing about the controversy that would surround our impending visit and our request for loans of artifacts from their museums.

One article from *Asahi Shimbun*, which Dr. Thiessen sent us in translation, carried the headline "American Museum Director Coming to Hiroshima to Request Loan of A-Bomb-Exposed Materials—Mayor Shows Reluctance to Exhibit with Bomber," and ended with a comment from the mayor:

> Hiroshima City is lending out A-bomb-exposed materials to convey to various parts in Japan and abroad our wish for peace. There is concern, as might be expected, that the materials, if loaned for display in the planned American exhibition, may be used to show off the impact of the bomb. I cannot say anything until I have their views fully explained to me.[19]

A CHANGE IN PLANS

By now the visit to Japan had been set for early April. I had originally thought that I should go to a first meeting with Michael Neufeld, the chief curator for the exhibition. But on March 16, I met with Neufeld, the chairman of his department; Tom Crouch, who would oversee the exhibition; and F. Michael Fetters, in charge of the museum's public affairs office. We discussed the matter, and I said that I now felt I should take Tom Crouch with me instead, because his presence might reassure the Japanese. Tom had built up good relations with the Japanese through an exhibition he had mounted at the National Museum of American History during the years he had been there. His "A More Perfect Union" exhibition on the internment of Japanese Americans during World War II had shown great sensitivity, and had won him many friends among Americans of Japanese descent. We were getting indications in the Japanese press that our exhibition would be controversial. I wanted to make sure we made a good first impression.

If we succeeded on this first mission, I suggested that Crouch, Neufeld, and the exhibition designer, William "Jake" Jacobs, go on a second visit, some weeks later, to select artifacts we might wish to borrow.

Admittedly, this was not the entire story. I had previously expressed disappointment to Neufeld about the lack of time he was devoting to the project. I had found myself making all the preliminary arrangements for the trip myself, where he should have been taking the lead, given that it was his project.

In Neufeld's defense I should say that all this was now suddenly surging forward at an awkward time for him. He had long-standing obligations to a publisher to adhere to a deadline for submitting the final draft of his book *The Rocket and The Reich*.[20] For some weeks he was unable to devote himself fully to the exhibition or to preparations for the trip to Japan. I was annoyed. I felt that Neufeld was not sufficiently well prepared, and thought that Tom would better help with public relations.

I emphasize this, in part, because a year later, when Neufeld came under enormous pressure to make major changes in the exhibition script, he showed himself levelheaded, effective, and ready to put in all required effort. My initial reservations then gave way to genuine appreciation.

THE EMBASSIES

On March 17, 1993, the day after I had decided that Crouch, rather than Neufeld, would accompany me to Japan on this first trip, I wrote to Ambassador Kuriyama, at the Japanese embassy in Washington. The significant parts of the letter read,

> I am writing to ask whether it might be possible for me to meet with you, at a time convenient to you, for perhaps half an hour, to ask for your advice:
>
> Last summer, you and Mrs. Kuriyama were so kind as to invite me and my wife to a dinner held at your residence in honor of Ambassador Shinichiro Asao and Mrs. Cheney, at the time head of the National Endowment for the Arts.
>
> On that occasion I had an opportunity to speak with Ambassador Asao about an exhibition the National Air and Space Museum is contemplating for 1995, in remembrance of the atomic bombings of Hiroshima and Nagasaki. Such an exhibition will undoubtedly be difficult. But the events of August 6 to 9, 1945, had an enormous impact not only in accelerating the end of World War II, but also in casting a shadow that persisted for forty long years of Cold War between world powers. Unless the public is willing to understand events that led to the bombings, and the terrible destruction that they wrought, the most valuable lessons that can be learned from history will be lost.
>
> Ambassador Asao['s] . . . deputy at the Foundation's Center for Global Partnership, Jun'etsu Komatsu, wrote to suggest a number of Japanese

> scientists and scholars whom we might wish to visit in Japan to obtain their views on how best to organize our exhibition....
>
> The reason for writing you now is my recognition that the exhibition we are planning cannot be carried out, with the kinds of aims we have in mind, unless we are able to enlist thoughtful commentary from Japanese scholars and scientists, medical and technical experts, statesmen and perhaps military specialists. The exhibition will not be fully effective if the sole views expressed are those of the American side.
>
> In order to succeed, we will undoubtedly need to earn the trust of senior figures in Japan, including the Mayors of Hiroshima and Nagasaki, representing the two populations most severely affected.
>
> I am planning to visit Japan, around the first of April ... and hope to meet with the mayors of both Nagasaki and Hiroshima, with the Directors of the two cities' Memorial Museums, ... medical scientists from the Radiation Effects Research Foundation, ... and a number of scholars of Japanese wartime and post-war history whom Prof. Akira Iriye of Harvard University recently recommended....
>
> Questions on which I would like to ask your advice center on the best ways to bring about the understanding by Japanese authorities that this is not meant to be a celebration, nor an exhibition about the power of technology. Rather, we would like to bring about an awareness by the public of the complex chains of events that can lead to enormous destruction with catastrophic consequences perhaps capable of affecting generations to come.
>
> I sincerely believe that we will only succeed in this attempt if we can bring to a focus the different views that thoughtful scientists, scholars and statesmen have discussed during the past half century, and that will require a partnership between our museum and Japanese colleagues.
>
> Your advice on these matters would have profound influence on our thinking and could lead to a deeper, more thoughtful, analysis. I would greatly appreciate an opportunity to hear your reaction and your thoughts.[21]

I attached the proposal we had circulated to all others, and hoped I would receive a response. For reasons that would become evident half a year later, none came.

I also called Paul Blackburn at the embassy in Tokyo and arranged to meet him at the embassy on April 9, after our visits to Hiroshima and Nagasaki, so we could inform him where matters stood.

Most surprising was a final telefax from Yoshitaka Kawamoto, director of the Hiroshima Peace Memorial Museum, who notified me on March 31, 1992, the day before our departure for Japan, that this was his last day before retirement. Our appointment at his museum would be with his successor, Mr. Harada, "who will arrive here tomorrow." I had heard that Japanese officials are fully active until the day of retirement and then, overnight, permanently leave office, relinquishing all responsibilities. We would be meeting Mr. Harada before he had even had an opportunity to get to know his own museum.

KEEPING ADAMS INFORMED

On March 29, just before leaving for Japan, I wrote to Bob Adams:

> I had promised to keep you informed about the Museum's activities on the exhibition of the *Enola Gay* we are planning for 1995.
>
> Late last summer, I mentioned the exhibition to Ambassador Shinichiro Asao, whom I met at a reception at the Japanese Ambassador's residence. Ambassador Asao was returning from his post in Italy to assume the Presidency of the Japan Foundation, headquartered in Tokyo. Through Ambassador Asao and the Japan Foundation we were put in touch with a number of contacts in Japan, and I will be meeting with many of those individuals next week. Tom Crouch, who will be accompanying me, will also be in on all the meetings.
>
> In preparation for these meetings we have sent an outline of the proposed exhibition to everyone involved, in order to make sure that our plans were understood ahead of time. Having seen that script, the following officials have assented to see us in the course of the week:
>
> In Hiroshima, Takashi Hiraoka, Mayor of Hiroshima,
> Yoshitaka Kawamoto, Director, Hiroshima Peace Memorial Museum,
> Dr. Itsuzo Shigematsu, Chairman, Radiation Effects Research Foundation
>
> In Nagasaki, Hitoshi Motoshima, Mayor of Nagasaki,
> Tatsuya Itoh, Director, Nagasaki Peace Memorial Museum,
> Dr. J.W. Thiessen, Vice Chair, Radiation Effects Research Foundation,
>
> In Osaka Dr. Kazuyoshi Otsuka, National Museum of Ethnology. He has favorably commented on the exhibition in the Japanese press,
>
> In Tokyo Paul Blackburn, Public Affairs Officer, USIS, at the U.S. Embassy
>
> I also have made Ambassador Takakazu Kuriyama, at the Japanese Embassy here in Washington, aware of the exhibition and expect to stay in touch with his staff.
>
> In addition, I have also had a chance to discuss the exhibition with Ambassador Mike Armacost's wife, who was in Washington without her husband last month. She said she'd tell the Ambassador about the exhibition, and also put me in touch with the embassy's cultural affairs officer, Robin Barrington, whom I then called in Tokyo.
>
> We have been working closely with Mrs. Hanako Matano, the Smithsonian liaison officer in Tokyo, and she has been helpful in setting up some additional meetings to introduce us to people whom we might not otherwise have approached.
>
> On the U.S. side, we have just had a first meeting with the Executive Director of the 50th Anniversary of World War II Commemoration Commit-

> tee. The Committee was established by Congress, works out of the Pentagon, and is largely staffed by active duty military personnel. Its Executive Director is Lieutenant General C.M. "Mick" Kicklighter (Retired). We approached Gen. Kicklighter's office recently. He and his senior staff then visited us and gave us a presentation on their efforts. In turn, I briefed them on our proposed exhibition and gave them the same script we sent ahead to the officials we will be meeting in Japan.
>
> The Committee is running an ambitious program, and would like us to work with them. They would like to use materials based on video footage we have on the Tuskegee Airmen. They'd also like to make use of footage we have on the *Enola Gay* and the atomic bombings of Hiroshima and Nagasaki. They asked us to put them in touch with competent historians who could make sure that the committee's products—pamphlets, public service video spots, posters—are accurate and balanced. And they are interested in the possibility of our preparing a pamphlet on the *Enola Gay* and the atomic bombings that could be used with educational materials they are sending out to schools nationwide.
>
> Aside from this, we have also begun to work with a number of scholars of World War II and post-war Japanese-American relations. I had mentioned to you that we had met with Prof. Akira Iriye of Harvard, whom you said you knew when you both were at Chicago. Iriye has put us in touch with a number of Japanese scholars, and I expect that Mike Neufeld, the exhibition's curator, will be meeting with them later in Japan.
>
> As you can see, we are working with a wide spectrum of interested parties. So far, we have been received well by all sides.
>
> I'll keep you informed on how our discussions in Japan, our collaborative efforts with the Pentagon, our long-standing exchange of ideas with the B-29 veterans' association and our liaisons with other groups evolve.[22]

JAPAN

I had been to Japan twice before, once in 1956 and again in 1969, when I encountered a completely different country, so rapid was the restructuring of Japan and its economy after the war. This time, however, additional surprises came in the way business was conducted, and particularly in the ubiquitous presence and intrusion of the press in Hiroshima and Nagasaki. Where we had anticipated private talks, we were invariably involved in discussions held in public, or tape-recorded if held in private.

After arriving in Tokyo in the late afternoon of April 2, Crouch and I briefly met with Mrs. Hanako Matano, the Institution's special representative for Japan. I had talked with her frequently on the phone, but it was a pleasure to meet this gentle dynamo of a woman in her fifties, who was helping me understand many of the Japanese customs I would need to recognize.

The next day, Crouch and I flew to Hiroshima, arriving in the early afternoon. After checking into our hotel, we took a tram to the Hiroshima

Peace Memorial Park and Peace Memorial Museum. We went through the museum, looking at everything carefully, at our leisure, to familiarize ourselves with its contents. Both of us were impressed with some of the objects, powerful photographs, and excellent videotapes of survivors' stories with good English captions. Any of these would make valuable additions to our exhibition.

The following Monday, April 5, 1993, we were picked up at 9:30 A.M. and driven to the city hall. On arrival we were introduced to some of the attendees who were to be present at the meeting with the Mayor: The first was Mr. Minoru Ohmuta, chairman of the board of directors of the Hiroshima Peace Culture Foundation, a former journalist and the senior man apparently of all of the Peace Memorial activities. Next in rank, as suggested by their seating, were senior officials attached to the mayor's office, followed by the new director, appointed April 1, of the Hiroshima Peace Memorial Museum, Hiroshi Harada.

A large contingent of local print and television journalists was also in the room. Even before the discussions with the mayor began, and throughout the meeting thereafter, lights and cameras were going and reporters were busily taking notes.

Mayor Takashi Hiraoka entered the large meeting room at 10:00 A.M. We shook hands. I sat next to him on his right at the head of the room. The Japanese officials sat in order of rank on his left. Facing them and sitting by himself on our right was Tom Crouch. The press was all over.

The mayor began with an apology for not having responded to our request that we be able to work with the Peace Memorial Museum and borrow artifacts for our exhibition, but he felt he needed to assure himself first of our intentions. Quite evidently a charismatic personality, he spoke forcefully about the need for peace and full abolition of all atomic weapons. He wanted to know would that be the thrust of our exhibition as well? I responded with an outline of why we had come to Hiroshima, why we were putting on our exhibition, and what we hoped it would contain. I emphasized the pivotal place of the bombings in the annals of twentieth century history, and the need for understanding how the decisions leading up to the bombings had come to be made. Unless such historical decisions were better understood, the future of civilization was going to continue to be insecure. Nevertheless, I said, we were not an organization that made political statements. We could not make calls for the abolition of all bombs, but we could include in our exhibition points of view of important constituencies, and the city's concerns about the abolition of bombs could certainly be included in that way. The suffering brought about by the bombings would receive understanding treatment. Tom Crouch, chairman of the department that would be mounting the exhibition would assure that. I mentioned Tom's earlier exhibition "A More Perfect Union," which appeared to be well known in Japan, and gave the mayor a brochure describing the exhibition. The mayor again forcefully stated the city's concerns, but also told me to further discuss our plans for exhibiting the *Enola Gay,* with the staff of the Peace Memorial Museum—a city museum that came under his jurisdiction.

The meeting, which had been scheduled to last twenty minutes, went on for a total of forty. Afterwards, the press remained and asked many questions.

We were then driven to the Peace Memorial Park. At the museum we were shown around by Mr. Harada and a curator. Again, television cameras followed us. I was asked to write an entry into the museum's guest book, again with cameras running, and then we retired to a conference room where discussions began in earnest.

Mr. Ohmuta led off. Occasionally, Mr. Harada added comments, though being new in his job, he could not speak with as much knowledge. Most of the statements were similar to those voiced by the mayor, and I repeated much of what I had said earlier. But positions did become clearer as the conversations went on. The discussion was tape-recorded at their request.

After lunch, Crouch and I were driven to the Radiation Effects Research Foundation (RERF). There we met with Dr. Shigematsu, Dr. Thiessen, and the four other directors on their staff. Also present were two young university professors from Hiroshima.

We outlined our reasons for coming to the RERF, and received literature that summarized more than four decades' worth of medical efforts to identify all symptoms displayed by atomic bomb casualties. We met for about two hours, after which Thiessen gave us a tour, showed us archival material in the form of painstakingly compiled dossiers on individuals, depicting exactly where in a building they had been at the time of the explosion and estimating how much dosage the individual had received. There was an interesting optical device from the 1950s that permitted slant angles to be calculated for exposure calculations; it might have been worth exhibiting. We were also shown around the laboratories where patients receive checkups at two-year intervals.

We were then returned to City Hall for one more press conference before the day was over. Clearly there was a great deal of concern in Hiroshima about an American exhibition dealing with issues they considered very much their own.

The following morning discussions resumed at the Peace Memorial Museum, where the others already were waiting in the conference room. With our permission, the tape recorder again was switched on. Over the next three hours we reached agreement that we should continue discussions with the museum staff, while putting together a detailed exhibition script by the end of the summer. There would be no commitment by the mayor or the museum until they had that script, but they would help us with information about materials that might be exhibited, so that the script could be specific. Even if they agreed to work with us, I assured them, we would continue keeping them informed, and if they or the city ever felt that they needed to withdraw, no questions would be asked. We would simply agree that we were unable to reach agreement.

One of the main Japanese fears was that the *Enola Gay*, by virtue of size alone, would dwarf any exhibit on the Hiroshima aftermath. We said we understood that, but felt that even so we could make a powerful exhibition of the catastrophic effects of the bombings.

In summary, we agreed that Mr. Harada would be their contact point; suggested that Mayor Hiraoka be invited to videotape a message on behalf of his city; said we'd like our curator and designer to visit Hiroshima to discuss details at a working level; and asked about the use of their videotapes of survivors' recollections. They explained that those tapes belong to the Peace Culture Foundation but might be available.

After lunch we continued to meet with Mr. Ohmuta and Mr. Harada, who said they now could talk more freely in the absence of the tape recorder, and that they were reasonably optimistic about our prospects of working together, though of course no guarantees could be given.

We talked for perhaps another forty minutes.

Late in the afternoon, Crouch and I took a train to Nagasaki.

The next morning, we went to the Nagasaki museum, the International Culture Hall, to look at their exhibits by ourselves. The photographs here were more striking than those in Hiroshima and the display of artifacts was better in having the pictures of the object's owners next to them—school children and others—making the exhibition come far more alive.

Shortly before 1:30 we went to City Hall, where we were ushered into a conference room. First to arrive after us was Brian Burke-Gaffney, a native of Canada, long a resident in Japan, who introduced himself as the translator for the session. Two dozen press and television people then began to crowd into the room, which was much smaller than the conference hall in which we had met with the mayor of Hiroshima. At 1:30 P.M., Mayor Hitoshi Motoshima appeared.

The mayor is a remarkable man. In December 1988, when Emperor Hirohito was dying, he was asked, during a regular session of the Nagasaki City Assembly about the emperor's war role. His response caused an immediate uproar. Speaking to reporters after the session, he had said,

> It is clear from historical records that if the emperor, in response to reports of his senior statesmen, had resolved to end the war earlier, there would have been no Battle of Okinawa, no nuclear attacks on Hiroshima and Nagasaki. I myself belonged to the education unit in the western division of the army, and I instructed troops to die for the emperor. I have friends who died shouting "banzai" to the emperor. I am a Christian, and I had difficult moments as a child when I was pressed to answer the question, "Who do you think is greater, the emperor or Christ?"[23]

These remarks caused such outrage among Japanese right-wing forces, and particularly Japanese veterans of World War II who felt it demeaned their service to emperor and nation, that sound trucks blaring demands for his death hounded the mayor for weeks. Threats arrived with the more than seven thousand pieces of mail that now inundated him. For months he and his family were under constant protection by the police. He could not go outdoors unattended. Alone in his position at first, he was subjected to the most vitriolic attacks. Then, slowly, public opinion shifted in his favor. But in January 1990, about a month after his police protection had been lifted, the bullet of a would-be assassin passed through his lungs.

By the time of our visit, three years later, he looked frail at seventy-one, but also at ease.

We shook hands and sat down at a long table. The mayor sat at its head. I sat to his right. To the left of the mayor, facing me, sat Burke-Gaffney, translating for us. Also present was Dr. Tatsuya Itoh, the director of the Nagasaki International Culture Hall (atomic bomb museum), Nagasaki Foundation for the Promotion of Peace, to whom we were later introduced.

The mayor began by greeting us and then began a long series of penetrating questions. He quite clearly also wanted the exhibition to advocate the abolition of all nuclear weapons. Would we be advocating that? I explained that we took no political positions, but that we would be glad to invite him to deliver a videotaped message representing the city's views on the subject. He said he felt Japan owed the United States an apology for Pearl Harbor, and that we owed one to Japan for the atomic bombings. Would our exhibition do that?

I said that apologies were not as important as trying to learn lessons. I mentioned that I had only come to the United States after the war. Many of the members of my family who had not been able to leave Czechoslovakia, the country where I was born, had been killed by the Germans in concentration camps. Should I expect an apology from the Germans? It made more sense to try to see how the Holocaust had come about, rather than to expect an apology. In the same way, I felt, our exhibition would serve a need by coming to a clearer understanding of the background for the atomic bombings. We had been scheduled to meet for half an hour, but the mayor's questioning continued for a full hour. In the end he said he would consult with members of the museum and his staff before reaching a decision on whether we could expect their help with our exhibition.

After some more press and television interviews, we took a cab to the museum. There we talked with Dr. Itoh and several members of his staff for the rest of the afternoon, occasionally interrupted by the press. In this session, we reached tentative agreement along the same lines as in Hiroshima. Again, we encountered great concern that the artifacts we wished to borrow and display would be lost next to the huge *Enola Gay* bomber. Crouch and I assured them that we had seen any number of powerful photographs in their museum that showed the human tragedy on the ground far more eloquently than words. We also hoped to borrow a watch or clock stopped at the precise hour of the devastating explosion; a scorched clog, the only remaining trace in the rubble, of a woman never found again; melted roof tiles, testimony to the awesome heat generated by the explosion; bent steel girders, evidence of the magnitude of the blast; and a little girl's lunch box, the only trace of her ever found by her parents, to be displayed with a small photograph of her. The rice and peas in the box are charred, recognizable only from their shapes.

We assured our hosts that their objects, though small, would not be lost. They would be displayed in a separate part of the gallery, out of sight from the *Enola Gay*, and would be impressive in their own quiet way.

The following morning, we were picked up just after 9:15 by an RERF staff car and brought to Dr. Thiessen's office. There we met with Dr. Kenjiro Yokoro, a survivor from Hiroshima. Though not present at the bombings, he had arrived there to help out, three days after the disaster, when still a freshman medical student. Yokoro expressed misgivings about our chances for success, stating that the survivors would probably protest, but he offered to help in any way he could.

In the evening, we flew back to Tokyo.

The next morning, Mrs. Matano, Crouch, and I visited the U.S. embassy to meet with Paul Blackburn, director of the United States Information Service for Japan and minister-counselor of embassy for public affairs, and his aide, Robin Barrington. We talked for about an hour and a half, and Blackburn seemed pleased that we had made so many contacts and covered so many bases. He suggested, however, that we should try to have ongoing updates on the moods in Hiroshima and Nagasaki through our contacts there, saying that moods could very quickly shift in Japan. Blackburn also said that he had briefed Ambassador Michael Armacost on our proposed exhibition, and that the ambassador would have liked to have met us, but had a previous engagement that morning.

Mrs. Matano, Crouch, and I then went off to have lunch with Mr. Shunichi Shibohta, executive secretary of the Japan Airlines Foundation; Hikaru Takeuchi, senior staff writer at the Mainichi Newspapers; and Susumo Miyoshi, chief producer, Cultural Affairs Department in the Project Planning and Development Division of Dentsu, Inc.

Mr. Takeuchi, who spoke excellent English, showed us his Air & Space membership card and wore a Wright flyer tie tack in honor of Crouch's biography of the Wright brothers, which he had read.

Mr. Shibohta was very much interested in sponsoring some form of conference in preparation for the exhibition. He was concerned to obtain U.S. press coverage of it, not just Japanese. We mentioned a list of potential American and Japanese participants Crouch and I had worked up ahead of time. We discussed a variety of conceivable formats and said we were greatly interested in following up the suggestion of a conference.

Unfortunately, for reasons I will recount later, nothing ever came of these plans.

In the two weeks after my return I sent Bob Adams a detailed report on the trip and wrote many letters of thanks to Japan. On April 15, I also wrote Ambassador Kuriyama a second letter to keep him informed.

This initial trip was followed by two more. In late May and early June, Tom Crouch, Michael Neufeld, and William "Jake" Jacobs, the exhibition's designer, visited Japan to begin discussions on the selection of artifacts we might borrow from the two museums. They came back with a long list of options, and began to decide on specific objects they would request. The atmosphere on this visit to Hiroshima and Nagasaki was friendly and cooperative. We appeared to be going forward with a strong exhibition that would meet American requirements without offending Japan.

JAPAN REVISITED

Causing offense was a constant worry. Cultural differences can easily lead to serious misunderstandings. I had arranged to return to Japan in early August to make sure that all potential problems had been solved. On this second trip I brought along, at their own expense, my wife Marianne and our son Eric, who speaks relatively fluent Japanese. Five days before our departure, Hanako Matano called me, worried about an apparent oversight. The Japanese Foreign Ministry's director-general for North American affairs, Yukio Satoh, had recently met with General Kicklighter, executive director of the 50th Anniversary of World War II Commemoration Committee, and Kicklighter had informed him about our intended exhibition. Mr. Satoh was very disturbed that the Smithsonian had not officially informed Japan.

I told Matano that I had, of course, written Ambassador Kuriyama, here in Washington, and had assumed he would forward the information to his Ministry of Foreign Affairs. Since Mrs. Matano's brother-in-law, Ambassador Makato Watanabe, was a highly respected member of the ministry, I also told her that I would appreciate any advice he might be willing to give us. Then I telefaxed her a copy of the letter I had written Kuriyama on March 17. I forgot that I had also written him on April 15, but that did not matter so much. By the time she took me to see Mr. Satoh on the afternoon of August 3, he was clearly in a friendly mood.

My wife and I arrived in Tokyo on Sunday evening, August 1. The next morning, Mrs. Matano came by to pick me up at the hotel for an appointment with Sakutaro Tanino, chief cabinet councilor for external affairs in the prime minister's office.

Mrs. Matano seemed to know everyone. On the way over to see Tanino I asked her whether she knew the newly announced prime minister designate Morihiro Hosokawa. Her reply was, "Since he was a little boy."

On arrival at Tanino's office I was struck by the cluttered but Spartan appearance of the corridors in the building and the crowded outer office, where five officials were busy with paperwork. We were ushered into a corner with sofa and coffee table and served iced tea in small glasses. At 10:15 Tanino's secretary, a friendly young woman who had set up the appointment, ushered us into Tanino's office.

Sakutaro Tanino, a man of medium height, with black, thinning hair, spoke excellent English. His position, like that of many of the most senior government officials, was a civil service appointment. In Japan, only the cabinet ministers and one or two deputies in each ministry are political appointees. All others stay on from one administration to another, including all ambassadors.

Mrs. Matano had sent Tanino the planning document I had circulated to the museum directors in Hiroshima and Nagasaki. Tanino asked a few questions, and showed Matano an internal ministry document, which referred to the meeting between Yukio Satoh and Mick Kicklighter, at which Satoh, much to his surprise, had learned about our exhibition.

Ahead of time, Mrs. Matano had provided Tanino a copy of my March 17 letter to Kuriyama, and Tanino said there evidently had been a mixup at their embassy; but it was now all cleared up. He was pretty sure that the new prime minister would encourage our exhibition and that his (Tanino's) office would tell the mayors of Hiroshima and Nagasaki that our exhibition was to be encouraged. I expressed hope that they would not feel pressured. He smiled quietly and assured me they would not.

At lunchtime Matano and I met with Atsuyuki Sassa, one of Japan's leading security experts. At one time he had been secretary general of the Security Council of Japan attached to the prime minister's office. With his background in antiterrorism, antihijacking, and management of similar national crises, he was deeply interested in the security measures we were preparing for our exhibition. He clearly realized some of the dangers it posed.

In anticipation of this meeting, I had asked Dr. Charles Hines, the Smithsonian's head of security to prepare an analysis for me. Hines, who as a general in the U.S. Army had served in a variety of security capacities, had long worked with us at the museum on plans to assure the safety of visitors as well as artifacts when the exhibition opened. He had provided me with sufficient background documentation to assure Sassa that security was a high priority.

The next afternoon, Mrs. Matano took me to the Ministry of Foreign Affairs, where we met with director general of North American affairs Yukio Satoh. He greeted her warmly, as a longtime friend. We were soon joined by Toshi Ozawa, director of the First North America Division, who had read our planning document, while Satoh, who had just returned from vacation, had not. I described the proposed exhibition. Satoh then mentioned the other anniversaries for 1995, namely the fiftieth anniversary of the creation of the United Nations and the twenty-fifth of the nonproliferation treaty. He told us he wanted to coordinate those activities, and then said he wanted to give us all possible support. I should stay in touch with Ambassador Kuriyama on my return and keep Tokyo informed through him. He was aware of the mixup that had occurred with my letter to Kuriyama.

Satoh was pleased that we were in touch with Akira Iriye and other Japanese scholars. He wanted to make sure that we had the best advice and did not inadvertently antagonize Japan's right wing.

I also had an opportunity to see Paul Blackburn at the American embassy again, to solidify our relations and assure a continuing exchange of information. Blackburn advised that since we had so much material in our exhibition, a gallery catalogue was essential to avoid confusion and later to have adequate documentation on the actual contents displayed in this exhibit. Since we expected to keep the exhibition open for only seven or eight months, such a catalogue would provide a lasting documentary basis for the debate certain to linger long past the exhibition's closing. Mrs. Matano, with whom I also discussed this issue later, similarly thought that our exhibition would raise enormous interest in Japan, and that a catalogue translated into Japanese would be desirable.

After these meetings in Tokyo, my wife, our son, and I took the train to Hiroshima and a few days later went on to Nagasaki.

On arrival at our hotel in Hiroshima, I found an urgent message from Mrs. Matano to call her in Tokyo. Apparently, Mr. Tanino, in the prime minister's office, had followed through on his request that the two mayors meet with me. I was to see Mayor Takashi Hiraoka of Hiroshima the next day, though Mayor Hitoshi Motoshima of Nagasaki was fully booked and would be unable to see me. Matano had already called Nagasaki to tell them not to worry, it was not so urgent that I see the mayor. I fully agreed with her, even though I appreciated Mr. Tanino's kindness in trying to arrange the meetings.

Over the next two days, I had a chance to meet with the Hiroshima Peace Memorial Museum's director, Hiroshi Harada, who had received our planning document and was greatly puzzled. Unfortunately, a large peace conference was in session because of the August 6 anniversary, and so most of the translators were occupied there. We tried to get along on the English of one of his staff members and on my son's knowledge of Japanese. But with a sensitive document like ours, where nuances mattered, we did not make sufficient headway. I became intensely aware of the extensive effort we would have to place on proper translation if the Japanese wanted to assure themselves about our exhibition.

The problems started at once with the title. Our tentative title, "Crossroads," I learned, does not have its English connotation when translated into Japanese. Mr. Harada could not make sense of the title "Traffic Intersection" for an exhibition on the mission of the *Enola Gay*. It took some time to explain that, and I decided we should seriously think about a more appropriate name.

We attended the August 6 memorial ceremony in Hiroshima and the corresponding August 9 commemoration in Nagasaki—both highly solemn occasions.

In Nagasaki I had an opportunity to speak again with Brian Burke-Gaffney, who had translated with so much ease during my April meeting with Mayor Motoshima. Aware now of the need for far more careful translations, I asked him whether we might secure his services. He agreed to help us, though he advised that we should also engage someone who could translate better from English to Japanese. His own expertise was in translating from Japanese into English. Translators are generally most skilled at translating from a foreign language into their own. Translating into a foreign language is more difficult. Burke-Gaffney recommended the names of a number of native Japanese translators who would be able to assure us of high-quality translations from English to Japanese, so that the cities of Hiroshima and Nagasaki could better understand the thrust of our exhibition in order to decide whether to lend us artifacts from their museums.

The Japanese museums were following normal practice in asking to be shown the particulars of our exhibition. Every museum asked for loans of materials first investigates how its artifacts will be displayed before granting the request.

In Nagasaki I also held further discussions on the exhibition with Dr. Itoh, the director of the Nagasaki museum. Again, our son helped with the translations, as did a member of Itoh's staff. Itoh had a number of questions about the script, which I tried to answer. Why, for example, had we dropped a second atomic bomb, the one that had devastated Nagasaki. Itoh commented, "The Japanese people think that one was quite enough." I responded that our side was partly trying to make it appear that we had a large number of these bombs, so as to pressure the Japanese into a quick surrender. We talked for about an hour.

In the course of my visit, Dr. Itoh and Mr. Harada had worked out opportunities for me to speak again with the mayors of both cities, this time much more informally than in the media-packed audiences they had held in April. Both mayors were friendly and encouraging. It was an upbeat mood.

This was the last visit to Japan. But Mr. Harada and Mr. Ohmuta, from Hiroshima, and Dr. Itoh and Ms. Takana, a member of his staff in Nagasaki, paid several visits to Washington. We were consistently plagued by translation problems, but work was proceeding as well as could be expected given the differences in Japanese and American perspectives.

IN WASHINGTON

The discussions in Tokyo had one further effect. I was given an appointment with Ambassador Kuriyama in his Washington offices on September 23. We had a cordial meeting, at which Mr. Kuriyama designated his aide, Mr. Seiichi Kondo, to act as liaison for our exhibit so that in future, communications would not fall between the cracks. A year later, when pressures generated by both the American and Japanese press were reaching a breaking point, these contacts would play an important role.

By the end of the year, we also had been assured that both mayors were consenting to lend us artifacts, subject only to the agreement we had worked out in April, basically permitting them the option to withdraw if they felt that the exhibition violated their beliefs. Mayor Hiraoka wrote on December 1,

> 1. The citizens of Hiroshima, the city that experienced the first atomic bombing in human history, have constantly appealed for the abolition of nuclear weapons and the realization of lasting world peace, based on their tragic experience of the atomic bombing. Keeping this in mind, the exhibition should be organized in such a way that "the spirit of Hiroshima" will be fully reflected. If it becomes difficult to make the exhibition with this in mind, the city of Hiroshima may refuse to lend out the materials.
> 2. At the time of transport, storage, and exhibition of materials, please handle them with utmost care, realizing that these materials are the ones left by deceased A-bomb victims. In addition, your museum should take full responsibility for all the expenses that will be incurred.
> 3. Concerning any other miscellaneous matter, it may be necessary to proceed with planning after coming to an agreement with the director of the Hiroshima Peace Memorial Museum upon sufficient deliberation.[24]

The letter from Mayor Motoshima of Nagasaki differed slightly, asking us also,

> 1. [To be] accurate [about] the heat, blast and radiation effects generated by atomic bombs.
> 2. Not to fail to include . . . human damage, including psychological damage due to radiation exposure.

3. [To give] special regard to the opinions of Nagasaki and Hiroshima concerning the exhibition of atomic bomb relics.
4. To give special regard to such opinions [also] . . . in editing peace messages from the mayors of the two cities.[25]

The Japanese press was not quite as sanguine. The *Japan Times* carried an article in English, reporting, "Yoshio Saito, secretary general of the Confederation of Atomic and Hydrogen Bomb Sufferers Organizations (Hidankyo), said he doubts the exhibition will back the call for a ban on nuclear weapons, as the U.S. 'continues to talk about nuclear deterrence.' The exhibition must go beyond being an apology to A-bomb sufferers for the inhuman A-bombing."[26]

With the mayors in tentative agreement, the directors of the two museums, Mr. Harada from Hiroshima and Dr. Itoh from Nagasaki, arrived early in January 1994 to confer on our planning document, which soon would be replaced by a much more detailed exhibition script.

INTERNATIONAL RELATIONS

In February 1994, Japan's prime minister Hosokawa paid his first official visit to the United States. My wife and I were invited to the dinner held at Ambassador Kuriyama's residence, which Vice President Albert Gore and several cabinet members also attended.

I had an opportunity to talk with United States ambassador to Japan Walter Mondale, who had accompanied the prime minister back to Washington. Paul Blackburn, on his staff, had already briefed him, but it was useful to talk directly with Mondale. Yukio Satoh, the Foreign Ministry's director general for North American affairs, whom I had met in Tokyo, was also there, and I was able to assure myself that the mood in Japan had not changed since the previous August. At dinner, I sat with Hiroshi Fukuda, Japan's deputy minister for foreign affairs, and Ambassador Winston Lord, assistant secretary of state for East Asian and Pacific affairs, and was able to acquaint them with our project as well. I had found it better to inform officials who eventually might become concerned about our exhibition as early in the process as possible, so they would not later be surprised by hearing about it from others. The consternation that can be raised by surprise had been well demonstrated by Mr. Satoh's reaction the previous summer on hearing about the *Enola Gay* from Kicklighter.

Later, as we departed, Ambassador Kuriyama told me that he had briefed the prime minister about our exhibition. I was pleased. We had informed everyone in Japan, up and down the line, and no serious difficulties had yet arisen with the exhibition. That was a satisfying sign.

15

Funding and Approval

SPECIAL EXHIBITION FUNDS

By now, funding had become a critical factor. Nadya Makovenyi who headed the museum's exhibits department, figured that she and her staff could mount this exhibition for $600,000, if all the work was done by museum staff rather than with an outside contractor. We would need no expensive computer interactives, and most of the exhibited items would be artifacts from our own collection or loans from the museums in Japan. For a gallery that eventually would expand to nearly ten thousand square feet, this was a very reasonable price, coming to $60 a square foot. More typical costs for new galleries were $200 per square foot.

Our problem was that even $600,000 was a great deal of money for the museum to find in its own budget. Normally, we did not need to fund our exhibitions from our federal budget, because virtually all major galleries opened in recent years had been sponsored by industries. In return for their support, the corporate donors would be prominently acknowledged in the gallery, giving them favorable publicity. With millions of visitors passing through the museum each year, a corporation could gain significant name recognition.

For several reasons, the *Enola Gay* exhibit could not be funded in this fashion. We all knew that we were dealing with a difficult exhibition that would inevitably be controversial. If the display were sponsored by a corporation or individual, critics would certainly accuse the museum of having caved in to the donor's demands. That would harm both the museum's reputation and the sponsor's. We would be better off handling this exhibition on our own and taking full responsibility for it. The controversial aspects would, in any case, also make any potential donor shy about sponsorship. And finally, the long moratorium on proceeding with the gallery, which had only been lifted at the end of 1992, had made us far too late to attempt raising funds for a gallery whose design and budget had to be assured right now.

With this in mind, my deputy Wendy Stephens and I went in late April 1993 to see Assistant Secretary Tom Freudenheim, who controlled the Institution's Special Exhibition Fund, SEF. Moneys from that fund were distributed annually, in a competition open to all Smithsonian museums. A maximum of $25,000 could be requested for planning an exhibition, and up to $250,000 could be obtained toward mounting it.

We now proposed to Freudenheim that he guarantee us $300,000 for the exhibition at once. This was half its cost. The museum would contribute the other half from its own budget. A letter I sent him the following week fleshed out the arrangement:

> . . . because we feel we should not accept funds from parties that might be construed as having a vested interest, our main difficulty will be to pay for all the expenses.
>
> We would like to propose to you, therefore, that the Museum and your Special Exhibition Fund equally split the expenses, not to exceed $600,000. The Museum will, one way or another, come up with half this amount from unrestricted Museum funds. And we suggest that the SEF contribute $25k for this year, $250k from the spring 1994 competition, and a final $25k from the spring 1995 competition. In return, we would submit no other proposals for the spring 1994 and 1995 competitions. If some other moneys did become available from third parties, those would be used toward the $600k and would reduce, in equal amounts, the contributions the SEF and the Museum had to make.
>
> I realize that this is a large request, but I believe that we will be mounting the kind of exhibition the Museum and the Institution should occasionally mount for the American public.
>
> I hope you will approve.[1]

The letter served merely to put our agreement in writing. At our meeting, Freudenheim had already told us he thought this was a reasonable proposal, but said he wanted to seek the approval of Secretary Adams before giving us a final OK. Soon we heard that Adams also approved in principle, but that he was not going to commit himself until he saw exactly what the exhibition would entail. Though I always assumed that he would eventually give us the release, his proviso did hold us hostage.

In the meantime, I had decided that going ahead now was not an excessive gamble. We desperately needed to get started. Wendy Stephens took the precaution of circulating a memorandum to the principals in charge, urging caution and restraint in spending. But at last we were under way.[2]

Secretary Adams's final approval did not come until ten months later, when Tom Freudenheim notified me that the museum would be receiving the requested $250,000 for 1994.[3]

ADAMS ENTERS THE FRAY

Within a few weeks of my return from Japan, Bob Adams independently visited Tokyo on other Smithsonian business. On returning to Washington, he asked for a meeting, and on June 4, 1993, Adams, Connie Newman, Tom Freudenheim, and I met in the secretary's office. Just before the meeting, however, I had received a call from Hanako Matano, who had heard from an extremely upset Mr. Shibohta about an interview Adams had given *Asahi Shimbun* while in Japan. He had told the newspaper that we were not doing an exhibition on the *Enola Gay* at all, but rather one on strategic bombing in World War II. Shibohta felt betrayed, and told Matano he was withdrawing his support for the symposium on Hiroshima and Nagasaki we had discussed.

So I was not in the best of moods when the meeting with Adams began. And it didn't get better when he started by telling me he thought concentrating on the *Enola Gay* was resented in Japan.

I countered that I had just learned that he had torpedoed the Japan Airlines Foundation's intention to fund a United States–Japanese symposium on Hiroshima and Nagasaki by telling *Asahi Shimbun* that our exhibition would deal not with those topics but with "strategic bombing in World War II," which was not our intention at all. I told him that I resented his having said that, and that we had been very cordially received in Hiroshima and Nagasaki and had talked with the mayors of those two cities. We also had talked with the head of the Japan Airline Foundation in Tokyo. I didn't understand how Adams could feel the Japanese would oppose the exhibition.

Adams responded he thought concentrating on the *Enola Gay* was a "lousy idea." He switched sides and now told me we'd have trouble with it in the United States. To this I replied we'd already been in touch with General Kicklighter's commemoration committee and had worked for years with veterans of B-29 groups.

Adams then conjectured that I wouldn't know where any attack might come from and that we'd probably get heavily attacked by someone. To safeguard against that, he wanted the museum to add material on the war in Europe, so that veterans of the European conflict would feel included as well. I told him that in an exhibition of limited size, I couldn't see how we could incorporate all that. Besides, the museum already had a gallery totally devoted to World War II that people could visit. Nevertheless, Adams wanted broader coverage of the war than just the *Enola Gay*. I said I'd talk to the curators to see what we could do.

We went on to discuss the title for the exhibition. Adams had strong feelings about what title we might use, and wanted to be involved in the choice.

When I left after nearly an hour, the one certain fact was that this had been the worst meeting with the secretary in all my years at the museum. Fortunately, we never had another like it. Perhaps both of us realized that this was an important exhibit and we would have to work together at it.

Six days later, on June 10, 1993, I sent Jim Hobbins, executive assistant to Adams, a new proposal draft the curators had prepared. I had heavily edited and rewritten it to conform to the agreements reached with Adams, and I proposed a new title, "Fifty Years On." My cover letter to Hobbins stated that this was a description of the exhibition for the secretary's use, and that

> I would greatly appreciate having the Secretary's comments, especially if negative, as soon as possible.... After some discussion, we decided we should use just one version that all of us could agree on, and I have accordingly modified the description to incorporate Bob Adams's suggestions.[4]

I attached the revised proposal which now read,

Fifty Years On

World War II ended in 1945. In Europe the war came to an end on May 8; in the Pacific on August 15. But in these two theaters the Allies had fought two very different conflicts. The men inducted in Europe in 1939 battled an enemy remarkably similar to the foe their fathers had met, with weaponry initially quite similar. As in World War I, battles might end with whole armies surrendering, and troops passing the remainder of the war in prisoner of war camps. And though wonder weapons, like long-range rockets and jet aircraft, were brought on line, none except radar had a major impact on the outcome of the war.

The war in the Pacific was different. Opposing forces battled an unfamiliar enemy. Racial and cultural differences fanned fears and inflamed hatreds. Savage battles ended with horrible losses on both sides, as men fought to their deaths, taking no prisoners and afraid to surrender for fear of consequences worse than death. And in the war against Japan there was indeed a vastly superior new weapon.

While all the major participants in World War II had conducted rudimentary experiments on nuclear fission, only the United States completed construction of an atomic bomb. Its employment against Hiroshima and

Nagasaki may not have influenced the ultimate outcome of the war, but it set in motion events that have had consequences of unparalleled proportions for our times.

In the first section of the exhibit, the necessary historical context will be given: the last phase of World War II in Europe and the Pacific and the increasingly brutal character of that conflict. This section will touch upon a variety of topics, including strategic bombing in Europe, the firebombing of Japan, and contrasting racist perceptions about the different enemies. (For a more complete discussion of the war in Europe, and battles at sea, visitors will, at this point, be directed respectively to the Museum's existing World War II and Naval Aviation galleries, while the new exhibition will continue with a focus directed at the war against Japan, inadequately discussed elsewhere in the Museum.)

A video of selected newsreel footage from the period will give a sense of the atmosphere in the last months of the war in the United States and Japan, as will posters, cartoons and other visual images of this period. The fighting on Okinawa and other islands and the kamikaze campaign will highlight the desperate character of the Pacific conflict. Among the artifacts that may be included are a Japanese "Okha" suicide bomb and leaflets and bomb canisters dropped by American B-29s.

This introductory section will prepare the ground for the next major unit: After a quick overview of the decision to build the atomic bomb and the history of the Manhattan Project, this unit will offer the visitor a nuanced picture of the way decisions were made in the American and Japanese governments in 1945, to provide visitors a deeper understanding of this complex topic: Quotations from major participants and key documents, such as the 1939 Einstein letter to President Roosevelt and the July 1945 Air Force order to drop the atomic bombs, will be used in this context. A *Little Boy* atomic bomb casing will be used here to indicate the reality of the bomb that was becoming available to American decision-makers. A fuller discussion of the topic, however, will be left to outside reading suggested in a brochure that will be available to visitors.

A key component of the next part of the exhibition will be the forward fuselage of the B-29 *Enola Gay*. The entire aircraft has been in the Museum's possession for many decades and will be fully restored by 1995, but it is too large to fit in any of the Museum's galleries. Therefore, the exhibit will be using only the portion comprising the aircraft's cockpit and bombsight, the evocative name painted on the port side of the aircraft, and the bomb-bay with a reconstruction of the special sway braces and latch. In the vicinity of this massive artifact we will treat the development and manufacturing of the B-29, the firebombing of Japan, the development of the bases in the Marianas, the 509th Composite Group (Col. Tibbets' special atomic weapons unit) and the final preparations for the missions. Beyond the fuselage itself, artifacts that can be used in this section will include an engine, propeller or other pieces of the *Enola Gay* (to further convey its massive scale), plus documents, pictures and memorabilia of the crews.

> The next major section of the exhibit will treat the bombings of Hiroshima and Nagasaki themselves and the ensuing surrender of Japan. The dimensions of the destruction and suffering in the two cities will be shown here through pictures and film of the victims, however upsetting that may be to some visitors. We will include, as far as possible, bomb-damaged artifacts from the two cities and other documents and artifacts from the missions. A *Fat Man*–type plutonium bomb-casing will also be shown here. A video or movie in this section would include footage of the missions and their aftermath, plus interviews with survivors on the ground and crew members of the attacking and accompanying aircraft.
>
> The exhibition would conclude by noting the debatable character of the atomic bombings, as well as their important role as one of the starting points of the nuclear age and Cold War. The closing video will include the perspectives of a whole range of people—historical actors, survivors, scholars and ordinary people, both Japanese and American. At the very end, visitors will be able to ponder what they have seen and record their own reactions and thoughts in comment books. They will leave the exhibition, it is hoped, with a deeper appreciation for the importance and the complexity of these watershed events in modern history.
>
> In setting this exhibition in motion, we have had the cooperation of a large number of organizations, both in the United States and in Japan. A consensus appears to have emerged that events, discussions of which were so often shunned in the past, should now, fifty years after their occurrence, be aired—perhaps as much as anything for the healing effect that such a debate might finally have. Discussions with the mayors of both Hiroshima and Nagasaki have taken place for the loan of items from these two cities' museums, whose directors have been most helpful. Japanese historians, humanists and artists have promised their cooperation in a search for accuracy of fact and balance to the presentation. The Radiation Effects Research Foundation, jointly administered by the United States and Japan, has volunteered to provide clarifying data on long-term health issues. Similar support has been obtained from U.S. veterans' organizations, which aided the Museum in its restoration of the *Enola Gay;* from the Executive Director and staff of the Congressionally mandated 50th Anniversary of World War II Commemoration Committee; from the staff of the American Embassy in Japan working for Ambassador Michael Armacost; and from a variety of other interested organizations.
>
> Fifty years after the atomic bombing of Japan, the National Air and Space Museum, with its unique collections of historic artifacts relevant to the events—most of them kept in storage and inaccessible to the general public for half a century—has an opportunity and, many would maintain, an obligation to mount an exhibition that will help visitors understand this pivotal moment in the history of World War II and the twentieth century.[5]

This outline best summarizes my intentions for the exhibition. I believe it also to be the best capsule description of the exhibition, as called for in production plans, eighteen months later, in January 1995, only days before the exhibit was finally canceled.

Adams passed copies of the proposal to former president of the World Bank Barber Conable and Chief Justice Rehnquist, both of whom served on the board of regents' executive committee. On June 21, Jim Hobbins sent me a note saying,

> Barber Conable today told me that he was not altogether pleased with the paper "Fifty Years On" because it focuses too much on the cost of dropping the bomb and too little on what the costs would have been of *not* dropping it. In Conable's view it pays insufficient attention to the number of American lives which would have been lost in the planned and imminent direct invasion of Japan—he cited official estimates of 100,000 killed in action (and he notes parenthetically, that "GI's are currently the beneficiaries of the bomb," as significantly fewer soldiers died than anticipated, leaving a greater amount of financial benefit per veteran than otherwise). On balance, he argues, dropping the bomb saved more lives than not dropping it.
>
> Conable thinks the exhibition should pay more attention to what the options were at the time—"not just the guilt that is felt in hindsight." And he adds that it was the *creation* of the bomb which introduced a major threat of terror for the world, and that same sense of terror would have beset the world whether or not the bomb had been dropped.
>
> Conable admits that his perspectives have been dictated at least in part by his personal experience—he was slated to be among the first of the American soldiers to invade Japan, and was, in effect, spared by the bomb. But he is quick to add that there are a lot of other Americans who share these views.[6]

I passed a copy of this along to Tom Crouch, with a note saying, "Tom—this is a view that, whether in hindsight correct or incorrect, should be strongly underscored in the introductory section. People *did* think this at the time."

Interestingly, Ben Nicks, who had flown thirty-five missions to Japan's home islands as pilot in command of a B-29, responded quite differently.

Ben had been in Washington for the Armed Forces Day weekend in mid-May, and we had invited him to a ceremony in which General Curtis LeMay's daughter, Jane LeMay Lodge, had officially donated her father's decorations and memorabilia to the museum. General Russell Dougherty, who had followed in LeMay's footsteps in heading the Strategic Air Command, had given a brief speech at this gathering, attended by many former Air Force officers, colleagues of LeMay. The day after this event, Nicks had come over for lunch, and we showed him our plans for the exhibition. He now wrote to Neufeld,

> As discussed previously at our meeting in May, I will circulate [*Fifty Years On*] through my acquaintances among the directors of the 20th Air Force

> associations with the confidence that they will in turn pass it on to members for their information and comments. The exhibit of the *Enola Gay*'s cockpit will certainly please them.
>
> On a personal note I find this version improves on the last one I have seen, inasmuch as it acknowledges without passing judgment that controversy exists on the morality and the necessity of strategic bombing in WW II.
>
> I think that when the display is finalized it would be fair if for every authority cited condemning WW II's strategic bombing (which of course includes both nuclear and fire bombing), another authority would be quoted pointing out the War Department's rationale and justification for the policy our military forces pursued during the conflict. General LeMay himself had no doubts on this score, as was attested to by your excellent speaker at the LeMay memorabilia presentation, General Russell Dougherty.[7]

Given Ben's membership in the 20th Air Force Association, the American Legion, and the Veterans of Foreign Wars, this might be considered as representative an opinion on veterans' attitudes as any one individual could have given us at the time.

Toward the end of June 1993, Neufeld and the exhibition team were drafting the "Planning Document" for the exhibition, which would go into greater detail than the proposal. Rather than having the display in an unenclosed space at the west end of the museum, it now placed the exhibition in a gallery of its own, where previously the museum had exhibited helicopters.

I met with the exhibition team and we discussed some of the major images that might greet the visitor on approaching the gallery. Most of the pictures the curators proposed were too graphic, as Gregg Herken pointed out:

> I'm not sure I would put the little boy at the centerpiece of the exhibit; besides the obvious fact that you'll catch hell for showing a burnt child right off, there is the substantive point that it will look like you plan on telling the events from the viewpoint of the victim. . . .[8]

Gregg's memorandum did not lead to an immediate response, although the image in question did eventually get removed. Neufeld's second draft, dated June 30, now bore the title "Ground Zero: The Atomic Bomb and the End of World War II." Its first sentence read, "Few events have had a more profound impact on our times than the creation of nuclear weapons and their employment against Hiroshima and Nagasaki." As I read on, Neufeld still had visitors first encountering "a central image of a survivor, such as one of a small boy taken immediately after the bombing of Nagasaki. He is about three years old, streaks of blood are visible on his cheek and he is staring straight at the visitor."[9]

But there were other problems as well. I responded to Tom at once, with a blunt memorandum, dated July 2, 1993. The highlighted passages in boldfaced print were part of the original. I wanted to make certain that my points were not lost:

> I am quite concerned by this draft. It seems to me that the revisions I made to the earlier, shorter draft after discussions with the Secretary have been eliminated.
>
> I am absolutely convinced Mike's new draft, as written, will be rejected out of hand by the Secretary. Its title, subheadings and description of the entry to the exhibit have little resemblance to the script which the Secretary, Barber Conable and the Chief Justice found acceptable. I cannot submit it to the Secretary in its present form, since I know he would reject it. And, because of that, we also cannot send it to our Japanese colleagues.
>
> The consistent problem with Mike's headings, subheadings, and introductory paragraphs, is that they do not do what the Museum always claims it intends to do: **To let visitors judge**.
>
> Mike appears at each stage to prejudge what is to be seen by the visitor and what the visitor will be told. Though the description of many of the artifacts and archival materials to be exhibited reads well throughout, his headings consistently emphasize only the most dramatic. A central image in the opening section will show "a small boy taken immediately after the bombing of Nagasaki." Does one have to add "streaks of blood are visible on his cheek." Why not let the visitor see for himself what the boy looks like?
>
> The opening paragraph again reverts to the form Mike has had all along. The context of Europe and Japan in World War II has been eliminated, after I had painstakingly inserted it at the Secretary's suggestion. The broader picture has been erased. Again, Mike concentrates on just one thing—the bomb and Hiroshima and Nagasaki. They all are right there in the first sentence.
>
> **Where is it that a visitor ever has a chance to formulate an independent opinion? Where does a visitor have a chance to see for himself whether the war in the Far East differed from that in Europe, or for that matter from other wars throughout history?**
>
> I am all for being specific with clear-cut examples. And I think that the more dispassionate description in the main parts of the text do that sufficiently well. But the headings seem to be overly dramatic and one-sided. We must insist on context, on balance, on providing a broader background to important issues, even in the section headings, so that the visitor indeed can judge.
>
> Of course, drama and striking examples are part and parcel of good exhibits, but only so long as they do not distort the message so that only one point of view comes across. And this particular exhibition, more than any one can think of, does not need any added drama. If anything, the labels must be dispassionate, perhaps even bland. In that respect I am convinced that normal exhibition practice needs to be deliberately violated. Instead of looking for striking labels and titles that encapsulate an idea or vision in a few simple words, we should go in the opposite direction of

> making the titles as broad as possible, so that visitors will not be forced into one particular line of thinking.
>
> Let us not be forced into conforming with standard practice. This is not a standard exhibition.
>
> A further concern is the final section. It simply does not do justice to the dangers to which the world's population now is exposed through the proliferation of fusion and fission bombs after World War II. The differences in power of fusion and fission bombs (a factor of 100), the differences in the sheer numbers (a factor of ten thousand) make today's world totally different from the world of July and August 1945 when only three fusion bombs were in existence. Today's abilities by even the smallest nations to produce nuclear bombs should also be emphasized. The visitor will undoubtedly want to have us present the facts on this, to complete the picture of World War II's main legacy.
>
> I would like to discuss all this with you as soon as possible, because time is running short and we need to make these changes.[10]

Four days later, on July 6, 1993, Tom Crouch, Gregg Herken, Nadya Makovenyi, Mike Neufeld, and I met to discuss these concerns, and to initiate revisions. Neufeld had a revised draft of the planning document, dated that day, which took into consideration some of my criticisms. It was once again titled "Fifty Years On."

The same day I also sent a paragraph on the proposed exhibition to be included in the newsletter periodically circulated to Smithsonian regents. It outlined my intentions to the regents:

Fifty Years On

> World War II ended in 1945. In Europe the war came to an end on May 8; in the Pacific on August 15. But in these two theaters the Western Allies had fought two very different conflicts. The men inducted in Europe in 1939 battled an enemy remarkably similar to the foe their fathers had met, with weaponry initially quite similar. The war in the Pacific was different. Opposing forces battled an unfamiliar enemy. Racial and cultural differences fanned fears, and inflamed hatreds. Savage battles ended with horrible losses on both sides, as men fought to their deaths, taking no prisoners and afraid to surrender for fear of consequences worse than death. And in the war against Japan there also was a vastly superior new weapon. From May to December 1995, the National Air and Space Museum intends to mount an exhibit featuring the restored front fuselage of the *Enola Gay,* the aircraft that dropped the first atomic bomb on Hiroshima on August 6, 1945. The exhibit will discuss the mood in the U.S. and Japan toward the end of the war, the Manhattan Project, the decision to drop the bomb, the modification of the B-29 bombers required for delivery, the actual operations launched from Tinian Island, the devastation and suffering in Hiroshima and Nagasaki, the long-term medical consequences, and finally the legacy of nuclear bombs—their proliferation during the Cold War and an uncertain future as increasing numbers of states learn to produce them.[11]

If I have emphasized these documents of late June and early July so heavily, it is because so much of the discussion in the two years to follow dealt with alleged intentions. The June 10 planning document and the paragraph for the regents clearly state mine.

On July 7, we had a further meeting of the exhibition team. Tom Crouch; our filmmaker, Patti Woodside; Mike Neufeld; project coordinator Victor Govier; and I discussed other potential titles besides "Fifty Years On." Gregg Herken also participated.

Since he had asked to be included in the decision, I wrote to Bob Adams the next day, listing for him some of the options we had discussed, and asking his opinion on different titles for the exhibition:

> ... There is some concern that the present title is somewhat clumsy. And, if we do keep it, we will probably need to augment it with a subtitle. But a number of other possibilities have been suggested, and I thought I should give you an opportunity to see which, if any, you could support.
>
> My own preference is to keep the title as understated as possible. But that does make it more difficult to be specific. Suggestions that have been made include:
>
> "Whirlwind: The Atomic Bomb and the End of World War II."
>
> Alternatives to "Whirlwind" as the main title are
>
> "The Long Shadow,"
>
> "The Crossroads,"
>
> "Ground Zero," and
>
> "Operation Alberta,"
>
> the last of these being the official code name for the military use of the atomic bomb. It is also possible to juxtapose the two parts of the subtitle, making it
>
> "The End of World War II and the Atomic Bomb"
>
> Another suggestion was:
>
> *Little Boy* and *Fat Man*: The Atomic Bomb and the End of WW II"
>
> If you wanted to stay with the main title I had submitted to you last time, we could add a subtitle, making it
>
> "Fifty Years On: The End of WW II and the Atomic Bomb"
>
> or, alternatively, something like
>
> "Fifty Years After World War II: Recollections on *Enola Gay, Little Boy,* and *Fat Man*"
>
> Of all these, my own preference would be for
>
> "The Crossroads: The End of World War II and the Atomic Bomb"
>
> The main title was suggested by Gregg Herken who, however, lists the bomb first in the subtitle. Please let me know your thoughts.[12]

Bob Adams called me a few days later, on July 12, and we discussed the possible titles I had suggested. That same afternoon I met with the exhibits team to decide on the best title. We settled on "The Crossroads," but the subtitle remained a problem. The following day I sent Adams a copy of the just-completed planning document:

> I enclose a copy of the planning document for our exhibition which, some weeks ago, I had promised to send you by mid-July.
>
> After your call, yesterday, I met with the exhibits team, to discuss some of your concerns about the title, "The Crossroads: The Atomic Bomb and the End of World War II." We had earlier considered "The Turning Point," which you also suggested, but felt that "Crossroads," just like "Turning Point," did leave uncertain the directions in which those roads led, and was a somewhat more familiar term. The question of juxtaposing the subtitle to read "The End of World War II and the Atomic Bomb," an option I had suggested in my letter to you, has two problems which I had unfortunately missed when I wrote you. First, "The End" only refers to the War, not to the bomb. And second, the bomb sequentially came before the end of the war, not after it. On both counts the subtitle reads less well when juxtaposed.
>
> Because of these questions, we decided for now to keep the title tentatively as "The Crossroads: The Atomic Bomb and the End of World War II." If that still faces you with difficulties, we will be glad to see if there is a way to find a variant that avoids awkward phrasing.
>
> Please do let me know what you think, both of the title and the planning document. It would help us a great deal to resolve any substantive issues as they crop up. Later, such changes become more difficult and costly.[13]

Adams read the document but remained concerned. On July 17 he responded:

> I've read your planning document with interest, and find much of it compelling. There could be an exhibit here that would do the Smithsonian credit. On the other hand, there are some fairly fundamental aspects of it with which I am no more in agreement now than when we have discussed them on previous occasions.
>
> 1. Title. Crossroads or Turning-Point is a fairly trivial matter, but the order that follows is not. I think it has to be "The End of World War II and the Atomic Bomb." This is, in fact, the underlying question of the overall stance of the exhibit that pervades all my other difficulties.
> 2. With my ordering of the title, there is clearly a missing section in the document after the first paragraph. In a document of this length and degree of detail, you cannot turn to the atomic bomb so quickly and abruptly. I assume you would want to treat the contrastive endings of war in the European and Pacific theaters more fully, with in all likelihood a discussion of the psychology

here on the home-front as people impatiently and fearfully awaited the war's winding-down.

3. The third paragraph is similarly affected ... I cannot accept the wording that this will be "an exhibit about the wartime development of the atomic bomb, the decision to use it against Japan and the aftermath of the bombings." This should be an exhibit commemorating the end of World War II, taking appropriate note of the atom bomb's central role in one theater, and seeing that decision-point as a decisive determinant of decades of strategic and political thinking and action that followed. The shift of stance is, from my perspective, subtle and vital.
4. ... You speak appropriately of the suffering of the bomb victims, but make no mention in this introductory section of the issue of prospective American losses if there were to be an invasion. I trust you have seen Barber Conable's letter, reflecting the strong feelings of one who was slated to participate in that invasion. He would rightly feel that the present discussion at this point lacks "balance."
5. Continuing along exactly the same line, I do not think it is appropriate that "Upon entering the exhibit.... The central image will be of a mushroom cloud...." The different ordering of the title has the clear implication that the initial set of images should concern the ending of the war in first one theater and then another—the return of some troops, at least a passing view of the discovery of the full horror of the Holocaust, etc. That then indeed sets the stage for the decision to drop the bomb on Japan, and for the appearance of the mushroom cloud as a commanding image further into the exhibit.
6. If I begin the planning document at page 4 ... I find that most of what is noted in the first five points above is well attended to. But your ordering of the title means that this follows rather than precedes the introduction of the atom bomb as the central subject. It is precisely my point that such an ordering greatly—and I think unacceptably—increases the risk to SI [the Smithsonian Institution]. By *beginning* with Unit 1 instead of your present Introduction we can reply to critics concerned about the atom bomb as the subject of an exhibit from any direction that this is essentially an exhibit commemorating the end of World War II and naturally also examining its sequels.
7. I continue to be uneasy that later sections of the planning document ... treat fully and sympathetically the horrors of the bombing—the fire-bombing as well as the atom bomb—but do not present in adequate depth what were perceived as the horrors experienced by Americans during all of the island invasions culminating with Okinawa. This is the Conable point, of course, and I urge you to treat it more fully and seriously. But this is more a change of emphasis and what will be perceived by some as "balance."

> Basically, I think the planning document from Unit 1 on is in very good shape.
>
> I do hope you will find these remarks helpful.[14]

The secretary's memorandum did not sit well with Tom Crouch. He wrote me,

> . . . It is clear the Secretary is not focussing on what really concerns him. . . . In item 3 he notes that: " . . . I cannot accept the wording that this will be 'an exhibit about the wartime development of the atomic bomb, and the decision to use it against Japan and the aftermath of the bombings.' " That is, of course, precisely what the exhibit outlined in the document *is* all about. Fiddling with the title and the introductory panel will not change that.
>
> In a nutshell, the Secretary is not consistent. On one hand he says: "Basically, I think the planning document from unit 1 on is in very good shape." On the other hand he identifies the exhibition outlined in that document as presenting an 'unacceptable risk to the SI.' What are we to make of that? You can't have it both ways. . . .
>
> I think that what really worries the Secretary is the fact that any morally responsible exhibition on the atomic bombing of Japan has to include a treatment of the experience of the victims. He knows we cannot escape that, and would not think of suggesting that we do so. At the same time, he knows that any exhibition including an honest discussion of that topic is most certainly going to upset a lot of visitors.
>
> You cannot solve the problem by obfuscation, by attempting to misdirect the attention of the visitors, or by discussing the very different end of the war in Europe and in the Pacific. None of that will fool our visitors . . .
>
> You and the Secretary are the ones who will have to accept responsibility for whatever we do. Do you want to do an exhibition intended to make veterans feel good, or do you want an exhibition that will lead our visitors to think about the consequences of the atomic bombing of Japan? Frankly, I don't think we can do both.[15]

Tom's letter required a decisive response. To me it was clear that we *could* do an exhibition that both dealt squarely with the mission of the *Enola Gay and* properly honored the veterans. Of course we would honor the men who had been willing to sacrifice so much. How could we not? I saw no reason why realistically exhibiting the *Enola Gay* had to interfere with that aim.

Historians have a strong sense of professional ethics and integrity. I could not force Tom Crouch or Mike Neufeld to express views in which they did not believe. If Crouch and Neufeld strongly disagreed with the views I had expressed in my July 2 admonition, or now, only two weeks later, with the thoughts the secretary had conveyed, I needed to know that so I could find other curators willing to mount the exhibition I had in mind.

I wrote Crouch and Neufeld the next day. Again, the underscoring of the final sentence was part of the original:

For two months now, I have been presenting to you the Secretary's point of view on *Crossroads*. I believe that his own letter now substantiates that I gave you a fair representation of his needs for the exhibition.

During this period, your response has consistently been the same as that given in the letter Tom wrote me yesterday. You talk of "obfuscation" and of "fooling" our visitors—strong words indeed. And the attempts I have made, in several drafts, to show how one might comply with the Secretary's wishes, are invariably erased in the next version you return, which essentially has kept your document at a standstill all this time.

I cannot, and do not wish to force you to do this exhibition if its contents will violate your conscience as professionals. I never asked the two of you specifically to do this exhibition. You volunteered. But time is running short. We have not quite 22 months before an opening.

I am writing to ask each of you, individually, to think about whether the Secretary's requests, taken literally and fully, will violate your professional ethics. If so, I will understand. But I need to know whether the problem is so serious, for you, that you wish to withdraw from the project.

I am fully committed to doing this exhibition. If neither of you can go on, I will need to find someone else to continue. I have no idea whom I could ask, and it will take me time to bring the right person on board. Hence the urgency of the situation. We have just lost two months; I cannot delay matters any further without seriously threatening the opening date and letting down all the many people who said they would work with us.

I now need a clear statement from you, which I hope will be an individual response, not a team, or departmental reply. It must be based on your conscience as an individual, on which honorable people can differ, and should not represent departmental cohesion:

If you decide to stay on with the exhibition, I will expect that you will work with full energy and enthusiasm, and take as a challenge the clear and dispassionate presentation of the different views on which so many of our honest critics insist.

That does not mean that we may not argue. But once a decision is made, as quickly as possible, I will expect you to again fall in behind the project and work with full vigor, rather than adding to our delays.

I will ask you to work to such a schedule that, should you ever feel you cannot in good conscience go on, you will at least have provided me with a cushion of some two months, to find someone else who might continue to pursue the project to a May 1995 opening. For now, and for the duration of the project, I need only to have that assurance from you, not a commitment to stay on at all cost.

I need your reply by the end of work tomorrow.[16]

The same day I also wrote to Bob Adams and Tom Freudenheim that I was sending them a revised version of the last draft I had sent them.

. . . I hope that the title we now have will meet your main concerns. Portions of the first three pages of the text, the ones you felt most strongly about, have also been rewritten. The others, which you felt were all right have remained as they were.

We will be proceeding now with more detailed planning. And when a full label script is in place, we will make it available. That will not be for several more months.

Please let me know if there are any further comments.[17]

The next day, July 23, Crouch came in and said he wanted to continue on with the exhibition, and he and Neufeld wrote me notes to that effect: Neufeld responded,

> The claim that we have held up the exhibit in any way is unfair. I threw myself into the process of formulating the exhibit and I promise my continued full commitment in the future. I very much want to stay on the exhibit team. I also promise to continue to be an advocate for what I consider to be the best, most intellectually substantive and most defensible exhibit that we do on this difficult subject. Tom's position and my own has been that the Secretary's demands are not that large, nor a violation of our professional ethics, they are just inconsistent with the content of the exhibit and his own concerns about it.[18]

Crouch wrote,

> 1. The present exhibition plan does not violate my professional judgment. The discussions to which you refer have had more to do with the way in which we should communicate with visitors than with the content of the exhibition. I once told you that if we ever arrived at a point where content became a problem for me, I would tell you. That still stands.
> 2. I very much want to continue as a member of the exhibition team. Moreover, I am puzzled and dismayed by the fact that you would question my level of energy, enthusiasm and commitment. That is neither fair nor true. I have given everything I have to this exhibition from the outset. I could only continue on that basis.
> 3. You seem troubled by my blunt and straightforward talk. I am sorry about that, particularly in view of the fact that I feel compelled to point out that the charges expressed in your memorandum are simply not true. We have not held the project up for two months—or two days. It is precisely on schedule—if not a bit ahead.
>
> Nor have we refused to incorporate your expressions of the Secretary's ideas into the various documents. You have seen every document and had every opportunity to comment and insist on change. You approved the planning document after lengthy discussions that resulted in important changes. If you did not think that document was responsive to the Secretary's concerns, you should have said so. We have argued, sometimes strenuously, for alternative strategies, but your decision has always been final. We would simply ask that you make yourself clear. If you believe a change is in order, say so. For my part, I will try to be a better listener.
>
> As noted above, I have no desire to leave the exhibition team. In view of the fact that your confidence in me seems to have been eroded, how-

> ever, you may prefer to involve someone else. If that is the case, I will certainly understand. My only goal has been to provide you with the best exhibition possible on this very difficult subject.[19]

These replies still indicated that we had been talking past each other. But the will was there to go forward with an exhibition that conformed to history and met the needs of the Institution, the museum, and the public.

All of us agreed that the principal object on display would be the *Enola Gay*. We also agreed that we could not concentrate on the aircraft as a piece of technology alone; we had to tell visitors about her historic mission, and that meant including the destructive effects and the casualties on the ground.

But we differed on how this story was to be told, and how it would be perceived by museum visitors.

As I saw it, Tom Crouch, who had done many excellent exhibitions before, hoped once again to enthrall the visitor with human interest stories. This is what had made his exhibition "A More Perfect Union," on the two hundredth anniversary of the United States Constitution, so compelling. He had shown how disregard for the Constitution had brought tragedy to Americans of Japanese ancestry. Now he saw an opportunity to show the tragedy that the atomic bombings had wrought. Indeed, when it came to writing a full gallery script over the next few months, Crouch would write the section that dealt with that tragedy, illustrating it with heart-rending vignettes.

For Mike Neufeld, the objective appeared to be to bring to our everyday visitors the latest scholarship on the decision to drop the bomb, the alternatives that might have existed, and the morality of its use. To be sure, we all agreed that the museum could only offer visitors a range of views and their rational foundations, but that already was likely to trouble many of our visitors.

Bob Adams clearly saw the dangers of dwelling on the mission of the *Enola Gay*. He thought that by casting that mission in the larger context of the entire war, a tacit justification of the bombings would appear. That, in fact, was how those involved in the war had seen it. Hiroshima and Nagasaki for them were just a logical consequence of all that had preceded.

I agreed with Adams about the dangers he saw. But I felt there was an alternative approach that would be easier to implement. My July 2 memorandum to the curators had emphasized that this was decidedly not a conventional exhibition. We should avoid all drama. We should refrain from pointing out moral issues. They would jump out at the visitor without our prompting. I wanted bland labels that simply dealt with facts, and where necessary, the evidence on which they were based. This did not mean obfuscation or fooling the visitors. It merely meant deliberate understatement where the temptation might be to embellish.

From my point of view, the very statement that we were not going to deliver a judgment on the morality of the bombing already meant that we had prejudged the existence of a moral problem.

To men like Burr Bennett, Donald Rehl, and William Rooney there were no moral dilemmas at all. Truman had merely chosen to save their lives instead of those of some Japanese. To them this made obvious sense. That most of the Japanese lives had been civilian made no difference. The distinction between military and civilians didn't really matter in an all-out war, where almost everyone with a weapon in his hand had been called away from civilian life. It was Japanese lives or American. Nothing could be simpler. Where was the moral dilemma?

As a practical matter, I also differed with Adams on enlarging the exhibit to include far more on the context of World War II. I did not see how we could accommodate a major addition to our gallery in the space available. The context of the war must, of course, be made clear, and we would do that in a brief introduction. But I felt that most visitors to the museum would already have learned about World War II in school, and that we would not need to do much more than refresh their memories.

A year later, I would come to understand that I had overestimated our visitors' knowledge about the war, and that we would need to add an entire gallery dedicated to the antecedents of the War in the Pacific and the bitter fighting throughout. That was an important change of emphasis; but it was still in the future.

For the moment, I believe, all of us realized that these issues would have to be resolved before the exhibition could open. We had twenty months in which to do that, and I had no doubt we would reach a sensible compromise. We had done that on all the other exhibitions we had previously mounted at the museum.

It was not to be!

Soon a public brawl would break out. The issues would be largely the same as those that we were debating internally; but the political pressures would escalate as competing interest groups sought to impose their views and perspectives on the exhibition, the museum, the Institution, and ultimately, the visiting public.

To justify their claim that they needed to apply all possible pressure, our critics would soon accuse us of inflexibility. But in the two years to date, we had drafted and redrafted the proposal half a dozen times, had gone through as many different titles for the exhibition, had redrafted the planning document three times, and had listened to dozens of people in and out of the Institution. We had also set up an exhibition advisory committee encompassing expertise in all the requisite disciplines and wide-ranging political views. That committee would meet to critique the first full draft as soon as it was ready.

I did not see any lack of flexibility on our part. We welcomed others' views, but we did not welcome attempts to force those views upon us.

16

Losing Friends

AIRPLANE ENTHUSIASTS

On June 28, 1993, Lin Ezell and I met with Michael Coup, a World War II "Flying Fortress" enthusiast, who had just come into town. Over lunch, he unfolded a number of ways that he and his B-17 restoration group might help the museum's efforts. Coup, chubby-faced and friendly, was much too young to have served in World War II. But he had a keen interest in World War II aircraft, particularly in Boeing Flying Fortresses. These were the Army Air Forces' B-17s, which had carried out most of the American raids on Germany in the last years of the war. Though not able to carry as much of a bomb load as the earlier B-24 Consolidated Liberator, the B-17 was equipped with far more armaments for its crew to defend itself against enemy fighter planes.

We had met Coup months earlier, when he and a group of colleagues, all of them B-17 restorers, had come to the Garber facility to talk with our restorers. They wanted us to do more with the two Flying Fortresses the museum owned, and suggested that they would be able to help us restore our historic *Swoose*, a B-17 that had flown in the Pacific theater and become the topic of a celebrated book. They wished to see it restored more rapidly than our small dedicated staff, busy with other projects, could manage.

Organizations of enthusiastic restorers of World War II aircraft can be found all over the United States. The Confederate Air Force is perhaps best known, because they fly a lot of air shows. But there are many others, including the Yankee Air Force, the Kalamazoo Air Zoo and the Collings Foundation. They restore airplanes to flying condition, paying more attention to their exterior appearance than to historical accuracy, and they don't worry about replacing an original engine with one of a different make in order to get their planes into the air. They are spirited groups that do a lot for aviation by giving rides in these wonderful old planes and getting youngsters enthusiastic about aircraft and flying. The air shows and rides bring in the revenue that keeps the groups able to restore and fly the aircraft.

These organizations' aims are different from those of a museum, which seeks to preserve the original technology for future generations and refrains from making convenient substitutions of modern parts for authentic components. Many of the enthusiasts also resent the National Air and Space Museum's refusal to fly any of its restored aircraft.

The museum doesn't fly its planes because so many museums that do, year after year, crash them in accidents—a loss to history. Even when the airplanes don't crash, they get heavily damaged. Just months before our meeting with Mike Coup, the Collings Foundation had toured its B-17 *Nine-O-Nine* and B-24 *All American* in the company of two vintage World War II fighters, a P-51 Mustang, and a German Messerschmitt 109, throughout Florida, visiting a dozen cities and "blowing" an engine on both the B-17 and the B-24. As a national repository, the Air and Space Museum could not afford to blow the engines on aircraft placed in our care, and so we never flew them.

Because the museum never flew the aircraft in its collections, it could also benefit from preservation rather than outright restoration. While the museum's small staff could only restore one or two airplanes a year, a tiny fraction of our large collection, it was possible to preserve a much larger number to prevent further deterioration beyond that already suffered. And by halting the deterioration, we were preserving the airplanes until we were able to find the resources to restore them. The concept of preserving airplanes is almost unknown to the groups bent on flying, since mere preservation does not permit them to take off and fly. But the museum had increasingly shifted emphasis to preservation because that provided us the best long-term strategy in safeguarding the nation's aerospace treasures.

In recent years, many of the National Air and Space Museum's restorations had been done by the staffs of other museums in exchange for a long-term loan. That way, museums all over the country were able to show the public some of the most historic airplanes, while doing a service to the nation by undertaking the restoration. An added benefit was that these museums could provide better housing for these aircraft until the National Air and Space Museum could build its planned extension.

While our meeting with Coup was friendly and constructive, it did not lead very far. There were not too many ways his B-17 restoration group could help us, short of adopting the same careful restoration techniques we employed, and they seemed reluctant to do that.

We did not mind that reluctance; it was natural given their aims, with which we had no quarrel. But we did mind the sharp criticism, not only of the National Air and Space Museum, but also of the Air Force Museum, in Dayton Ohio, criticism launched by the Collings Foundation that spring. Bob Collings, of Stow, Massachusetts, who wrote a periodic newsletter every couple of months, had attacked both museums in his January/February 1993 issue:

> Over the past year we've had the opportunity to review in-depth the restoration activities at the Air Force Museum at Dayton, Ohio, and the Smithsonian's restoration and storage facilities at Dulles Airport and Silver Hill, Maryland. To say we are *shocked* and *disappointed* would be gross understatement. This disappointment stems from the **low level of restoration activities** relative to the size, budget and reputation of these two institutions and secondly to the **priorities and policies** of the organizations.[1]

We understood Bob Collings's viewpoint, and of course, the National Air and Space Museum would have liked to restore our collections at a more rapid pace. The museum's annual budget request for Federal funding for many years listed this area of activities—management of our collection—as the museum's most urgent need. Year after year, however, we were overruled—by the Institution, the Office of Management and Budget, or Congress—and given money instead for activities we had assigned a lower priority. There was little we could do except occasionally transfer staff from other areas into hands-on work with airplanes and spacecraft. But this was difficult, since members of the staff were highly specialized, and could not be easily retrained. A historian cannot substitute for a skilled restorer, and an exhibit designer cannot be cognizant of the latest preservation techniques.

While the Collings Foundation's first newsletter expressed disappointment, it was followed in the March/April 1993 newsletter by a more severe attack:

> The Smithsonian is presently being misled by a top-heavy administration preoccupied with matters not related to its role of preserving historical artifacts but rather their personal interest in social and environmental issues.... The NASM needs a major redirection "back to basics" of its stated purpose. It's very doubtful that Smithsonian Secretary **Robert Adams** or the Director of the Air and Space [Museum] **Martin Harwit** are going to provide that redirection. If you're concerned enough you might want to contact **Ron Stroman**—Staff Director Human Resources and Intergovernmental Relations or Congressman **Edolphus Towns**—Head of the Subcommittee....[2]

The community of airplane enthusiasts is a tightknit group. In the July 1993 issue of *Aviation*, a magazine that caters to enthusiasts, its editor, Arthur Sanfelici, took up the cause. He criticized the museum's low restoration rate, and the message that had come through in the World War I gallery. Sanfelici gave us anything but the critical approval the gallery had received in the nation's press when the exhibit opened in November 1991. He argued that we should have done more in that gallery to display aircraft than to provide commentary on such matters as their effectiveness and the carnage in the air that characterized a war in which slow, fragile craft flew low-level flights to survey the enemy's lines and were easily and routinely shot down. The World War I airplanes that we had so painstakingly restored for this exhibition received no praise.[3]

The massive restoration we had undertaken on the *Enola Gay* was taking its toll. The roughly twenty man-years we had already spent on this one project had taken away time during which we could have restored a substantial number of smaller airplanes. In recent years, about half our efforts at the Garber facility had gone into just this one aircraft. While those wishing to see the *Enola Gay* restored might be pleased, we were losing many friends who wanted to see other aircraft given equal care. We did not have the resources to please all the enthusiasts and normally would have shrugged off their disappointments as temporary. But disaffections were adding up. A year later they would result in a request for an investigation by the General Accounting Office, which reported,

> In December 1994, Senator Kay Bailey Hutchison was contacted by a historic aircraft organization, which said that NASM was not properly managed and in particular was not restoring a sufficient number of aircraft, thereby allowing its collection to deteriorate. We were asked to assess the rate of aircraft restorations; examine the adequacy of facilities for preserving aircraft; and if preservation problems exist, identify options to better care for the aircraft collection.

The fifty-five page report covered many points, summarized the many difficulties the museum was attempting to overcome with limited resources, gave advice on how that might be accomplished, but also admitted,

> . . . even if NASM were to increase its restoration efforts, the museum would not have adequate space to properly display or store the aircraft.[4]

But let us return to the spring of 1993.

MICK KICKLIGHTER

On May 10, Mick Kicklighter had written me an enthusiastic letter recapitulating our March 29 meeting:

> We are particularly excited about the educational possibilities from our association. Your support with public service announcements, fact sheets and other educational resources will greatly assist us in one of our most important missions, that of educating our nation's youth on the impact of World War II.

The letter also recalled Kicklighter's earlier promise to try to help us with funding:

> We will keep an eye out for potential sponsors to support your exhibition on the *Enola Gay*. One possibility is the Air Force Association, headed by General Monroe Hatch, Jr. (USAF retired), whom we briefed on April 14. Both General Hatch and his directors were enthusiastic about getting involved in our committee's World War II programs. . . .[5]

I would have not have minded receiving money from the commemoration committee, because the funds they disbursed came from the Congress, and thus the American people as a whole, a source nobody would find controversial. I did not know much about the Air Force Association at the time. I had not had occasion to work with them. But accepting money from most organizations was problematic, since they generally had their own agendas. I explained our reluctance on this point to the general at our next discussion, a week later.

At that meeting, on May 19, 1993, several members of the museum staff and I met for a second exchange of ideas with General Kicklighter and his team—this time in their Pentagon offices. I reported on my trip to Japan, mentioning Mr. Shibohta's then still open offer for support from the Japan Airlines Foundation for a symposium on Hiroshima and Nagasaki to be jointly held with the Japanese. Kicklighter thought the United States Information Service also sponsored such conferences at the East-West Center in Hawaii, and that they might be willing to share costs for such a colloquium, possibly in a year—on the forty-ninth anniversary of the bombing, in 1994.[6]

We discussed a variety of other joint activities and materials Kicklighter's team wanted the museum to prepare for a newsletter and calendar they would put out; and they thought we might be able to help the National Building Museum with a project. The general's team again asked whether we could prepare a fact sheet on the atomic bomb, and also a scholarly piece for teachers. And Kicklighter also asked me to send him my report on my recent trip to Japan.

Most gratifyingly, Kicklighter offered help in defusing attacks that might be launched at us from an "extremist right wing"—some of our respondents who acted as though World War II with Japan had not yet ended. We also talked more about possible funding for the exhibition.

Kicklighter advised that we get in touch with General Monroe Hatch, executive director of the Air Force Association. I did not know Hatch, but called his office to see whether I might invite him over for lunch so we could talk about the project. Normally, this was an easy way to meet for an informal discussion. But in several attempts at contacting him I was told that Hatch would be traveling and unavailable.

Somewhat dismayed, I talked to retired Air Force Brigadier General William Constantine, who served as a docent at the museum—a volunteer guide to our exhibitions. Constantine said he knew Hatch well, and that he would speak to him. But the months passed and he was not succeeding either. In mid-August, Constantine told me he was leaving on a prolonged motorcycle trip across the country, and I took up the effort once again.

In the meantime, we were continuing to talk about funding that might be available directly from the 50th Anniversary of World War II Commemoration Committee. On July 13, 1993, I wrote to Commander Luanne Smith, on Kicklighter's staff, responding to her request for the planning document we were still modifying at the time and for the expected cost of the exhibition.

> . . . any portion of the exhibition costs that your office could cover would be greatly appreciated. . . . Through this exhibition, we share one of the goals of the WWII Commemoration Committee: to encourage the study of history, particularly by our youth . . . and to convey the message to all that the existence of atomic weaponry calls for even greater vigilance on behalf of peace.[7]

THE AIR FORCE ASSOCIATION FINDS FAULT

On August 20, I finally reached General Monroe Hatch by telephone. Our conversation was far from pleasant. He came straight to the point. He did not like our exhibition.

His hostility was puzzling. The AFA's magazine, *Air Force,* had just published a cover story on the restoration of the *Enola Gay* in its August issue. It had included many beautiful color pictures taken at the Garber facility, and was not at all critical.[8]

I was puzzled also that Hatch should be familiar with our intentions, given that we had not circulated the proposal or planning document widely, and that we had been revising it so frequently in the past two months. I told him I would send him the most recent copy of the planning document at once, and did so the same day, with a cover note that I hoped would reassure him.

What I did not know was that W. Burr Bennett, Jr., had also seen the *Air Force* magazine cover story and had written to the magazine's editor, John Correll, on August 6:

> I am one of a small group of B-29 veterans of World War II engaged in a struggle with the Smithsonian Institution to display the *Enola Gay* proudly.

> With the *Enola Gay* on my mind, I instantly recognized the B-29 head-on photo on the cover of the August issue of the magazine.... A small group of veterans who served in B-29s has become a *Committee for the Restoration and Proud Display of the Enola Gay.*
>
> ... I just received a summary of the current plans for the exhibit (enclosed). Please note the lack of any reference to Pearl Harbor and the savings of six million lives anticipated by the Japanese to be lost in an invasion....
>
> The statute under which military equipment given to the Smithsonian is to be exhibited includes the statement, "The valor and sacrificial service of the men and women of the Armed Forces shall be portrayed as an inspiration to the present and future generations of America." Somehow, the enclosed plans for the exhibit do not reflect that purpose.
>
> Our Committee has collected over 5,000 signatures from around the world asking the Smithsonian to display the plane proudly ... if there is any way in which you could lend your support towards the proud display of the *Enola Gay*, our small group would be most grateful.[9]

The enclosure was the proposal draft "50 Years On," cited in its entirety in Chapter 15.

On receiving this letter, Correll at once wrote Monroe Hatch,

> I am aware—as you are—of the controversy about the Air & Space Museum. I've been carrying background material on the situation in my briefcase for some time, but it hasn't risen above more pressing priorities, so I haven't really studied it yet. My quick take is that the Air & Space Museum isn't quite as guilty as it's said to be, but I'll have a more informed opinion later.[10]

Perhaps, the "50 Years On" draft did not alarm Correll, since he wanted to study the situation more carefully, and it was this draft that best defined our intentions at the time. But, as noted by Correll, both he and Hatch already were aware of a brewing controversy, and he may have wished to assess the mood among his readers more thoroughly before deciding how to address the issue. In that regard, he might have wished to verify the five thousand signatures on the veterans' petition.

Having received the planning document and my letter of August 20, Monroe Hatch responded on September 10:

> I thank you for letting me see the revised planning document for the *Enola Gay* exhibition. I wish I could give you a favorable reaction to it, but the new concept does not relieve my earlier concerns and, in some respects, it seems even less balanced—possibly because of the details now given—than the earlier concepts were.
>
> The paper says the Smithsonian is non-partisan, taking no position on the "difficult moral and political questions" but the full text does not bear out that statement. Similarly, you assure me that the exhibition will "honor the bravery of the veterans," but that theme is virtually nonexistent in the proposal as drafted.

> The concept paper dwells, to the effective exclusion of all else, on the horrors of war. We agree that war is horrible, which is why the Air Force Association has always set such great store by the deterrence of war and has held that the nation should not enter armed conflict until other means of resolution are exhausted. Once war begins, casualties are inevitable. It is less than honest to moralize about the casualties unless one also claims the war to be immoral, and I don't believe many people are ready to say that about World War II. . . .
>
> Furthermore, the concept paper treats Japan and the United States in the war as if their participation in the war were morally equivalent. If anything, incredibly, it gives the benefit of opinion to Japan, which was the aggressor. . . . Artifacts seem to have been selected for emotional value (the schoolgirl's lunch box, for example) in hammering home a rather hard-line point of view. . . .
>
> It is not just a matter of fairness to veterans—supposedly achieved by giving them the chance to say a few words on videotapes at the end of the exhibit. Balance is owed to all Americans, particularly those who come to the exhibition to learn. What they will get from the program as described is not history or fact but a partisan interpretation.

Hatch's letter included a two-page description of what he and the AFA wanted the exhibit to be. It differed from our concept in emphasizing far more strongly Japanese aggression and atrocities. But it also gave a perspective that nobody but Air Force sources ever cited:

> Roosevelt relentlessly pressed Marshall and Arnold to bomb Japanese cities. The President, outraged at Japanese military operations against China, had expressed a desire to see Japan bombed as early as December 1940, one year prior to the Japanese attack on Pearl Harbor. As commander-in-chief, Roosevelt, through General Marshall, gave Arnold and his operational commanders carte blanche to do whatever was necessary to defeat Japan as quickly as possible with the least loss of American lives. This was also President Harry S. Truman's overriding objective.
>
> Determined to prove that a conventional bombing campaign could defeat a modern nation, General Arnold at Potsdam in July 1945 stated his position that it was not necessary to drop the atomic bomb. Arnold's position was based on his fervent desire to drive Japan out of the war by conventional bombing without a ground invasion.[11]

CONSULTING THE AIR FORCE

I did not know about it until two and a half years later, but on receiving the exhibition planning document, Hatch had sent a copy to Air Force assistant vice chief of staff Lieutenant General Tom McInerney at the Pentagon. McInerney in turn had passed it on to Air Force historian Richard Hallion, director of the Center for Air Force History. Hallion's reply to McInerney was dated September 8, 1993, and consisted of a four-page critique of the document, with two pages of addenda.

Hatch had based his response to me on Hallion's critique. Hatch's two-page description of an alternative exhibition that the Air Force Association favored was identical word-for-word to the addenda appended to Hallion's analysis, with a handwritten postscript explaining, "I've just received what seems an excellent [alternative] outline for this exhibit prepared by Herman Wolk of our staff—suggest it be sent to Hatch for his consideration!"

Hatch had simply copied Wolk's lengthy outline. The two paragraphs I've quoted above that deal with Roosevelt and Arnold are taken directly from Wolk's proposal.

I will return to the significance of these two paragraphs later, after showing how often and how fervently they were claimed to represent an essential element in atomic bombing history.

Hallion's letter to McInerney began,

> As disturbed as we might be by some of the things we read here, this is actually a much better product than what the Smithsonian originally was going to do!
>
> Harwit's letter, at least, pays homage to the veterans. I feel, however, that [it] holds a potential flaw—namely that the exhibit be "appreciative of courage, but also sympathetic to suffering." I read this as "appreciative of American courage but sympathetic to Japanese suffering." In fact, as you are aware, one of the major reasons we dropped the bomb was to end unnecessary allied and Japanese suffering, particularly given the casualty tolls people assumed would be encountered.
>
> I encourage General Hatch to meet with Martin, Tom Crouch, and Michael Neufeld . . . a Canadian with strong antiwar/anti-AF prejudice. I think that Monroe could certainly help establish a productive dialogue here. . . .

On what basis Hallion should have labeled Neufeld in this fashion I do not know. That Neufeld is Canadian cannot be disputed, but to claim him to have prejudices that would preclude him from mounting an honest exhibition seems scandalous. Over the years of working with him, I gained great respect for Neufeld's skill and integrity as a historian; the honors and positive critiques his historical works have won confirm that judgment.

Having set the stage in this fashion, Hallion's letter turned to more detailed commentary, much of it reflecting his personal views:

> I personally believe that conventional air attacks by the AAF and USN/USMC air elements won the war, making it impossible for Japan to supply and defend itself. But if we had continued to pound Japan from the air by conventional attacks, the civilian toll would ultimately have been a lot higher than was inflicted by the two atomic bombings—after all, raids were already producing Nagasaki-like casualty tolls. Is it the kind of weapon that determines whether a war is "humane" or not? Put another

way, did it ultimately matter to the Japanese victim whether he/she was a) peppered by shrapnel b) blown apart by a HE [high explosive] bomb c) incinerated by napalm or d) vaporized by an atomic bomb? Given these options, many would opt for d) as more merciful. . . .

In view of this statement, Hallion's very next paragraph seems surprising:

Unit (400) of the exhibit, [dealing with the aftermath of the bombings at Hiroshima and Nagasaki], strikes me as possibly tasteless. I am not opposed to selective photographs that show graphic injury or damage—but the power of a broken person is overwhelming, and may leave visitors with the mistaken impression that nearly all victims of the Pacific were Japanese.[12]

Hallion was right in worrying that the museum had to be careful not to give the impression that most victims of the Pacific war were Japanese. All of us agreed on that. But if we did not show the destruction and suffering atomic weapons had caused, we would also be misleading the public. We needed to consider both these concerns in mounting the exhibition.

A FIRST CASUALTY OF THE AFA

Most of Hatch's other suggestions were pretty much what we were planning to exhibit anyway, except that he ended with "... the exhibit should note the enormous technological import of the nuclear age." Given that nuclear energy plays a relatively small role in the U.S. economy, the museum's planning document had concentrated mainly on the military importance of nuclear power.

I responded to this letter briefly, on October 7, and again tried to reassure Hatch:

The views you expressed do not seem to me to greatly differ from those we thought we, at the National Air and Space Museum, had expressed in our planning document. Some of the fears you voiced about our portraying Japan and the United States as having positions morally equivalent in the War really surprised me. Such a position would clearly be outrageous and never entered any of our minds.

I believe that the exhibition can be compassionate to the human sufferings that resulted from the bombing of Hiroshima and Nagasaki without also contending that the use of the bomb was morally right or wrong. Its use simply is historical fact. Our exhibition would attempt to show how President Truman faced the question of whether or not to use the bomb and the factors that shaped his decision.

It would help us at the Museum a great deal if we could have the benefit of a more detailed discussion. I would like to invite you to come over for lunch, sometime this month, so that you could speak not only with me but also with Tom Crouch, the chairman of our aeronautics department, and Michael Neufeld, the curator in that department who is coordinating the exhibition. All three of us would like to hear your views in greater detail and tell you more about our plans. It would be important to understand particularly those places where we have differences, so that we could better come to grips with areas that potentially would lead to controversy.[13]

Unfortunately, however, we had already suffered our first casualty in our dispute with the AFA. Throughout the late summer, the museum staff had been working with General Kicklighter's aides, who thought their committee could help fund the exhibition. But toward the end of August, the tone turned quite tentative, and on September 24, Sandy Rittenhouse-Black, from the museum's exhibits department, distributed a memorandum:

I have just gotten off the phone with Commander [Luanne] Smith. She told me that the funding for the exhibition this year does not look good. General Kicklighter is uncomfortable with approving funds for the exhibition without prior review of the project by the Air Force Association and the Air Force historians. General Hatch of the Air Force Association has some reservations with the project and has voiced those to General Kicklighter. Commander Smith said the only way to change General Kicklighter's mind for funding this year would be for Dr. Harwit to talk with General Hatch and convince him of the value of this exhibition in order that he convince General Kicklighter. Commander Smith also indicated this would probably not be possible by the end of today.[14]

Kicklighter had consulted with the Air Force and the Air Force Association. In response Hallion had telefaxed Kicklighter just that day, September 24, 1993, and had included both his letter to Hatch and Herman Wolk's proposal for an exhibition the museum should mount. At the end of the handwritten cover note, Hallion had penned,

Bottom line: Be careful dealing with the SI
Will be happy to discuss, Dick Hallion[15]

Two years later, Navy Commander Luanne Smith, who had worked with Kicklighter at the time but now was retired, commented,

At one time, our Committee explored the idea of funding the exhibit. General Kicklighter, however, is a very cautious person, and I believe, Dr. Hallion's comments may have prevented us from doing so.[16]

RUSSELL DOUGHERTY

Since I had for so long tried in vain to have a discussion with Monroe Hatch, I had in the meantime asked General Russell Dougherty, who had at one time also been executive director of the AFA, to meet with me. I figured he as well as anybody might be able to alert me to the association's fears about the exhibition. Since I did not know Dougherty well, I had invited him to come over for lunch with his law partner, Carrington Williams, a strong supporter of the museum's efforts to build an extension at Dulles. Before coming over, however, Dougherty also expressed reservations:

> You were thoughtful to provide such a comprehensive letter concerning your plans for displaying the *Enola Gay* in 1995 . . . General Monroe Hatch and several others have kept me advised of your plans for the display. . . .
>
> I must say that I regret your decision to display the *Enola Gay* with the panoply of the moral issues of war set as a counterpoint. It is my opinion that emotional scenes and subjective scenarios such as those you sent Monroe are sure to create great angst and endless controversy—all without resolution. An "objective" script for such a conflicting display is, in my judgment, impossible.
>
> You have the material for a comprehensive display of the artifacts of the attack, and the historical facts surrounding this dramatic period. They make a powerful display without moralistic judgement, pro or con. What happened—the anatomy of the causes underlying major decisions in war, particularly *this* decision, are far beyond a museum's ability to analyze and portray in an exhibit of the tools of the atomic attack. Many of us would urge you to rethink the portrayal of the controversial, moralistic aspects of the Hiroshima bombing as you are now contemplating.[17]

General Dougherty's view, that this was too complex a topic for the museum to handle, and that we should not involve ourselves in moral issues, was one that we would hear repeatedly over the next fifteen months. Given that we had some outstanding historians on our staff and many of the world's leading experts on the exhibition's advisory committee that would be going over our script before it was finalized, I felt no such qualms concerning complexity. As far as moral issues were concerned, I too felt that the museum should avoid passing judgment; but that was very different from pretending that moral issues were not in the minds of those who had discussed the morality of the bombings in 1945, before deciding to use atomic weapons. I responded to Dougherty on October 20:

> I especially appreciated your frank comments on the proposed exhibition. At several points in your letter you mention the associated moral issues. I would, therefore, like to assure you that the Museum does not intend to pass moral judgment. In fact, our more detailed planning document specifically states "These are difficult moral and political questions and the Smithsonian Institution can take no position in that regard." All of us here agree with you that we are not an organization suited for that purpose.
>
> Our intention is to provide the public as accurate a historical picture as we can, and to provide documentation to support whatever we show.
>
> The issue we face is that the American public is both well educated and thoughtful. Unless we provide broad coverage that respects the intelli-

gence of our visitors, and fully explore the many questions that naturally arise, they will wonder why we leave certain topics unmentioned.

I believe that we can mount an exhibition that will be informative; will honor our veterans who risked and often sacrificed so much; will provide an accurate portrayal of the mood in the United States and in Japan toward the end of the war; will document the factors that influenced President Truman in his decision to use the bomb; will describe the bombing and its effects at Hiroshima and Nagasaki; and will show the legacy of atomic weaponry in a world that had to learn to live with an ever present nuclear threat.

All these are issues commonly discussed by Americans today, in connection with the *Enola Gay*. I do not think we should sidestep them. They can be answered truthfully and honorably, and that is our intention.

I look forward to an opportunity to hear your views on this. For us at the Museum it is important to listen, to fully understand, and then take into consideration the carefully formulated views of individuals like yourself, who have long pondered the issues and understand them so well.[18]

VETERANS OF THE 509TH READY TO HELP

Joanne Gernstein was a young curator working with Michael Neufeld on the exhibition. On October 27, 1993, she wrote me a note countering the AFA's concerns, and pointing out how much support we were obtaining from veterans of the 509th Composite Group, who had been directly involved in the atomic bombings, and in particular from veterans who had flown the Hiroshima and Nagasaki missions.[19] By that time, she had already talked with and had favorable responses from Dutch Van Kirk, the *Enola Gay* navigator; George Caron, the tail gunner; and Dick Nelson, the radio operator. She had also worked with Tom Classen, 509th deputy group commander; Cecil King, head of all their B-29 maintenance; Fred Olivi, copilot on *Bockscar* on the Nagasaki mission; Charles Levy, copilot of *Great Artiste*; William "Pappy" Hulse, flight engineer, and William "Locke" Easton, copilot of *Next Objective*, many of whom she thought would lend us objects.

Four days later, on November 1, Joanne updated her memorandum and added to her list of 509th members willing to help the names of Charles Sweeney, who piloted *Bockscar* to Nagasaki, and Fred Bock, the regular *Bockscar* pilot, though not on that mission. She added,

> Friday afternoon I spoke with Paul Tibbets. Although reluctant at first, by the end of the conversation he had invited me to have lunch with him and his crew when they visit in February. However, he still has reservations. His main concerns seemed to revolve around these issues. First, he wanted to make sure that we did not say that strategic bombing was not successful during World War II. Second, he wanted to make sure that we did not blame the 509th. He seemed to be pleased with the plans for Unit

> Three, which will not only discuss B-29 development (introducing Tibbets as a key player), B-29 production, the creation of the 509th, the multifaceted nature of the 509th (its unusual combination of flight and service crews within the same bomb group) and both atomic missions (Nagasaki is often downplayed).[20]

FACE TO FACE WITH THE AFA

The November 1, 1993, lunch with General Russell Dougherty and Carrington Williams was friendly and noncommittal. The November 19 lunch scheduled with Monroe Hatch was not. Hatch had asked whether he could bring a colleague, who turned out to be *Air Force* magazine editor-in-chief John Correll, whom none of us knew at the time. Tom Crouch and Mike Neufeld also joined us. Hatch and Correll immediately lit into us on the *Enola Gay* exhibition, for which, as far as we knew at the time, they had only seen the planning document.

Throughout lunch, we kept saying that none of the issues they were fearing were as they claimed, and tried to tell them what we actually intended to do. After lunch we took them to Crouch's office to show them sketches. They asked to become part of the exhibition's advisory committee, but I said that we would not do that. The advisory committee was made up of various experts we had selected to cover different areas. However, we were willing to send them the script when we mailed it out to the advisory committee and would appreciate their comments in return. I thought they left appearing less upset, but on returning to his office, Correll wrote a memorandum to file, which I only saw eighteen months later in mid-1995.

> We said the concept paper was not balanced, and that it did not provide adequate background or accurately depict the context in which the decision to drop the bomb was made. All three museum people (especially Crouch) said the concept included all of these things. We said there is a huge difference in impact between a few words in the script and an emotion-grabbing artifact like a little girl's burned lunch box.
>
> We also said the concept goes out of its way to spotlight Japanese suffering.... Correll asked if it would show GIs *dead*.... We made an issue of the emotional impact of the school child's lunch box and pointed out that there was nothing on the other side for balance. Harwit asked what we had in mind. We mentioned several possibilities of Japanese behavior.
>
> Harwit dismissed those suggestions, saying the exhibit should not show Japanese atrocities because that would make *Enola Gay*'s mission appear to be one of revenge—i.e. unfair to the Americans! (This was one of two instances when the Air & Spacers rejected content that we would regard as balanced on the pretext that it was unfair to Americans.) Furthermore, Harwit (supported by Neufeld) said the airplane itself was a domi-

nating "militaristic" and "macho" element in the exhibit. We challenged that interpretation, but Harwit said the Japanese would see it as he said. (After lunch, Harwit introduced us to a staff member, [Joanne Gernstein], who has been assigned to keep frequent contact with crew members of *Enola Gay* and *Bockscar* both for general liaison and to seek memorabilia for the exhibit.)

Neufeld acknowledged that his low U.S. casualty estimate (20– 30,000) was for invasion of the southern island only, and only for the first month at that. He said higher casualty estimates—such as the often-cited 500,000—could not be used because veterans groups use a figure of 1–2 million (??!!) And would not be satisfied with anything lower. The solution, therefore, is not to use any casualty estimate—conveniently eliminating the impact of a key point in the decision to drop the bomb. This, like Harwit's reluctance on Japanese atrocities, just happens to tilt the balance toward the point we believe they are really trying to make, and to which we object. . . .

Harwit heaped praise on enlightenment of Japanese. He said one of the mayors (Hiroshima or Nagasaki) had almost been assassinated for saying the Emperor was wrong in not acting sooner to end the war. Harwit indicated the mayor may say something along those lines in his recorded presentation in the exhibit. Correll asked if either of the mayors intended to say that Japan had also been wrong in *starting* the war. (Of course not.). . . .

The word-for-word exhibit script will be ready around January. They will send us a copy for review and comment.[21]

The memorandum's last sentence shows that Correll was well aware that we were going to send him a copy of the exhibit script for review and comment only. This is significant, because later Correll was to violate this agreement.

The following week I wrote Hatch,

It was good to have a chance to meet with you and John Correll, last Friday, so that Tom Crouch, Michael Neufeld and I could tell you more about our planned exhibition and hear your concerns.

I came away from that meeting with a reinforced appreciation of the depth of your commitment and loyalty to the men who fought with the U.S. Army Air Forces in World War II. Your concern that our exhibition express the nation's appreciation to those who served, and that we properly honor what they did for our country, also was clear.

We do not differ with you on any of this, and I would like to believe that all veterans who served in the War will be pleased with our exhibition when it opens. We already are in close touch with many groups now, and they are helping us with their memories and with the memorabilia they are willing to let us use.

As promised, we will send you the complete script for the exhibition when ready, early in the coming year, and look forward to your comments when you receive it.

Please be assured that we take your comments very seriously. With the many interests the Air Force and this Museum share, it should be

> straightforward for us to accommodate the concerns you had expressed to us, most of which in any case paralleled our own thinking.[22]

I wrote to Hatch again three weeks later, to see whether he would be willing to be one of the commentators whose videotaped views would be presented near the exhibition's exit, where visitors would see a variety of representative opinions displayed:

> When you visited the Museum last month . . . we mentioned that a final section of the exhibition would feature a range of viewpoints on the events of August 1945.
>
> We are planning on videotaping representatives with well-articulated opinions, cogently presented so they will fit into something like a one-to-two-minute segment. A finished videotape including all of these segments will be part of our exhibition. It would be important to us to include a representative for the Air Force Association, speaking for the many veterans who flew in World War II, risking their lives, often failing to return home.
>
> I am writing now to see whether you, personally, would be willing to make a statement for the exhibition, to be included in these videotapes. In our discussions, it was clear to me that you had the veterans' interests deeply in mind, and that you would be an outspoken and persuasive spokesman on their behalf. Should you prefer to nominate some other spokesman, we would of course be glad to take your advice, but I do hope that you would be willing to take on this responsibility yourself.[23]

Hatch accepted this offer early in January 1994:

> I accept your invitation to make a videotaped statement. . . .
>
> As you know, I remain concerned about the balance and perspective of the exhibit. I still hope to convince you that greater objectivity is required throughout the exhibition, not just in statements like mine in the final section. Nevertheless, I do appreciate this opportunity to provide a viewpoint that I'm sure many Air Force Association members and military veterans will share.[24]

In the meantime, Correll had followed up on Burr Bennett's August 6 letter, where Bennett had again quoted from the legislation authorizing a National Armed Forces Museum (Title 20 of the United States Code §80a), "The valor and sacrificial service of the men and women of the Armed Forces shall be portrayed as an inspiration to the present and future generations of America." In a letter to Hatch, written on December 16, Correll omitted mentioning that this paragraph of the U.S. code, written in 1961, specifically governed the establishment and operation of a museum that had never been built. Though Correll noted and underlined the words "in no way rescinds §§77–77d"—meaning that the legislation had no bearing on the National Air and Space Museum—he argued, apparently without having obtained legal advice, that the quoted sentence made it "obvious—except to those who willfully misconstrue it—the spirit in which Congress intended these "contributions" to be displayed when it passed this law."[25]

From then on, it appears, the AFA, and eventually most of the other veterans' organizations, would tacitly assume that the National Air and Space Museum was "willfully misconstruing" the law governing its operations. Had they consulted legal opinion, they would have found that the cited legislation specifically ruled out any alteration of the legislation that applies to the museum.

THE EXHIBITION SCRIPT

By January 31, the museum had a first draft of the exhibition script ready to send out, and I sent a copy to Hatch, with a cover letter that again specifically asked him not to circulate it. I also thanked him for his willingness to provide a videotaped commentary for the exhibition. The entire text reads,

> I am sending you an advance copy of the entire script of the *Crossroads* exhibition. It still has quite a number of rough edges, which I expect will be removed as our advisory committee provides us with criticisms.
>
> Please let me have any comments you have. They are important to us.
>
> I would appreciate it, however, if you did not circulate the material at this time, since it is not yet in suitable form. We expect that a final draft should be ready for wider distribution and comment in a few weeks.
>
> In the meantime, I am really pleased that you have offered to participate personally in the exhibition with a video commentary. We very much wanted to have your viewpoint included.[26]

17

The Script

In late July 1993, the museum had reached tentative agreement on how the *Enola Gay* should be exhibited. The planning document tried to spell that out in sixteen double-spaced pages. But those pages conjured up different expectations in each of us—Bob Adams, Monroe Hatch, the curators, and myself, to say nothing of the Japanese.

The next step, preparation of a full first draft of a script, would crystallize those expectations and make possible a more meaningful exchange. To help the museum through a debate that represented as many thoughtful views as possible, we had already assembled an advisory committee of experts, whose opinions we would weigh carefully in preparing a second draft to be tried out on a wider circle.

Usually a complex exhibition script went through five or six drafts before a final document was ready. Not until then was the exhibits staff able to finalize a design and go into production. For an exhibition of the size we planned, the scope of construction was equivalent to putting up walls, electric lines, lighting systems, air-conditioning ducts, fire-extinguishing sprinkler systems, and all-handcrafted furnishings for a five thousand-square-foot home. In addition, the silk-screening of label text to accompany the display of artifacts was equivalent to typesetting a moderate-sized book. All this would take time, which was why we urgently needed to have a first draft of the script by the end of January 1994 in order to open the exhibition on time in late May 1995. We needed those fifteen months for refining the text, gathering artifacts to be borrowed from other museums, and constructing the entire gallery.

Once a script was drafted, half a dozen to a dozen members of the museum staff independently worked their way through it to verify facts and search for possible errors. Then, a dedicated advisory committee of experts was invited to go over the script and designs to provide broad advice on items or perspectives that might have been overlooked, and to pass judgment on balance.

PREPARING THE SCRIPT

Mike Neufeld, as lead curator for the entire exhibition, was responsible for coordinating the gallery's script. He was assisted by Tom Crouch, who acted as manager of the curatorial team, and by two young curators, Thomas Dietz and Joanne M. Gernstein. Work on the script could not begin until the planning document was approved, which occurred in late July 1993. With a January deadline just six months away and an anticipated script five hundred pages long, all four curators became totally immersed in writing the script.

Writing sessions were continually interrupted by meetings on the concurrently evolving gallery layout, the responsibility of the exhibition's chief designer, William "Jake" Jacobs. Ultimately the display of objects would take the upper hand, setting the mood for the gallery, particularly for the casual visitor who wanted to see only the *Enola Gay* and one or two other primary artifacts. For them the words would play merely a subordinate role. For more serious students, who might spend hours reading every word, hungry for details and nuances, the accompanying words—the script—would be paramount. The script and the display of artifacts had therefore to be carefully interwoven to make the gallery enjoyable and effortless for both these groups.

Right from the start, the exhibition, then bearing the title *The Crossroads: The End of World War II, the Atomic Bomb and the Origins of the Cold War,* was designed to have five sections, respectively designated, *100, 200, . . . , 500.* Before entering these, the approaching visitor would first be greeted by a large, mood-setting montage "Today is V-E Day," declaring the war in Europe over, but with newspaper headlines indicating that only half the war was won. A difficult fight in the Pacific still lay ahead.

On entering the gallery proper, the visitor would be immersed in section *100,* titled "A Fight to the Finish," which introduced the antecedents of the war—the naked Japanese aggression in Asia in the 1930s; the attack on Pearl Harbor; the early years of the war in the Pacific, marked by terrible losses for the Allies; and the enormous casualties suffered by both sides at

Iwo Jima and Okinawa as the Japanese initiated their suicidal attacks. An Okha piloted bomb would point at a towering photograph of the aircraft carrier *Bunker Hill* burning fiercely after two kamikaze aircraft dove into her off Okinawa on May 11, 1945. Numerous smaller artifacts, film footage, and still photographs would be augmented by a historical text. This introduction would be followed by a description of the development of strategic bombing in World War II, escalating to General Curtis LeMay's incendiary bombing of Japan.

Next—in section *200,* "The Decision to Drop the Bomb"—the visitor would see the origins of the Manhattan Project, the technically staggering crash program to design and construct an atomic bomb, and the perspectives on the bomb among President Truman's advisors. The fanatically stubborn mood in an almost-defeated Japan would be made understandable in terms of a nation unwilling to give up an emperor who symbolized the country's seven centuries of independence and self-sufficiency. And finally, the factors President Truman considered in reaching his decision to use the bomb would be enumerated, to the extent they have become known through papers declassified in the intervening years. Here archival material would play a prominent role, but a major artifact would be the huge *Fat Man* plutonium-bomb casing of the type used at Nagasaki. It would reveal the immense size and weight of these early devices.

The visitor would then enter section *300,* "Delivering the Bomb," and come upon the fifty-six-foot-long forward fuselage of the *Enola Gay*, with a *Little Boy* uranium-bomb casing of the type used at Hiroshima beneath the open bomb bay doors. This would be accompanied by a history of the construction of the B-29 bombers, which were crucial to bringing the war to the Japanese home islands; the story of the training of the 509th Composite Group, which eventually carried out the missions on Hiroshima and Nagasaki; and a description of those missions. Memorabilia and photographs provided by members of the 509th would lend a human touch.

In a fourth unit—section *400,* "Ground Zero: Hiroshima, 8:15 A.M., August 6, 1945; Nagasaki, 11:02 A.M., August 9, 1945"—visitors would see scenes on the ground at Hiroshima and Nagasaki; artifacts borrowed from the two cities' museums; and film footage showing survivors sometimes vividly recalling their experiences. These would emphasize the immense scale of nuclear havoc and its effects on the cities' populations.

Finally, visitors would enter section *500,* "The Legacy of Hiroshima and Nagasaki"—a display on the postwar problems of nuclear proliferation humanity now faces.

While the script was being written and the gallery layout designed, Patti Woodside, the museum's filmmaker, was also at work on a fifteen-minute film in which visitors to the gallery could witness members of the crews of the *Enola Gay* and *Bockscar* recalling their thoughts during the missions to

Hiroshima and Nagasaki and their impressions fifty years later. Woodside was also starting to gather film footage from wartime newsreels, to set the mood for visitors through sounds and images almost forgotten after fifty years. In addition, she was screening short film segments we hoped to borrow from the Japanese museums. Here, older people recalled the day the bomb had exploded over their city. And finally, Woodside was beginning to plan the series of two-minute interviews, at the exhibition's exit, where perhaps half a dozen leading personalities would express a variety of views on the mission of the *Enola Gay,* the end of the war, and the nuclear age that ensued. General Monroe Hatch had agreed to do one of these, and the mayors of Hiroshima and Nagasaki had also shown an interest in adding their cities' voices.

The artifacts, the script, and the films formed an indissoluble triad that had to be melded into a complete exhibition. The Okha piloted suicide bomb, the two atomic bomb casings, the forward fuselage of the *Enola Gay,* and the photographs of battles and devastation would give visitors their first big impression. The label script accompanying the objects would provide added detail and a more complete historical setting. And the film footage would infuse a personal element to spark the exhibition to life through the words of ordinary people who had experienced extraordinary times and could bring home to visitors the essence of war.

The film footage had an additional purpose. The success or failure of a controversial exhibition can depend on whether a museum gets the communities having the greatest stake in the display to become active participants in its design—"to buy in." To do that the museum must enrich an otherwise purely historical display with the voices of those who have special knowledge based on having been there. We hoped that the inclusion of the newsreel film footage, and particularly the recollections of the *Enola Gay*'s and *Bockscar*'s crew members, would convince veterans that their voices were being heard. We were also trying to get General Tibbets to agree to be videotaped to express his thoughts, both because they would provide a historical perspective, and because they would bring added credibility to the exhibition.

A MEDIUM OF THE SENSES

Exhibitions are media of the senses. The most challenging problem the museum faced was to shape the exhibition in its entirety so that visitors would leave with reasonable impressions and perceptions.

Retired Admiral Noel Gayler, member of the museum's Research Advisory Committee, had worried as early as the fall of 1987, as others did later, that a massive, gleaming *Enola Gay* would give the impression that the museum was celebrating raw power.

To avoid this perception we needed to show that the bomb had caused unimaginable damage and suffering. This would be done by emphasizing the thoughtful debates that had preceded the decision to drop the bomb (section *200*) and by treating with understanding the effects of the bomb on the two cities and their populations (section *400*).

We could imagine that a visitor who had just seen the fifty-six-foot-long forward fuselage of the *Enola Gay* (in section *300*) might come into section *400* and walk right past the small, modest-looking, everyday objects we could borrow from the museums at Hiroshima and Nagasaki—a wooden clog, a small watch with its hands frozen at the exact hour of the bombing, some scorched clothing, a child's lunch box with charred contents. Because of the very modesty of these objects, we felt we had to call our visitors' attention to them. To do that, we resorted to deliberately compassionate text and photographic images. Whether this text hit just the right note, sounded excessively sympathetic or possibly too callous, would have to be tested on our advisory committee, focus groups, and others representative of our visitors.

An exhibition script, taken by itself, conveys no indication of the display's overall impact, largely dominated by impressive artifacts and film footage. To critics who read only the script and could not imagine the raw visual impact of the aircraft and the atomic bomb, section *400* might have appeared excessively sympathetic to the Japanese. But this would certainly not have been the conclusion visitors would have reached having gone through and seen the entire exhibition. The humanity of our fighting men and their sacrifices were so strongly conveyed in the exhibition's video-film footage, which could not be reflected in the script, that no visitor could have left that exhibition feeling anything but proud of the people who had served their country throughout the war or on the two atomic missions, even though the atomic bombings had inflicted terrible catastrophes on our wartime enemies.

The exhibition's overall impact critically depended on the interplay of this triad—the object, the written word, and the commemorative video testimony.

In contrast to books, exhibitions are largely visual and auditory experiences. The sights and sounds dominate the words. Why else bother? If words and small images alone sufficed, it would be far less expensive to produce just a book.

The museum's critics never bothered to think of that. Concentrating solely on the written word and Xerox copies of photographs whose relative sizes in the planned display they did not know, they drew conclusions about an exhibition medium whose main impact derives not from words on exhibit labels, but from apt juxtapositions of objects, documents, films, sound, and lighting. Nor was the text they analyzed a final script, but rather the first of many drafts that would normally follow—and that did, in fact, follow.

Immersed in this analysis, they never seemed to notice that in the midst of the words, phrases, and paragraphs on which they so intently focused sat a fifty-six-foot-long, brightly polished, brilliantly sparkling bomber, an atomic bomb casing immediately beneath its open bomb bay. That this enormous object, towering above the crowds, might affect how visitors were impressed by the exhibition seemed to escape notice. Nor did the critics take into account the emotionally powerful, fifteen-minute film the museum had expressly produced for the gallery. Here, *Enola Gay* and *Bockscar* veterans recalled their missions, the training that had preceded them, the thoughts on their minds as they flew to their targets, their impression on witnessing the horrendous upheaval, their conversations as they left the burning cities behind, and their thoughts in the fifty years since.

For the critics, the exhibition was only words. They seemed unable to properly appreciate what powered the display or what its prime components and true proportions were.

AN EVOLVING SCRIPT

The script's five chapters had been written by four different curators. Style and language differed markedly among chapters. That is not unusual in a first draft. The museum employs a full-time editor assigned to exhibits whose task is to take a script once it has undergone initial revisions and rewrite it in a style that is consistent throughout and also consistent with a house style common to all the museum's exhibitions.

Quite aside from this, considerable imbalance is quite normal in a first draft. Every curator brings some intended or unintended bias to his work. The museum's job is to test successive drafts of the script on as many different focus groups as possible, to hear their reactions, and eventually to reach a balance that is broadly, if not universally, accepted. This procedure is time-consuming but essential.

The first draft of the script usually will be ready a year or two before the anticipated opening of the gallery. Sometimes it is available far earlier. At this stage, if not sooner, an advisory committee of outside experts is convened to advise the museum on the script, to check it for accuracy and balance, and to convey how visitors might perceive it. Thereafter the script is revised, and it often goes through five or six successive drafts before it passes all tests for accuracy, balance, and perceptions.

To cite one example: For an uncontroversial exhibition on "How Things Fly," to open in 1996, the first draft was ready in spring 1993. It then had to be entirely rewritten that summer because it had taken on aspects of a treatise on aerodynamics instead of speaking simply about how things fly. It was repeatedly rewritten thereafter, and though then in good shape, it continued to be revised until close to opening day.

This was also the intended procedure for exhibiting the *Enola Gay*. For a year, revisions would continue. Production of materials for the exhibition would begin late in 1994, or at latest in January 1995, for an opening scheduled for late May 1995.

In this same fashion, the first *Crossroads* script also underwent major revisions. Later versions, bearing a new title, *The Last Act: The Atomic Bomb and the End of World War II,* were copyrighted by the Smithsonian. The final version, as it stood on January 30, 1995, is available only from the Smithsonian legal counsel by appeal to the Freedom of Information Act. The first script, which the museum had not copyrighted, has since been published, though without any images, and is available.[1] Readers should, however, bear in mind that a script concentrates primarily on the words on exhibit labels; they should try to imagine the greater impression that the actual exhibition, with all its objects, sounds, and images, would have conveyed.

THE ADVISORY COMMITTEE

For *The Crossroads* exhibition we invited an advisory committee of eight people.

Edwin Bearss was chief historian of the National Park Service and the man behind the fiftieth anniversary commemorations at Pearl Harbor. Those commemorations had encountered strong opposition from veterans' groups wishing to exclude anything that went beyond celebrating them and mourning their fallen comrades. Historical contemplation was to be rejected, even though its supervision by Bearss, a wounded and decorated veteran of the World War II Guadalcanal campaign, should have commanded the veterans' confidence. We hoped to obtain Bearss's guidance on how to avoid the same kind of controversy.

For similar reasons we had also invited Edward T. Linenthal, professor of religious studies at the University of Wisconsin, Oshkosh. He had extensively studied how communities and nations commemorate sacred places and events, and in particular, battles and battlegrounds. Linenthal had attended the fiftieth anniversary ceremonies at Pearl Harbor and written about the passions and controversies that surrounded those commemorations:

> There were many perceptions of the lessons that appropriate commemoration could bring, yet Pearl Harbor survivors—and, indeed, many Americans of the World War II generation—were adamant about the need to keep faith with patriotic orthodoxy, a faith that would tolerate no alteration.[2]

Fearing this same rejection of a historical approach to our exhibition in favor of long-accepted views that sometimes lacked factual basis, we had brought in both Bearss and Linenthal to advise us on how to proceed.

We had asked Barton J. Bernstein, professor of history at Stanford University, to join the advisory committee as a leading student on the decision to drop the atomic bomb. His research had documented the casualties and loss of lives that military analysts in 1945 had predicted for an invasion of Japan.[3] These figures were likely to prove controversial, and we wanted to be sure we provided visitors with the most authoritative figures.

Martin J. Sherwin, director of the John Sloan Dickey Center for International Understanding at Dartmouth College, was asked to join because of his widely cited book *A World Destroyed—The Atomic Bomb and the Grand Alliance,* in which he had analyzed the diplomatic issues involved in President Truman's decision to make use of the bomb—again a potentially controversial topic on which we hoped to obtain the most reliable guidance.[4]

We asked Victor Bond, an internationally recognized radiation physiologist at the Brookhaven National Laboratory, whether he would be willing to examine our script to make sure we had properly presented the medical effects suffered by the populations exposed to atomic radiation.

Stanley Goldberg, a historian of science, was asked for two reasons. He was the biographer of General Leslie Groves, head of the Manhattan Project and a controversial figure in the literature on the project. He had also successfully mounted an exhibition on the fortieth anniversary of the atomic bomb at the National Museum of American History, and we felt that he could give us guidance on how to handle our sensitive subject.

We also invited Richard Hallion, military historian in charge of the United States Air Force Center for Air Force History, for two purposes. First, he could speak with authority on behalf of the United States Air Force and its predecessor, the United States Army Air Forces; and second, he had for many years been a curator at the National Air and Space Museum, knew our procedures well, and could bring both an outsider's and insider's perspective. As an added member of this committee, Dr. Hallion brought with him Herman S. Wolk, a historian of World War II, also attached to the Center for Air Force History.

Akira Iriye, professor of history at Harvard University and past president of the American Historical Association, was asked because of his extensive insight into Japanese-American relations in and around World War II. Born and raised in Japan and educated in the United States, he was able to bring keen insight on both American and Japanese perspectives. He had already been very helpful to us in suggesting Japanese scholars to contact.

And finally, we had invited Richard Rhodes, arguably the world's most expert scholar on the origins and development of both the atomic and hydrogen bombs. As author of the Pulitzer Prize–winning book *The Making of the Atomic Bomb,* Rhodes would be immensely helpful on any questions dealing with the history of the bomb and able to put this knowledge into layman's terms.[5]

THE ADVISORY COMMITTEE MEETS

The committee met all day, February 7, 1994. Prof. Iriye, unable to attend, had thoughtfully telefaxed his comments that morning:

> I have gone through the document, and I have paid particular attention to the historical descriptions contained in the sections marked 200 and 500. I like what I find there. It seems to me that all the statements are carefully written and reflect the authors' obvious intention to present as judicious an interpretation of controversial events as possible. I have no problems at all with the statements concerning the origins of the war, wartime hatred in the two countries, fire bombing of Japanese cities, the decisions (although it is argued that there was no "decision" as such, and I agree) that led to the atomic bombings, their military, political, and moral implications, direct and indirect damages caused by the bombings, the long-range implications of the coming of the nuclear age, etc. Only irresponsible fanatics, who do exist in both countries, would take exception to the document.[6]

The last sentence is particularly significant because it represents the views of a respected scholar who has long studied and thoroughly understands both the United States and Japan.

Iriye's note was encouraging in that it focused on the more difficult portions of the exhibition, which dealt with widely debated issues.

Interesting statements in the day's discussions included Martin Sherwin's capsule comment that the exhibition struck him as a celebration of the bomb. Rhodes, Wolk, and Bond disagreed. Stanley Goldberg felt that we did not have enough interview material of people who shaped the atomic bomb program, and in a similar vein, Rhodes said he missed any emphasis on "scientific discovery" in our discussion of the Manhattan Project.

But these were among the very few comments regarding the exhibition as a whole. I was disappointed that most of the committee members concentrated on only one or two topics of current academic research interest, leaving whole portions of the exhibition untouched. Several times during the day I asked the committee to discuss other aspects, but the debate invariably gravitated back to the decision to drop the bomb and the necessity for dropping it.

Normally, the expert advisory committees the museum appointed gave us far broader, more dispassionate advice. This committee, perhaps because it was so intensely involved in current research on the topic, was far more narrowly focused. Committee members dwelled more on their own areas of expertise than on the broader aspects of the exhibitions that would be of interest to the ordinary visitor.

With this proviso, the comments we did receive from the committee were generally encouraging. I will quote them somewhat extensively, because they were so different from the criticism we would later hear.

Edwin Bearss wrote,

> It was an honor and a pleasure to participate in the panel that was convened on February 7. . . . The give and take that highlighted the discussions was intellectually stimulating and bodes well for the success of what will be a thought-provoking exhibition to mark the 50th anniversary of the development, deployment, delivery and explosion of the atomic bombs over Hiroshima and Nagasaki.
>
> As a World War II Pacific combat veteran, I commend you and your colleagues who have dared to go that extra mile to address an emotionally charged and internationally significant event in an exhibit that besides enlightening, will challenge its viewers.
>
> Enclosed please find the draft, which I have reviewed. In the margins of the draft I have identified a number of editorial suggestions. . . .
>
> The superior quality of the label texts and of the objects and illustrations identified by the exhibit designers and researchers sets a pattern of excellence that all aspire to, but few achieve.[7]

The other committee members struck similar notes and would continue to back us.

The Air Force historian, Dr. Richard Hallion and his colleague Herman Wolk commented,

> Overall this is a most impressive piece of work, comprehensive and dramatic, obviously based on a great deal of research, primary and secondary.
>
> While little is left untouched, no mean feat in itself, there are places and themes which perhaps require either more, or less, emphasis in order to give the exhibit a better contextual basis.
>
> They then summarized their main conclusions on President Truman's deep concern with potential American casualties, and President Roosevelt's determination to heavily bomb the Japanese. They also felt a need for more images focusing on the Japanese brutality to subject peoples, 1931–1945.[8]

Most of these points, and a few of lesser importance, were then addressed in a more detailed listing, two pages long, roughly fourteen items all in all. They included a request for more emphasis on Japanese atrocities in the war; a clear indication that Japan had been the aggressor; two more references to the pressure Roosevelt was exerting on Army Air Forces commanders to bomb Japanese cities; two notes on the numbers of American casualties President Truman expected in an invasion of the Japanese home islands; and eight minor technical notes, some merely typographic.

It ended with Hallion's handwritten message: "Again—an impressive job! A bit of 'tweaking' along the lines discussed here should do the trick."

Hallion and Wolk were right in their perception that repeating Truman's concern about the diplomatic implications of the bomb could give visitors an unintentionally slanted picture. This was the kind of criticism that was very helpful. And perhaps more emphasis needed to be given to Truman's concern about losses at Okinawa.

Interesting, in retrospect, was Hallion's request that we emphasize Roosevelt's insistence on bombing the Japanese "heavily and relentlessly." Months earlier, on September 10, 1993, Monroe Hatch had already made the same point, in his very first letter to me. At the time, I had thought nothing much of it, because obviously Roosevelt, as Commander in Chief, would have had to be responsible for the massive bombing of Japan. The Air Force could not unilaterally decide on such a bombing campaign.

The Air Force Association and the Office of the Air Force Historian would later repeatedly stress Roosevelt's role in the bombing of Japan. It seemed to become increasingly important to them.

Because Hallion's and Wolk's critique was so generally favorable, and because both men would later deny that they had praised the script, I have included a copy of the original. The penned-in notes are in Dick Hallion's hand. [Fig. 17-1]

From Dick Hall
& Herm Wolk
Office of the AF
Historian
7 Feb 94

Comments on Script, "The Crossroads:
The End of World War II, the Atomic Bomb
and the Origins of the Cold War"

Overall, this is a most impressive piece of work, comprehensive and dramatic, obviously based upon a great deal of sound research, primary and secondary.

While little is left untouched, no mean feat in itself, there are places and themes which perhaps require either more, or less, emphasis in order to give the exhibit a better contextual balance.

Through sheer repetition, the script gives the impression that the Truman administration was more concerned with the atomic bomb as a diplomatic weapon against the Soviet Union than as a route to shorten the war and avoid heavy American casualties [references to dropping the bomb to impress the Soviets are on: EG:200-L2, p1; EG:220-L2 p19; EG:240-L2, p32; EG:240-L5, p37, mentioned twice here; EG:240-L7, p39; and EG:270-L8, P66]. By this repetition, the impression is that the diplomatic rationale was the major reason the bomb was dropped although the script notes that "most scholars have rejected this argument."

President Truman's deep concern on June 18, 1945 with potential American casualties, based on the casualty rate on Okinawa, does not come out clearly and effectively in the script.

President Franklin D. Roosevelt was the leading American official advocating the bombing of Japanese cities "heavily and relentlessly." Outraged by Japanese brutality in China, Roosevelt indicated even prior to December 7, 1941, that he wanted to see Japanese cities bombed. Arnold and Norstad (and then LeMay in the Pacific) were under enormous pressure from Roosevelt and Marshall. Somewhere in the script there should be a sentence or two pointing this out.

There should be more images focusing on the Japanese brutality to subject peoples, 1931-1945.

EG:100-L2, p5

-- Four or five sentences take us from 1931 to 1941. Japanese aggression associated with the drive for a Greater East Asia Co-prosperity sphere, and the brutality against subject peoples, resulted in a severe reaction in the United States, among other nations. This brutality, especially in China, a country that Americans generally were sympathetic to, and the reaction to it, seem to be missing here.

"For most Japanese, it was a war to defend their unique culture against Western Imperialism." Is this relevant to 1931-1941? Who was the aggressor here? The U.S.? The western allies?

-- Good sections on Iwo Jima and Okinawa, detailing the ferocity of the fighting. The text also brings out especially well the scale of kamikaze operations.

EG:120-L2, pp28-29

-- The highest American official calling for the bombing of Japanese cities -- even before December 7, 1941 -- was President Franklin D. Roosevelt. FDR is quoted on p29 as condemning the bombing of civilian populations. However, Roosevelt during the war put intense pressure on Marshall and Arnold to bomb Japanese cities with incendiaries. Michael Sherry, in his book, asks how it was that the American Operational air commanders (LeMay) in the Pacific had carte blanche as to tactics and strategy. The answer is that the Commander-in-Chief, FDR and Truman, made clear to Marshall that everything should be done to end the war as soon as possible with the least loss of American lives. President Roosevelt's view was that Japan should be bombed "heavily and relentlessly" [FDR report to Congress, Jan 7, 1943, in War Messages of FDR, p32, Washington 1943].

EG:120-L6, p34

"General Curtis E. LeMay proposed a radical change in tactics after taking command... in January 1945." Actually, LeMay continued General Hansell's tactics until March 1945. He also got the same disappointing results as Hansell in January and February 1945.

EG:130-L3, p39

"By the spring of 1945... had brought the Great Depression to an end."

Why not merely state that World War II -- government spending for munitions, etc. -- brought the Great Depression to an end. Why nail it to Spring 1945?

EG:130-L7

"The Yellow Peril" -- Is this overdrawn as "anti-Asian racism," given the American sympathy and support for China?

EG:130-L9

"B-29s of the 20th Air Force" -- Air Forces are spelled out -- Twentieth -- Bomb Groups are designated by numerals -- 87th -- and commands by roman numerals -- XXI Bomber Command.

EG:260-L5, p54

The question of U.S. casualties should an invasion be necessary -- script notes that postwar estimates were too high and that "military staff studies" estimated thirty to fifty thousand casualties in OLYMPIC. However, Truman had in front of him on June 18,

1945 the Luzon (31,000), Iwo Jima (20,000) and Okinawa (41,000) casualty figures. And as the script notes, if OLYMPIC took the same rate of casualties as Okinawa, the casualty figure estimated was 268,000.

EG:260-L6, p56

"The Japanese and American lives that would have been lost in an invasion have often been used to justify the atomic bombing of Japan." The point is that Truman made the decision based primarily on the estimated American invasion casualties. He wanted to prevent, in his own words, "an Okinawa from one end of Japan to the other."

EG:270-L7, p64

Spaatz, Commander of the newly created U.S. Army Strategic Air Forces in the Pacific.

EG:270-L8, p66

Again, the comment that "even if the American deaths in an invasion of Japan would have been significantly lower that the post-war estimates." The numbers that counted were the ones Truman wrestled with in June 1945.

EG:320-L1, p12

"... the Twentieth Air Force would be directly under the command of the JCS," with General Arnold as executive agent of the JCS.

-- Good sections on "the Battle of Kansas" and crew training of the 509th.

EG:320-L13, p27

Note that General LeMay, in addition to being a veteran of the European campaign, had replaced K.B. Wolfe at XX Bomber Command.

EG:320-L13, p27

LeMay was under pressure from FDR, Arnold, and Norstad to attack Japanese cities with incendiaries.

Again — an impressive job! A bit of "tweaking" along the lines discussed here, should do the trick...

18

Once-Secret Documents

REMARKABLE INSIGHTS

As the exhibition was taking shape, I was especially fascinated by a handful of documents involving President Truman and his secretary of war. Reading them was like breaking through a fog of historians' opinions, to personally glimpse fragments of truth. Would not veterans, whose lives had been so greatly affected, feel that same thrill on finding them in our exhibition?

One of the most significant documents was Albert Einstein's 1939 letter to President Roosevelt that set the atomic bomb project in motion. Einstein was on vacation at the time and had been sought out by the young Hungarian physicist Leo Szilard, the first man to think about the possibility of a nuclear bomb. Szilard was desperate to bring to the attention of the president the possibility that German scientists might already be at work on an atomic bomb. But neither Einstein nor Szilard knew just how detailed a letter they could address to the president of the United States, and so they drafted two versions, both of which Einstein signed. Eventually the longer of the two was transmitted to President Roosevelt from Einstein's summer retreat at Nassau Point, Peconic, Long Island.

August 2nd, 1939

F.D. Roosevelt,
President of the United States,
White House
Washington, D.C.

Sir:

Some recent work by E. Fermi and L. Szilard, which has been communicated to me in manuscript, leads me to expect that the element uranium may be turned into a new and important source of energy in the immediate future. Certain aspects of the situation which has arisen seem to call for watchfulness and, if necessary, quick action on the part of the Administration. I believe therefore that it is my duty to bring to your attention the following facts and recommendations:

In the course of the last four months it has been made probable—through the work of Joliot in France as well as Fermi and Szilard in America—that it may become possible to set up a nuclear chain reaction in a large mass of uranium, by which vast amounts of power and large quantities of new radium-like elements would be generated. Now it appears almost certain that this could be achieved in the immediate future.

This new phenomenon would also lead to the construction of bombs, and it is conceivable—though much less certain—that extremely powerful bombs of a new type may thus be constructed. A single bomb of this type, carried by boat and exploded in a port, might very well destroy the whole port together with some of the surrounding territory. However, such bombs might very well prove to be too heavy for transportation by air.

The United States has only very poor ores of uranium in moderate quantities. There is some good ore in Canada and the former Czechoslovakia, while the most important source of uranium is Belgian Congo.

In view of this situation you may think it desirable to have some permanent contact maintained between the Administration and the group of physicists working on chain reactions in America. One possible way of achieving this might be for you to entrust with this task a person who has your confidence and who could perhaps serve in an inofficial capacity. His task might comprise the following:

a) to approach Government Departments, keep them informed of the further development, and put forward recommendations for Government action giving particular attention to the problem of securing a supply of uranium ore for the United States;

b) to speed up the experimental work, which is at present being carried on within the limits of the budgets of University laboratories, by providing funds, if such funds be required, through his contacts with private persons who are willing to make contributions for this cause, and perhaps also by obtaining the co-operation of industrial laboratories which have the necessary equipment.

I understand that Germany has actually stopped the sale of uranium from the Czechoslovakian mines which she has taken over. That she should have taken such early action might perhaps be understood on the ground that the son of the German Under-Secretary of State, von Weizsäcker, is attached to the Kaiser-Wilhelm-Institut in Berlin where some of the American work on uranium is now being repeated.

Yours very truly,
Albert Einstein[1]

This letter, written just weeks before Germany's invasion of Poland and the outbreak of World War II, and six full years before the atomic explosion at Hiroshima, conveys the scientific essentials known in the summer of 1939, and worries about uranium supplies and methods of delivering a bomb to an enemy. It also shows a remarkable innocence about the enormous scope and costs the project would ultimately entail, and the massive organizational structures required to bring it to fruition. But pre–World War II science was modest in scope. "Big science" as we know it today had not yet arrived. That would come with the establishment of the Manhattan Project and the realization of the atomic bomb.

The museum was fortunate. The National Archives promised us the loan for our exhibition of the original letter, signed in Einstein's own hand. Visitors would experience the thrill that only comes at seeing an original document, artifact, or machine that forever changed the world.

Another revealing document in the exhibition was Secretary of War Henry Stimson's unbelievably farseeing, three-page, April 25, 1945, report to the new President Truman.

Stimson was a remarkable man. Born in 1867, he had been secretary of war from 1911 to 1913 under President Taft, and had served as secretary of state in President Hoover's administration. With war looming in 1940, President Roosevelt turned to this longtime Republican and asked him to serve once again as secretary of war. Stimson continued to serve under President Truman following Roosevelt's death. His work done, he retired at age seventy-eight, shortly after the war's end. The depth of his perspective after so many decades of public service at the cabinet level were to serve Truman well.

Truman was thrust into office on April 12, 1945. The dying Roosevelt had never told his vice president about the atomic secret. And so twelve days later, Stimson sent Truman this note:

April 24, 1945

Dear Mr. President,

I think it is very important that I should have a talk with you as soon as possible on a highly secret matter.

I mentioned it to you shortly after you took office but have not urged it since on account of the pressure you have been under. It, however, has such a bearing on our present foreign relations and has such an important effect upon all my thinking in this field that I think you ought to know about it without much further delay.

> Faithfully yours,
> Henry L. Stimson
> Secretary of War.[2]

Scribbled underneath in Truman's hand is the message to his appointments secretary, Matt Connelly, "Matt—Put on list tomorrow, Wed. 25,—HST."

Significant in Stimson's note is his emphasis on "our present foreign relations." As the contents of the museum's exhibition became known to veterans and veterans' organizations, they objected to the notion that Truman or his administration might have considered the use of the atomic bomb in any diplomatic context. Yet here, in his very introduction to the atomic bomb, Truman was already alerted to the broader significance of this weapon.

HENRY STIMSON'S CRYSTAL BALL

The following day, Stimson prepared a three-page report marked "Memo discussed with the President, April 25, 1945. Top Secret: Original and 3 carbons made." I will cite it in detail here, because it amazingly foresees all the major atomic issues of the next fifty years—diplomatic, moral, and strategic—on which the veterans' associations soon would begin attacking the exhibition.[3]

1. Within four months we shall in all probability have completed the most terrible weapon ever known in human history, one bomb of which could destroy a whole city.
2. Although we have shared its development with the UK, physically the US is at present in the position of controlling the resources with which to construct and use it and no other nation could reach this position for some years.
3. Nevertheless it is practically certain that we could not remain in this position indefinitely.
   a. Various segments of its discovery and production are widely known among many scientists in many countries, although few scientists are now acquainted with the whole process which we have developed.
   b. Although its construction under present methods required great scientific and industrial effort and raw materials, which are temporarily mainly within the possession and knowledge of US and UK, it is extremely probable that much easier and cheaper methods of production will be discovered by scientists in the future, together with the use of materials of much wider distribution. As a result, it is extremely probable that the future will make it possible to be constructed by smaller

> nations or even groups, or at least by a large nation in a much shorter time.

Stimson clearly recognized the enormous destructive power soon to be in American hands, and also the inevitable danger that other countries and even "groups"—"terrorists" in today's language—could sooner or later build such bombs. All this was relatively straightforward, but he foresaw a great deal more. His report continues:

> 4. As a result, it is indicated that the future may see a time when such a weapon may be constructed in secret and used suddenly and effectively with devastating power by a wilful nation or group against an unsuspecting nation or group of much greater size and material power. With its aid even a very powerful unsuspecting nation might be conquered within a very few days by a very much smaller one, although probably the only nation which could enter into production within the next few years is Russia.

These sentences ring remarkably modern fifty years later, as we try to determine the extent to which Iraq, North Korea, or Pakistan may have developed a nuclear arsenal. They recall that South Africa has acknowledged building, and subsequently dismantling, six atomic bombs—twice the number the United States had built by the end of World War II. And of course, Stimson was right again in assuming that the Soviet Union would be the first to independently build an atomic bomb. However, he saw further still:

> 5. The world in its present state of moral advancement compared with its technical development would be eventually at the mercy of such a weapon. In other words, modern civilization might be completely destroyed.

This sentence is important in defining prevailing attitudes in 1945.

In its earliest proposals to exhibit the *Enola Gay,* the museum had already mentioned the moral dilemmas posed by the atomic bomb. We had also stated that we could only depict that debate, and would refrain from taking a position.

For Secretary of War Henry Stimson, morality and "moral advancement" were serious matters as he tried to help his new president in guiding the world to a safe future:

> 6. To approach any world peace organization of any pattern now likely to be considered, without an appreciation by the leaders of our country of the power of this new weapon, would seem to be unrealistic. No system of control heretofore considered would be adequate to control the menace. Both inside any particular country and between the nations of the world, the control of this weapon will undoubtedly be a matter of the greatest difficulty and would

> involve such thoroughgoing rights of inspection and internal controls as we have never heretofore contemplated.

This perhaps is the most amazing paragraph of all. It took the world fifty years of arms negotiations between the United States and the Soviet Union to finally establish "such thoroughgoing rights of inspection and internal controls as we have never heretofore contemplated." Yet Henry Stimson clearly foresaw their necessity and inevitability. Otherwise, as emphasized in his paragraph 5, "modern civilization might be completely destroyed."

Now Stimson returned to the importance of foreign relations, moral responsibility, and potential disasters to civilization:

> 7. Furthermore, in the light of our present position with reference to this weapon, the question of sharing it with other nations and, if so shared, upon what terms, becomes a primary question of our foreign relations. Also our leadership in the war and in the development of this weapon has placed a certain moral responsibility upon us which we cannot shirk without very serious responsibility for any disaster to civilization which it would further.
> 8. On the other hand, if the problem of the proper use of this weapon can be solved, we would have the opportunity to bring the world into a pattern in which the peace of the world and our civilization can be saved.
> 9. As stated in General Groves' report, steps are under way looking towards the establishment of a select committee of particular qualifications for recommending action to the Executive and legislative branches of our government when secrecy is no longer in full effect. The committee would also recommend the actions to be taken by the War Department prior to that time in anticipation of the postwar problems. All recommendations would of course be first submitted to the President.

How clearly Stimson foresaw the ties between the atomic bomb and the development of the postwar world! And yet, this would become a point veterans' organizations would violently oppose. The veterans would stridently resist the museum's attempts to show the mission of the *Enola Gay* as a watershed, marking the end of World War II while also ushering in the nuclear age.

For the postwar era, the Cold War, the arms race, the veterans would argue along two different lines. One was that this was a separate issue. We should not muddy the waters with such irrelevancies, which would only serve to diminish the glorious victory our men had won and cast doubt on the morality of President Truman's decision. The museum should simply show the atomic bomb bringing about the end of a terrible war. Period! The difficulty with that approach was that it clearly glossed over an issue many visitors would have in mind.

The alternative was that rather than portraying them as weapons of war, atomic bombs should be hailed as the peacekeepers of the world. They had maintained a fifty-year-long standoff without any major wars. The trouble with that assertion is that fifty years is but a brief moment in human history, and the correctness of the peacekeeper notion remains to be tested over future millennia—not half a century.

Instead of editorializing, the museum had chosen simply to state the problem posed by nuclear weapons. Veterans would revile this approach as "politically correct" and "revisionist," and ask us, as Paul Tibbets publicly did, to show the *Enola Gay* with a 1945 frame of mind:

> I suggest to you that few have ever attempted to discuss the missions of August 6th and August 9th, 1945, *in the context of the times. . . .* The *Enola Gay* has become a symbol to different groups for one reason or another. I suggest that she be preserved and given her place *in the context of the times* in which she flew.[4]

Here we were trying to do just that, by presenting that mission in precisely the way that Secretary of War Henry Stimson saw it in 1945, and in just the way he presented the issues to President Truman in his very first briefing on the subject, and yet this was labeled as revisionist, rather than as a portrayal in context.

The tragedy responsible for these unfortunate accusations is that Stimson's eloquent report remained top secret, hidden for twenty-eight years until declassified on May 14, 1973. Its suppression and subsequent neglect by all except a handful of historians permitted a myth to develop among veterans that Truman and in fact all responsible people at the time cared about one thing only, to save lives and bring American boys home.

Had they known of this report, they would have seen that Stimson gave no advice to the president on whether the bomb should or should not be used. He described the bomb's destructive power. Independently of its use against Japan, which he never even mentioned in the document, Stimson enumerated the dangers and moral problems that the very existence of the bomb would pose for the postwar world. Those dangers could be surmounted only if sovereign nations consented to sweeping changes.

The April 25, 1945, report is not Stimson's sole mention of diplomacy and the atomic bomb. His diary entry for Sunday, May 13, 1945, takes up a State Department request for his comments on agreements to be reached with the Soviet Union in view of their expansionist plans in the Far East. Among these is his query, "Is the entry of the Soviet Union into the Pacific war at the earliest possible moment of such vital interest to the United States as to preclude any attempt by the United States to obtain Soviet agreement to certain desirable political objectives in the Far East prior to such entry?" Specific issues listed involve Soviet aspirations for the future of China, Manchuria, Korea, and the Kuril islands. One question raised is,

"Should a Soviet demand, if made, for participation in the military occupation of the Japanese home islands be granted or would such occupation adversely affect our long term policy for the future treatment of Japan?" And Stimson comments, "These questions cut very deep and in my opinion are powerfully connected with our success with S-1." S-1 was the code name for the Manhattan Project.

Two days later, on May 15, his diary entry reads,

> . . . The trouble is that the President has now promised apparently to meet Stalin and Churchill on the first of July and at that time these questions will become burning and it may be necessary to have it out with Russia on her relations to Manchuria and Port Arthur and various other parts of North China, and also the relations of China to us. Over any such tangled wave of problems the S-1 secret would be dominant and yet we will not know until after that time probably, until after that meeting, whether this is a weapon in our hands or not. We think it will be shortly afterwards, but it seems a terrible thing to gamble with such big stakes in diplomacy without having your master card in your hand.[5]

The Potsdam Conference, to which Stimson was referring, eventually was postponed to start July 16, and Truman gave the order that the atomic bomb be tested no later than that day. Despite poor weather, the Trinity test at Alamogordo, New Mexico, took place in the early morning hours of the sixteenth. The plutonium bomb was successfully exploded, and Truman was immediately notified at Potsdam. Ten days later it led to the Allied ultimatum to the Japanese.

Truman, though new in office, appears to have fully shared Stimson's views. On the other hand, General MacArthur had been pressing for the Soviet Union to declare war on Japan and share the burdens shouldered by American troops. Truman spoke about this with Stalin at Potsdam and received the assurances he requested. The museum's exhibition planned to record that

> During the Potsdam Conference, Stalin promised to declare war on Japan by August 15. Truman wrote in his diary on July 17, "Fini Japs when that comes about." But a day later he wrote, "Believe Japs will fold up before Russia comes in. I am sure they will when Manhattan appears over their homeland.[6]

We also planned to display an entire page from Truman's diary for July 25, 1945. It reads,

> We met at 11 A.M. today. That is Stalin, Churchill and the U.S. President. But I had a most important session with Lord Mountbatten and General Marshall before that. We have discovered the most terrible bomb in the history of the world. It may be the fire destruction prophesied in the Euphrates Valley Era, after Noah and his fabulous Ark.
>
> Anyway we think we have found the way to cause a destruction of the atom. An experiment in the New Mexico desert was startling—to put it

> mildly. Thirteen pounds of the explosive caused the complete disintegration of a steel tower 60 feet high, created a crater 6 feet deep and 1200 feet in diameter, knocked over a steel tower 1/2 mile away and knocked men down 10,000 yards away. The explosion was visible for more than 200 miles and audible for 40 miles and more.
>
> This weapon is to be used against Japan between now and August 10th. I have told the Sec. Of War, Mr. Stimson, to use it so that military objectives and soldiers and sailors are the target and not women and children. Even if the Japanese are savages, ruthless, merciless and fanatic, we as the leader of the world for the common welfare cannot drop this terrible bomb on the old Capital or the new. He and I are in accord. The target will be a purely military one and we will issue a warning statement asking the Japs to surrender and save lives. I'm sure they will not do that, but we will have given them the chance. It is certainly a good thing for the world that Hitler's crowd or Stalin's did not discover this atomic bomb. It seems to be the most terrible thing ever discovered, but it can be made the most useful.[7]

The exhibition script would continue,

> On July 26, 1945, the three largest Allied powers at war in the Pacific, the United States, Britain and China, issued the Potsdam Declaration, which demanded that the Japanese Empire surrender immediately or face "prompt and utter destruction." ... On July 28, Prime Minister Suzuki announced that his government would ignore ("mokusatsu") the declaration. As a result the United States used the atomic bomb.[8]

A most revealing document is a letter Truman penned to powerful Senator Richard Russell on August 9, just hours after the bomb was dropped on Nagasaki. It portrays Truman as a far more caring man than conveyed in the statement often quoted by veterans' organizations, claiming Truman had "never had lost a night's sleep" over the decision:

> For myself I certainly regret the necessity of wiping out whole populations because of the "pigheadedness" of the leaders of a nation, and, for your information, I am not going to do it unless it is absolutely necessary. It is my opinion that after the Russians enter into the war the Japanese will very shortly fold up. My objective is to save as many American lives as possible but I also have a human feeling for the women and children of Japan.[9]

Our exhibition further recorded,

> On August 10, while discussing the Japanese surrender offer, President Truman ordered that no more atomic bombs be dropped until further notice. According to the diary of Commerce Secretary Henry Wallace, Truman told the Cabinet that "the thought of wiping out another 100,000 people was too horrible. He didn't like killing, as he said, 'all those kids.' " Although he had written in his Potsdam diary in July that the target for the first bomb would be purely military, Truman clearly understood after

Hiroshima that whatever the target, the atomic bomb could destroy whole cities.

Because of Truman's order, General Groves held up the shipment to the Pacific of the plutonium 239 core for another *Fat Man* bomb, which was to be available for a mission around August 20. Further plutonium cores could have been shipped to the Pacific about every three to four weeks thereafter. But no uranium 235 for a *Little Boy* bomb would have been available for some months.[10]

The documents we exhibited and the accompanying text could not have failed to convey the thoroughness with which questions about the atomic bomb were considered by Truman and his advisors in the context of a fast-moving war in which lives were daily lost. Truman may not have been shown every opinion offered. He was not, for example, shown a petition by a number of Manhattan Project scientists, dated July 17, 1945, the day after the successful Trinity test. That petition, signed among others by Leo Szilard, who six years earlier had so insistently urged Albert Einstein to write President Roosevelt, urged Truman to give Japan every incentive and opportunity to surrender, before deciding, and then only as a last resort, to use an atomic bomb. The exhibition displayed this petition in full—again a document reflecting the context of its times and the opinions of some of the leaders of the Manhattan Project:

A PETITION TO THE PRESIDENT OF THE UNITED STATES

Discoveries of which the people of the United States are not aware may affect the welfare of the nation in the near future. The liberation of atomic power which has been achieved places atomic bombs in the hands of the Army. It places in your hands, as Commander-in-Chief, the fateful decision whether or not to sanction the use of such bombs in the present phase of the war against Japan.

We, the undersigned scientists, have been working in the field of atomic power. Until recently we have had to fear that the United States might be attacked by atomic bombs during this war and that her only defense might lie in a counterattack by the same means. Today, with the defeat of Germany, this danger is averted and we feel impelled to say what follows:

The war has to be brought speedily to a successful conclusion and attacks by atomic bombs may very well be an effective method of warfare. We feel, however, that such attacks on Japan could not be justified, at least not unless the terms which will be imposed after the war on Japan, were made public in detail and Japan were given an opportunity to surrender.

If such public announcement gave assurance to the Japanese that they could look forward to a life devoted to peaceful pursuits in their homeland and if Japan still refused to surrender our nation might then, in certain circumstances, find itself forced to resort to the use of atomic bombs.

> Such a step, however ought not to be made at any time without seriously considering the moral responsibilities which are involved.
>
> The development of atomic power will provide the nations with new means of destruction. The atomic bombs at our disposal represent only the first step in this direction, and there is almost no limit to the destructive power which will become available in the course of their future development. Thus a nation which sets the precedent of using these newly liberated forces of nature for purposes of destruction may have to bear the responsibility of opening the door to an era of devastation on an unimaginable scale.
>
> If after this war a situation is allowed to develop in the world which permits rival powers to be in uncontrolled possession of these new means of destruction, the cities of the United States as well as the cities of other nations will be in continuous danger of sudden annihilation. All the resources of the United States, moral and material, may have to be mobilized to prevent the advent of such a world situation. Its prevention is at present the solemn responsibility of the United States—singled out by virtue of her lead in the field of atomic power.
>
> The added material strength which this lead gives to the United States brings with it the obligation of restraint and if we were to violate this obligation our moral position would be weakened in the eyes of the world and in our own eyes. It would then be more difficult for us to live up to our responsibility of bringing unloosened forces of destruction under control.
>
> In view of the foregoing, we, the undersigned, respectfully petition: first, that you exercise your power as Commander-in-Chief, to rule that the United States shall not resort to the use of atomic bombs in this war unless the terms which will be imposed upon Japan have been made public in detail and Japan knowing these terms has refused to surrender; second, that in such an event the question whether or not to use atomic bombs be decided by you in the light of the considerations presented in this petition as well as all the other moral responsibilities which are involved.[11]

Historians have often argued that President Truman should have been shown this letter. But basically it did not add much to what Stimson had already reported to Truman on April 25. The only difference was to explicitly urge the President not to be the first to use an atomic bomb except as a last resort. The scientists' request to make the terms of surrender public would have impracticably extended the war. It took months after peace was declared to come up with a new constitution for Japan. And that is what such a detailed plan would have entailed.

DIPLOMACY AND MORALITY

So secret was the atomic bomb in July 1945 that only a very few people knew of its existence. All the more remarkable, therefore, that those who were informed, had developed it, or would be responsible for deciding on its use so clearly recognized their moral dilemma and the bomb's significance for postwar diplomacy.

An appreciable number of leading scientists, including the bomb's first and most persistent proponent, Leo Szilard, understood both these implications—moral responsibility for first use and the bomb as a diplomatic weapon. Stimson had pondered these questions for months. Truman's diary shows his concern about killing women and children. His order to stop further atomic strikes immediately after Nagasaki, when he realized that this bomb would do more than disintegrate or knock over steel towers—as his diary had recorded—confirms the moral compunctions he felt. And Truman's diary shows that he clearly recognized the bomb as a diplomatic tool, "It is certainly a good thing for the world that Hitler's crowd or Stalin's did not discover this atomic bomb."

We already saw Chief of Staff William D. Leahy calling the bomb "This barbarous weapon," adding, "I was not taught to make war in that fashion ... by destroying women and children."[12] And Eisenhower recalled telling Stimson at Potsdam of his opposition to the bomb—though no independent record of the conversation exists—"I disliked seeing the United States take the lead in introducing into war something as horrible and destructive as this new weapon."[13] Letters and memoranda such as these, flashbacks through newsreel film footage, videotaped recollections of those who had been involved in the bombings, in the air and on the ground, not to mention the *Enola Gay* herself, the atomic bombs the 509th had delivered, and other memorabilia of the times, would have made this a great exhibition.

And yet, the museum and our script were about to come under the most intense criticism from veterans' organizations and the media for mentioning the bomb in a diplomatic context and even more, for explicitly showing the scale of destruction and suffering or referring to the existence of a long-standing debate on the morality of using the bomb—even though we, again explicitly, undertook not to take a stance in the debate.

RACISM

Equally vehement were the veterans' organizations' denials that racism played a major role in the war in the Pacific. While there is no evidence that the decision to use the atomic bomb was in any way racially motivated. John Dower's book *War Without Mercy* convincingly demonstrates that the Pacific war on the whole *was* mired in racial fears.[14] Where else did one find the atrocities the Japanese inflicted on their prisoners of war, or the disfigurement of American corpses by the Japanese. Where else would one find a picture in *Life* magazine of a young woman with the skull of a dead Japanese soldier on her desk—a memento sent home by her boyfriend in the Pacific.[15] At the meeting of the exhibition's advisory committee, Edwin Bearss, who had fought in the Pacific, also told us of his gunnery sergeant in the Pacific, who had a watch chain made of gold teeth he had pulled out of the mouths of Japanese corpses. The horrors inflicted by the two sides on each other in the Pacific war far outweighed those that Americans encountered fighting the Germans in Europe. Yet our mention of these attitudes so outraged the press and the veterans that we eventually withdrew that section from the script. The hatreds had not played an essential role in the use of the bomb, and there was no sense in needlessly detracting from matters more relevant to the mission of the *Enola Gay*.

19

The AFA Lobbies for Its Own Version of History

I was pleased with the advisory committee's response. Air Force Chief of Staff Merrill A. "Tony" McPeak, a member of the National Air and Space Museum Advisory Board, had worried about our producing a script that would be unacceptable to the military or to veterans. Dick Hallion's position as Air Force historian and head of the Center for Air Force History and his complimentary note at the end of the day allowed me to hope we had overcome a main hurdle.

My Deputy, Gwen Crider, was on vacation that week and so that Friday I left her a note to find on her return. The two first entries were

> 1. We had a good review of *Crossroads* on Monday. In particular, Dick Hallion, the Air Force Historian, was satisfied. He said he'd be glad to convey that to Gen. McPeak. I then called McPeak to let him know, and he sounded very pleased.

John Correll had called and asked for an interview. My note to Gwen continued:

> 2. On Tuesday, I gave John Correll, Editor of *Air Force* magazine an interview. . . . It's hard to know how well that went. Correll kept asking questions that sounded as though he would like us to repudiate what we'd done in the past, on the *World War I* gallery,

> in the early drafts for *Crossroads*, and so on. It was a little puzzling and we'll have to see how he writes it up.[1]

We did not have long to wait.

My January 31, 1994, letter of transmittal to Air Force Association executive director Monroe Hatch had specifically asked for comments and requested that he not circulate the script. Unfortunately, rather than letting me have their comments, Hatch turned the material over to his staff and set them in motion. By mid-March, John Correll, editor-in-chief of the AFA's *Air Force* magazine, had authored for his magazine's April issue a critique, which attacked all the script's shortcomings, sometimes by quoting passages out of context. By then also, Jack Giese, the AFA chief of media relations, had started a media blitz. And Stephen P. Aubin, director of communications, was initiating a campaign against the Smithsonian in Congress. Also involved in this campaign was Kenneth A. Goss, the AFA's director of national defense issues. And of course, Hatch himself directed these efforts, becoming involved openly only where overt representation by the AFA was needed.

While the spring and summer of 1994 were to make evident that some form of massive offense was in the works, none of us at the time knew how broad an effort had actually been initiated. Its scale became apparent only after the Air Force Association released some of the relevant documentation in May 1995. I will turn to that now.

JOHN CORRELL

On March 15, 1994, the Air Force Association issued a "Special Report," consisting of two parts, both authored by Correll.[2] The first, "War Stories at Air & Space," bore the heading "At the Smithsonian, history grapples with cultural angst." The second, "The Decision That Launched the *Enola Gay*," was headlined "In April 1945, the new president learned the most closely-held secret of the war."

This second article was a twenty-page typewritten summary of the last few months of World War II, five pages of which consisted of a chronology and references. Much of this paper was fairly standard, and most of the factual information agreed with the script we had just released. Correll cited the same figures for casualties expected in an invasion of Japan that we had quoted. For an invasion of the southern island Kyushu, we had cited the original figures that General Marshall, Admiral King, and Chief of Staff Admiral Leahy had given President Truman at their meeting on June 18, 1945. Marshall's and King's figures ranged from 30,000 to 50,000 for the first thirty days of operations—which, for all one knew at the time, might have sufficed to induce surrender. Leahy's figures, based on the entire operation on Okinawa, while not clearly stated, could be interpreted as anticipating casualties as high as 268,000 for the entire operation on Kyushu, with perhaps 50,000 killed. Correll had not gone to the primary sources, but had taken figures given in McCullough's biography of Truman and come up with similar numbers.[3]

I mention this for two reasons. First, the memorandum Correll had written for his files on November 11 had accused us of planning not to give any casualty figures for an invasion, whereas we had done just that and come up with figures he now verified. And second, the American Legion would later attack the museum for having used casualty figures that were too low—though they were the same as Correll's.

After the war Truman often spoke of half a million to a million American casualties, and sometimes of half a million to one million American dead. Those were the figures that the veterans also cited in letters they wrote to the museum—claiming that the United States would have suffered one to two million deaths had our troops been called on to invade the Japanese home islands.

Another interesting feature of Correll's article is his paragraph on a subject that would also become controversial later:

> In his memoirs, Truman said a consensus had been reached in July, during the Big Three meeting at Potsdam, by Secretary of State James Byrnes, Stimson, Leahy, Marshall and Arnold that the bomb should be used. In fact, the advice was not as clear-cut as Truman depicted it in his memoirs. Although Arnold supported the decision, he declared his view at Potsdam that use of the bomb was not a military necessity. Leahy had reservations about the decision also. And at a meeting with Truman July 20 during the Potsdam conference, Gen. Dwight Eisenhower, commander of allied forces in Europe, advised against using the atomic bomb (although he said later his reaction was personal and not based on any analysis of the circumstances).[4]

The Eisenhower statement, in fact had been made not to Truman, but to Stimson, but it is remarkable that Correll was willing to say that some of Truman's advisors had been against dropping the bomb. The museum's first script, which also cited Eisenhower's and Leahy's statements, later came under attack from other veterans' organizations and from military service historians, on the grounds that Leahy had been against all strategic bombing and that there was no record of Eisenhower's statement until his 1948 memoirs.

Correll's other article in this special report, "War Stories at Air & Space," began with a critique of the museum's script, but then digressed to criticize a whole range of other activities of the museum, ending up with criticism of the direction the Smithsonian as a whole was taking.

He noted that the Committee for the Restoration and Display of the *Enola Gay* had "collected 8,000 signatures on a petition asking the Smithsonian to either display the aircraft properly or turn it over to a museum that will do so." Then, referring to the planning document issued the previous July[5] and now superseded by the more detailed script, Correll described his vision of the planned exhibition:

> The restored aircraft will be there all right, the front fifty-six feet of it, anyway. The rest of the gallery space is allotted to a program about the atomic bomb. The presentation is designed for shock effect. The museum's exhibition plan notes that parents might find some parts unsuitable for viewing by their children, and the script warns that "parental discretion is advised."
>
> For what the plan calls the "emotional center" of the exhibit, the curators are collecting burnt watches, broken wall clocks, and photos of victims—which will be enlarged to life size—as well as melted and broken religious objects. One display will be a schoolgirl's lunch box with remains of peas and rice reduced to carbon. To ensure that nobody misses the point, "where possible, photos of the person who owned or wore these artifacts would be used to show that real people stood behind the artifacts." Survivors of Hiroshima and Nagasaki will recall the horror in their own words.
>
> The Air and Space Museum says it takes no position on the "difficult moral and political questions" involved. For the past two years, however, museum officials have been under fire from veterans' groups who charge that the exhibition plan is politically biased.

The museum staff and I soon came to agree with Correll that it was more important to have children come to the exhibit than to retain some of the more explicit photographs showing the horror of the atomic bombings. We removed both the graphic pictures and the parental discretion sign.

Correll's critique was not all negative, although somewhat puzzling, since what he called the "latest script" was actually our first full script. The comparisons he drew must have referred to his own reading between the lines of that brief planning document of July 1993:

> The latest script, written in January, shows major concessions to balance. It acknowledges Japan's "naked aggression and extreme brutality" that began in the 1930s. It gives greater recognition to U.S. casualties. Despite some hedging, it says the atomic bomb "played a crucial role in ending the Pacific war quickly." Further revisions of the script are expected.

Quite sensibly, Correll continues,

> The ultimate effect of the exhibition will depend, of course, on how the words are blended with the artifacts and audiovisual elements. And despite the balancing material added, the curators still make some curious calls.
>
> "For most Americans," the script says, "it was a war of vengeance. For most Japanese, it was a war to defend their unique culture against Western imperialism."[6]

This last sentence ultimately became the slogan cited over and over again by newspaper columnists and veterans' organizations, who quoted it for more than a year after we had already removed it from the script, recognizing that it was clumsy and easily misunderstood. But it makes sense to show the full litany of Japanese atrocities within which this sentence had always been embedded. Taking it out of that context gave it an entirely different shading. The exhibition label that included the cited words read,

> A FIGHT TO THE FINISH
>
> In 1931 the Japanese Army occupied Manchuria; six years later it invaded the rest of China. From 1937 to 1945, the Japanese Empire would be constantly at war.
>
> Japanese expansionism was marked by naked aggression and extreme brutality. The slaughter of tens of thousands of Chinese in Nanking in 1937 shocked the world. Atrocities by Japanese troops included brutal mistreatment of civilians, forced laborers and prisoners of war, and biological experiments on human victims.
>
> In December 1941, Japan attacked U.S. bases at Pearl Harbor, Hawaii, and launched other surprise assaults against Allied territories in the Pacific. Thus began a wider conflict marked by extreme bitterness. For most Americans this war was fundamentally different than the one waged against Germany and Italy—it was a war of vengeance. For most Japanese, it was a war to defend their unique culture against Western imperialism. As the war approached its end in 1945, it appeared to both sides that it was a fight to the finish.[7]

This label was intended as an introduction to a more detailed presentation that followed. The sentence about Japan's unique culture tried to explain why Japan in 1945 refused to surrender, in the face of obvious defeat, unless guarantees were given that the emperor could be retained. For Japan, the emperor, a demigod figure, represented seven centuries of autonomy. Their troops' suicidal kamikaze tactics could only be understood as the response of a culture that feared loss of its heritage more than the eradication of its youth.

OTHER CRITICISM

At this point, Correll stopped discussing the script and turned to a shotgun barrage of quotations from anyone unhappy with the museum.

He cited a three-year-old letter to *Air & Space* magazine by Ben Nicks, who had complained about a much earlier plan to discuss the *Enola Gay*'s mission in terms of the "controversial issue of strategic bombing." Correll quoted Nicks's complaint that this was "Simply a transparent excuse to moralize about nuclear warfare. A museum's role is to present history as it was, not as its curators would like it to be."[8] All of us at the museum fully agreed with that last sentence. We also were proud of the sixteen-month-long series of talks, panel discussions, and symposia we had held, in which so many leading figures of strategic bombing had participated, including General Curtis LeMay, in whose Twentieth Air Force Ben Nicks had served as a B-29 command pilot.

Correll also cited his own boss, Monroe Hatch, who had criticized the planning document for treating "Japan and the United States as if their participation in the war were morally equivalent. If anything, incredibly, it gives the benefit of opinion to Japan, which was the aggressor. " Given the litany of Japanese atrocities cited above, which the script had listed right in the opening statement on the war in the Pacific, I don't see how Hatch could maintain this viewpoint after reading the script.

Correll further faulted the museum's World War I exhibition that had been critically acclaimed in the nation's press three years earlier when it opened in November 1991. The *Washington Post*'s Hank Burchard had headlined his review "Plane Truths During WWI" and praised the museum for its "quantum leap forward into historicity." The *Chicago Tribune*'s Michael Kilian, emphasizing this same dedication to authenticity, had headlined his piece "Grounded in reality—Exhibition finds the mythic WWI ace was a flight of fancy."[9] Correll, however, was offended by such straight talk, with its portrayal of the young airmen whose life expectancy upon reaching the front was only three weeks.

He quoted a three-year-old article by *Washington Post* correspondent Elizabeth Kastor:

> Dr. Harwit said he would like the museum to have an exhibit "as a counterpoint to the World War II gallery we now have, which portrays the heroism of the airmen, but neglects to mention in any real sense the misery of war. . . . I think we just can't afford to make war a heroic event where people could prove their manliness and then come home to woo the fair damsel.

With a penchant for quoting only parts of a statement, Correll had left out the full *Washington Post* quotation in which this last sentence was embedded:

> We really are at a point now where, inadvertently or deliberately, it would be possible if not to destroy the population of the Earth, at least to drive us back to very primitive conditions. I think we just can't afford to make war a heroic event where people could prove their manliness and then come home to woo the fair damsel. If you survived such a war now as a warrior, your fair damsel would probably have been pulverized by the time you came back.[10]

I stand fully behind that complete statement. If a national museum portrays warfare, it needs to show youngsters more than gleaming machines described as able to fly higher, faster, and farther, with more maneuverability and "fire power capable of inflicting heavy punishment." People die in war! Children should realize that.

Correll now moved on to Arthur H. Sanfelici, editor of *Aviation* magazine, who had criticized the museum a year earlier for not sticking to the business of restoring airplanes and displaying them without comment: "a new order is perverting the museum's original purpose from restoring and displaying aviation and space artifacts to presenting gratuitous social commentary on the uses to which they have been put."[11] Sanfelici's "gratuitous social commentary" was just the straight talk that had pleased the *Washington Post*'s Hank Burchard so much.

I want to emphasize that these were not dominant themes anywhere in the museum. All of us on the staff were fans of aviation and spaceflight. Many of us were pilots. I certainly enjoyed piloting a plane, though I was nowhere near as proficient as the former Air Force and Navy aviators on the staff who had made a career of combat flying. My policy for the museum's new galleries was simply a matter of stating the obvious, rather than deliberately suppressing it.

Next, Correll brought up the museum's charter and again quoted Title 20, §80a of the United States Code: "The valor and sacrificial service of the men and women of the Armed Forces shall be portrayed as an inspiration to the present and future generations of America," without mentioning that this referred to an Armed Forces Museum that had been neither funded nor built in the thirty-three years since the enactment of the legislation. Nor did he mention that the wording of that legislation included a paragraph specifically stating that its intent was not to affect the National Air and Space Museum's activities in any way.

And finally, not content with all this, Correll took on Smithsonian secretary Adams for the direction he had set for the Institution as a whole and criticized a variety of exhibitions that had been mounted since he took office—at the Museum of American History, at the Museum of American Art, and at the Museum of Natural History. The Smithsonian five-year plan in Correll's view was "laden with politically correct goals and lumpy language."

The two articles constituting Correll's special report were printed almost verbatim, though minus their lists of references, in the April 1994 issue of the Air Force Association's official magazine, *Air Force*.[4,6]

My letter to Monroe Hatch had specified, "Please let me have any comments you have. They are important to us. I would appreciate it, however, if you did not circulate the material at this time, since it is not yet in suitable form." No doubt, Hatch's colleague Correll had let us have his comments. However, by the third week of March, even before the April issue of *Air Force* magazine had been printed, the AFA had already gone into high gear, copying and circulating the exhibition script to other veterans' organizations, to the media, and to Congress. Steve Aubin, the AFA Director of Communications, recalls sending out massive packages of background information,

> ... I can tell you many boxes went out of here, four or five inch high boxes with documents in them—and we said, "Come to your own conclusions." Obviously from our point of view, if you want to talk to Hallion, he's an Air Force historian....[12]

Boxes that thick might contain more than a thousand pages of documents. Asked how many reporters plowed through all this, Giese adds,

> I guarantee they went through it for specific areas. If there was a question about "the war of vengeance" we'd tell them where to go. We had analyses and brief outlines. We made it as simple for them as possible to get the information out that we wanted. They had everything at their disposal.

Pressed on whether the reporters should have plowed through the script to make up their own minds, Giese responds,

> That's their job, but they don't have the time to do that.

In short, Aubin and Giese were swamping the media and congressional aides with more material than any of them would have time to read, and then telling them "where to go" in these thick packages to "get the information" the AFA wanted to have emphasized. When the *Wall Street Journal*'s editorial page writer Dorothy Rabinowitz, asked for information, Giese says,

> Anything she needed was provided. I called it "the package from hell," or our "kitchen sink package."[13]

THE CONGRESS

By March 24, 1994, the Senate Committee on Rules and Administration, had sent me a letter signed by its chairman, Democrat Wendell Ford, and Republicans Ted Stevens, Robert Dole, Thad Cochran, Jesse Helms, and Mitch McConnell:

> We are concerned about information we received recently regarding a proposal for an exhibit entitled "Hiroshima and Nagasaki: A Fiftieth Anniversary Exhibit at the National Air and Space Museum."
>
> Information provided to us indicates that the exhibit may focus on very specific issues such as the decision to drop the atomic bomb and avoid other facts that place that decision and prior events in the context of a war that lasted for over three years and cost the lives of over 290,000 American servicemen and women. We are very concerned that any analysis of the atomic bombings does not inadvertently lead to a revised view of the events that led up to the difficult decisions that were made in a time of war. Out of respect for the deceased American servicemen and women as well as historical integrity, we believe that the bombing of Hiroshima and Nagasaki should be presented in the context of WW II, and must be careful to reference the many factors that contributed to the United States' decision to drop the atomic bombs.

> Thank you for the opportunity to share our concerns regarding this exhibit. We are hopeful that the Museum will be sensitive to the issues we have raised.[14]

And on March 30, Republican senator Nancy Kassebaum wrote to Bob Adams,

> It has come to my attention that a number of B-29 veterans and Air Force organizations are unhappy with the Smithsonian Institution Air and Space Museum's planned 1995 exhibition of the *Enola Gay*. . . .
>
> In order to resolve this situation, I suggest the famed B-29 be displayed with understanding and pride in another museum. Any one of three Kansas Museums—the Kansas Aviation Museum at McConnell Air Force Base in Wichita, the Combat Air Museum at Forbes Field in Topeka, and the Liberal Air Museum in Liberal—would be ideal for showcasing the *Enola Gay*. . . .[15]

These letters received carefully considered answers. The Committee on Rules and Administration has oversight of the Smithsonian, and the concern voiced by six of its most senior members, including the chairman and the ranking Republican, was a serious matter. My April 4 answer to the committee members read,

> Thank you for your letter of March 24, 1994, concerning the National Air and Space Museum's upcoming exhibition in conjunction with the fiftieth anniversary of the end of World War II. There is, unfortunately, significant misinformation and unfounded rumor circulating regarding the planned exhibition, whose working title is "The Crossroads: The End of World War II, the Atomic Bomb, and the Origins of the Cold War."
>
> A detailed script for the exhibition, which is to open in May 1995, has been carefully scrutinized for accuracy and balance by a committee consisting of some of the nation's leading scholars, including Dr. Richard Hallion, Chief of the U.S.A.F. Center for Air Force History; Edwin Bearss, Chief Historian of the National Park Service, a decorated veteran of the war in the Pacific and the organizer of the Pearl Harbor commemoration in 1991; Prof. Akira Iriye of the Department of History at Harvard University; the Pulitzer Prize–winning author on the subject, Richard Rhodes; and several other distinguished experts.
>
> "Crossroads" is only one of several exhibitions the Museum has mounted to focus on World War II as we commemorate the fiftieth anniversary of that conflict. It is a larger display than the others, reflecting both the innate size of the airplane, *Enola Gay* and the many requests from veterans the Museum has received in recent years to publicly display it. The aircraft, which is too massive to fit into the Museum on the Mall, has been undergoing restoration and conservation for nearly a decade, and the front 60 feet of the forward fuselage, including the cockpit and the evocative name of the aircraft, will be displayed as a central feature of the exhibition. Knowing what the *Enola Gay* represents to many World War II veterans, we are especially pleased that they shortly will have the opportunity to see

the artifact, together with memorabilia of her crew and other members of the 509th Composite Group, prominently featured in the exhibition.

While the exhibition will include a great deal of information on both the decision to use atomic weapons and the results of that use, its goal is to provide a thorough and objective presentation of an event which was a turning point in the history of the 20th Century. It places that event in a rich historical context, from a moving account of the bitter fighting in the Pacific, to the presentation of estimated casualty figures, both Allied and Japanese, that might have resulted from an invasion of Japan, to a factual description of the immediate destruction and long-term results of the use of atomic weapons at Hiroshima and Nagasaki. While I do not believe that the *use* of the atomic bomb is something our nation would want to celebrate, neither is it anything for which we should apologize.

To address the specific points outlined in your letter, let me assure you that the exhibition script treats our servicemen and women as it should: as skilled, brave, loyal, and dedicated members of the United States' Armed Forces who honorably served their country. Additionally, as touched upon above, the discussion of the use of atomic weapons is most certainly presented in the context of the entire war and, most importantly, makes no judgment as to the morality of the decision. If we are successful in our intent, our visitors will not have any preselected point of view imposed on them, but rather, will be provided with objective information on the basis of which they may attain a deeper understanding and formulate their own opinions. This is the goal of all our exhibitions.

The end of World War II was a pivotal time in our nation's, and indeed the world's, history. The beginning of the atomic age and the advent of the Cold War changed forever the way nations conducted both domestic policy and international relations. Whatever one's opinion about Hiroshima and Nagasaki, one factor cannot be denied—it was the line of demarcation between the end of one era and the beginning of another. . . .

I would welcome the opportunity, along with the exhibition curators, to brief the members of the Committee as a group or individually. . . .[16]

THE NATIONAL MEDIA

The media also were being drawn in. The conservative *Washington Times* responded first, on March 28, 1994, with an article headlined "Rewriting History," which quoted John Correll describing our planned exhibition as "politically biased" and Secretary Adams responsible for a "trend of politically correct curating." It also resurrected Hatch's two sentences that gave his views on the exhibition's earlier planning document: "treats Japan and the United States as if their participation in the war was morally equivalent. If anything, incredibly it gives the benefit of opinion to Japan, which was the aggressor." This pair of sentences, too, were to hound us for the next year, along with the "war of vengeance/unique culture" quotation, which also appeared in this same short article.[17]

The *Washington Times* gave me an opportunity to respond a few days later, where I was correctly quoted as saying "the exhibition describes the naked brutality of Japanese forces in concrete terms, calling attention to the rape of Nanking, the treatment of POWs, the use of Chinese and Koreans as slave laborers, and the conduct of biological and chemical experiments on human victims."[18]

Soon, however, the assault broadened to include the other media. On April 7, and again on April 13, John Correll appeared on the Jeff Kamen radio talk show, where listeners calling in were outraged by the same set of quotations: "war of vengeance/unique culture," and "morally equivalent/benefit of opinion to Japan."

In the May issue of *Air Force* magazine, I was given limited space to answer at least some of the points Correll had raised in his two feature articles. I took advantage of that, and wrote,

> At the heart of John Correll's impassioned critique of the National Air and Space Museum, and its parent institution, the Smithsonian, lie two stark and simple questions. The first of these is more readily answered: Is the function of the Museum to showcase the technology of aviation and spaceflight purely in terms of man's urge to go higher, further, faster? Or should we also show how America evolved in the twentieth century in response to opportunities these new technologies offered?
>
> The law which established the Museum is clear on this point. It directs that "The National Air and Space Museum shall memorialize the national development of aviation and space flight; collect, preserve, and display aeronautical and space flight equipment of historical interest and significance;" and "provide educational material for the historical study of aviation and space flight."
>
> Exhibiting airplanes to highlight their historical significance or mounting a display dedicated to a historical study of aviation would seem consonant with that directive. Now comes the more difficult question: If such a historic study is to be undertaken, should it be celebratory? Or if not actually celebratory, should it remain silent on matters that are undeniably central?
>
> Here Mr. Correll seems to opt for silence, at least in the context of the Museum's planned exhibition called "The Crossroads: The End of World War II, the Atomic Bomb, and the Origins of the Cold War." Early in his article he acknowledges that "The *Enola Gay*'s task was a grim one, hardly suitable for glamorization." But he also admonishes the Museum to display the aircraft in a patriotic manner that will instill pride in the viewer.
>
> Patriotism is one thing: Given the ferocity of the fighting and the high casualty rates in the last few months of the War in the Pacific, and given that the country had gone to enormous expense to perfect an atomic bomb, wielding enormous destruction, President Truman's decision to drop the bomb was met with widespread relief in August 1945. General Curtis LeMay and others who were broadly knowledgeable, were of the

opinion that the atomic bombing of Hiroshima and Nagasaki was unnecessary to end the war, but the fact is that it did take place.

The facts also are clear on the enormous destruction and human suffering that resulted.

Does this mean that the United States should apologize for its use of the atomic bomb to end World War II? Of course not! Should we show compassion for those who perished on the ground? As human beings, I believe we must. Should we take pride in the bombings? That's a tougher question. Pride in having found a way to end a terribly costly war? Yes. Pride in the extent of destruction? I can't see how.

Of course, every serviceman was jubilant and relieved as the war quickly ended in response to the two atomic bombs and the concurrent declaration of war by the Soviets. Even the tenacious Japanese had to admit defeat in the face of such adversity.

The exhibition that the National Air and Space Museum is planning for 1995 and the fiftieth anniversary of those events will seek to honor our veterans, who were willing to sacrifice their lives and often did. (The Museum asks veterans who have pictures or memorabilia they would be willing to loan us for this exhibition to contact us.) It will show the circumstances that led to the development and ultimate use of the atomic bomb. It will show the immediate consequences of the bombings, the rapid increase of nuclear arsenals over the forty-year-long Cold War, and the present-day dismantling of nuclear weaponry.

Is this an exhibition that is consistent with the legislation which established the museum? I believe it is, since its aim is to look at the historical significance of the *Enola Gay*. This is not solely my own opinion. Plans for the exhibition have been carefully scrutinized by a committee that includes some of the nation's leading military historians and experts on the War in the Pacific, among them (Air Force Historian) Dr. Richard Hallion and Edwin Bearss, Chief Historian for the National Park Service, a wounded veteran of the Pacific War and organizer of the fiftieth anniversary commemoration of Pearl Harbor in December 1991.

In closing, let me say that the museum invites and welcomes the comments of readers of *Air Force* magazine. Though my vision for the museum differs from John Correll's, his concerns are important to me. I hope the above response helps *Air Force* readers better understand our intentions.

Appended was a reply from editor-in-chief John Correll,

I asked Dr. Richard Hallion if the Center for Air Force History was indeed satisfied with the exhibition plan. He said, "The exhibit, as currently structured, is not one we would have done. We feel that though the museum has made considerable progress over its original concepts, it still needs to show that the central issue behind dropping the bomb was shortening the war and possibly saving upwards of 500,000 Allied troops."[19]

I thought Correll must surely be misquoting Hallion, who had written such a positive critique of our script in February. I also could not believe that Hallion should have claimed half a million lives could have been saved by the bombing. No evidence for that was ever cited by any World War II military leader, nor by professional historians since.

THE AFA LAUNCHES ITS ATTACK

I made another attempt to patch up relations with Monroe Hatch. On April 7, 1994, I wrote,

> I am glad we had an opportunity to talk on the telephone last night.
>
> I fully understand your feeling that the Museum's planned exhibition does not reflect your personal views, and that we will most likely continue to differ. When the Museum undertook this project, we knew it would be difficult and perhaps controversial even among our strongest friends and supporters. But we felt that if we took care to remain objective and presented the history as honestly as possible, we would be providing a service to the public. We continue along that road and hope that we will have your support, if not your endorsement.
>
> As I said on the phone, I think both your Association and our Museum are interested in basically the same things, though we may appear to be far apart. We all wish to honor the achievements of aviation, the men and women who took part in those developments, and the historic changes that flight through the atmosphere brought about in this century. Some of the stories are simple, others more complex. We try to bring them to the public as best we can.
>
> I do not believe that either the Air Force Association or the National Air and Space Museum can benefit from airing our differences through the columns of the *Washington Times*, or any other newspaper. Because of that I was somewhat puzzled by today's report that you had delivered a copy of our script to the *Times* with a challenge for them to see for themselves. I would like to think it went to them before you and I had a chance to talk on the phone. I don't particularly mind having the *Times* see the script, since it has been distributed quite widely by now, but it wasn't clear why you should wish to pursue the matter that far. More than likely it will reflect badly on the Museum, the Air Force Association and the U.S. Air Force which the Association represents.
>
> Let me propose that we agree to disagree, but do so with dignity. We at the Museum would certainly be pleased with that.[20]

In the meantime, the AFA team had swarmed all over: During the next nine months, John Correll, Jack Giese, and Stephen P. Aubin by their own count gave twenty-eight radio interviews and, with Monroe Hatch himself putting in an appearance for the international CNN coverage, took part in thirty further television interviews. This included one appearance by Richard Hallion on CNN America. Print coverage for the remainder of the year exceeded 330 articles. Simultaneously, an increasing number of letters was showering Congress.[21]

BEHIND THE SCENES AT THE AFA

On April 4, 1994, Stephen Aubin had written to Ed Bolen, legislative director on Senator Nancy Kassebaum's staff. He had been in previous contact with Bolen on the letter Senator Kassebaum had written to Secretary Adams on March 30; he had also provided Bolen a copy of the exhibition script. In return, Bolen had sent Aubin a courtesy copy of the letter the senator had written to Adams. Aubin now wrote to Bolen to express the AFA's thanks: "We greatly appreciated Senator Kassebaum's letter to Secretary Adams. Thanks again for all your help and your interest in this matter."

But Aubin's real concern was to neutralize the response I had been able to insert into the *Washington Times* after their first piece on John Correll. Wishing to assure that the Congress heeded the lines the AFA favored, he further wrote,

> I am writing to give you some information on the latest chapter in the *Enola Gay* / Smithsonian battle.
>
> You may have seen Dr. Martin Harwit's response to the *Washington Times* clip on our Special Report, which I have enclosed. In it, Dr. Harwit suggests that the Air and Space Museum has been wrongfully criticized for its planned 1995 exhibit. . . .
>
> I think you will agree that the 559-page script for the exhibit tells quite a different story. As I indicated in my letter to John McCaslin, the few references to Japanese aggression and atrocities pale next to the treatment of the U.S. role in dropping the atomic bomb and other "aggressive" actions. Note especially the use of photos and graphics; they will tell the story to the majority of visitors to the exhibit.
>
> Dr. Harwit's so-called balance is a slap in the face to all the Americans who fought in World War II and especially to the memory of those who died for freedom.[22]

Three days later, Aubin wrote to Ron Stroman, majority staff director, and Marty Morgan, minority staff director of the House Government Operations Subcommittee on Human Resources and Intergovernmental relations: "I have attached the analysis that you requested of the 559-page script, dated January 12, 1994, which we left with you this past Tuesday." Here Aubin referred to John Correll's special report, with the tacit implication that it represented an independent authority:

> The meticulous analysis completed by the author of our original report, "The Smithsonian and the Enola Gay," dated March 15, 1994, further confirms our previous conclusion.
>
> As we stated during our meeting, our main concern is balance, context, and fairness. After reviewing our findings in this analysis, I think you will agree that this script is seriously flawed as it now stands. We will be providing you with more material next week on the chronology of events with respect to AFA's involvement in this issue.[23]

A week after that, Aubin followed up with a three-page chronology of contacts between the AFA and the museum as seen through AFA eyes.

The following month, the AFA's Ken Goss wrote a letter to twenty-two members of the House, fifteen members of the Senate, and four key congressional staffers, including Dan Stanley, administrative assistant to Sen. Bob Dole:

> You may have seen material in the *Washington Post* and in other publications regarding the controversy over the proposed exhibit of the historic *Enola Gay* aircraft.... Because you also may have received information from the NASM staff and other sources, the Association wanted to share our position and supporting information with you.
>
> ... the proposed exhibit is politically biased and reinterprets history in a manner that distorts the role of the United States in World War II. The gravity of this distortion is exacerbated by the portrayal of the United States as irresponsible in the use of nuclear weapons to bring the War to conclusion thus saving both American and Japanese lives.
>
> ... unless major, substantive revisions are made to the exhibit, Americans will be viewed as waging a war of vengeance against the Japanese; the Japanese will be viewed largely as victims of American aggression.[24]

Meanwhile, John Correll had found another line of attack, which he implemented in a second AFA report in late June[25] and another article in *Air Force* magazine in September.[26] It involved statistics.

The exhibition the museum had planned all along, centered on the *Enola Gay* and her mission. At Secretary Adams's urging, we had included a small introductory section on the antecedents of the war, the end of the war in Europe, and the last months of bitter fighting in the Pacific, which reached its height at Iwo Jima and Okinawa in the suicidal resistance on land and kamikaze attacks at sea. But apart from that, befitting the museum's mission, we had largely concentrated on the air war against Japan. The Manhattan Project, the design and production of the long-range B-29 Superfortress, and the methods developed by Colonel Tibbets and the 509th Composite Group had changed the nature of that air war—in fact, changing the character of war forever. This was the theme of the exhibition built around the *Enola Gay,* the two atomic bombs, and the Okha kamikaze bomb we had planned to display, together with artifacts and images showing the enormous destructive power of the new atomic weapons.

With this concentration on the mission of the *Enola Gay*—whose crew had returned unscathed—the images of casualties we showed were almost entirely photographs of wounded Japanese, with a small number of those who had been killed. John Correll now published the statistics. The museum's most recent draft of the script contained thirty-two pictures showing Japanese casualties and only seven showing Americans. There were eighty-four pages of text and ninety-seven photos that Correll judged to show Japanese suffering. How he obtained those particular numbers was not clear.

Correll continued, "By contrast—and demonstrating our point about lack of context—the new script devotes less than one page and only eight visual images to Japanese military activity prior to 1945." He also complained that many of the pictures of the aftermath of the bombing showed Japanese women and children. This was not surprising, since so many of the able-bodied men were away, serving in the military; but Correll used his statistics to assert "The defining characteristics of the museum's plan include the unilateral emphasis on Japanese suffering in the war ... [and a] distinctive ideological tilt."

Nobody seemed willing to admit that the photographs we hoped to exhibit were not new to Washington. Many had already been displayed fourteen years earlier, in an exhibition Senator Mark Hatfield had mounted in the U.S. Senate itself, in June 1980. The *Washington Post* had favorably reviewed that exhibition at the time, though they now criticized the museum's intended exhibition.[27]

AUGUST 10, 1994

August 10, 1994, definitely was not a good day. At 3:30 in the afternoon, Connie Newman, Mark Rodgers, Mike Fetters, and I went up to Capitol Hill at the request of Congressmen Sam Johnson (R-Texas), Henry Bonilla (R-Texas), Tom Lewis (R-Florida), Robert Dornan (R-California), Duncan Hunter (R-California), and Joseph McDade (R-Pennsylvania). We met in the Longworth House Office Building, Room 4116. Johnson, Lewis, and Dornan all were former Air Force men. Johnson and Dornan had been fighter pilots in Vietnam. McDade was there in his dual role of Smithsonian regent and Republican member of the House.

While we were still mingling around, waiting for everyone to arrive, Congressman Johnson told me his son was an astrophysicist in Tucson. Knowing we were going to have a rough ride, I regretted not knowing this young fellow-astrophysicist so I could at least start the session with something pleasant to say to his father. But I had no such luck.

The room was small. A large oblong table dominated the middle, so that the congressmen's staffers, who had come along but were not invited to sit, had to stand pressed against the walls.

The session started. Hunter and Johnson were especially angry about the exhibition and quoted from early drafts. They were upset at what we had hoped would be seen as obviously ironic statements from 1939, quoting Hitler as against bombing civilians and George Marshall as for bombing Japan. They were appalled that we intended to display ground zero at

Hiroshima and Nagasaki, or people hurt on the ground. Lewis, who had been an eighteen-year-old Army Air Forces recruit in 1943 and had served as gunner and cannoneer in World War II and the Korean conflict, was particularly incensed, accusing us of casting kamikaze pilots in a heroic light, while showing nothing about the heroism of our own servicemen.[28]

In responding to this onslaught, I admitted that irony was presumably misplaced in the script if it was that easily misunderstood. I also tried to explain that we were attempting to show the kamikaze pilots as a symbol of Japanese fanaticism—but to no avail.

The congressmen insisted that the museum had no business getting involved in important historical matters. That was for the National Museum of American History, not for us. The National Air and Space Museum should stick to displaying the great technological achievements of aviation and spaceflight. One of them read me the legislation the Congress had enacted and the museum's mission as stated therein. I told them that this was exactly what I was doing: "displaying," "memorializing," and "providing educational material" on "aeronautical equipment of historical interest and significance." The congressmen countered that I was misinterpreting these words.

Bonilla didn't say much. McDade, a Smithsonian congressional regent, tried to be nice and calm the mood. Dornan talked more than anyone else, showing off his knowledge of historical trivia. He informed us that this day was the 149th anniversary of the enabling legislation that had established the Smithsonian Institution.[29] He intended to introduce a comment on this in the Congressional Record for the day. His loquaciousness provided welcome relief from the pounding I was getting from the three former airmen.

Toward the end of the ninety-minute session, Congressman Johnson leaned forward across the table and said, "Dr. Harwit, you participated in atomic bomb tests. Did they affect the way you are planning this exhibition?" I replied that one's past experiences have an inevitable effect on the work that follows, but that I did not think I was unduly influenced. He then had another question. "Could you," he asked "order a nuclear bomb to be dropped on an enemy?"

This is not a question I am asked every day. I thought I might evade a hasty answer by saying, "That is an order only the president of the United States can give."

I was not to be let off that easily. Johnson parried, "Never mind that. Could you give the order if you were in a position to do so?"

Johnson had not formulated his question precisely. But he looked so intense about this evident test he was administering, that I assumed he wanted to know whether I would order a first strike.

I had no time for a reasoned reply. My mind raced back nearly forty years to Eniwetok and Bikini. I recalled the hydrogen bomb that had turned night into day for us more than two hundred miles away. I remembered the day of the first hydrogen bomb drop, the intense flash of light and heat, and the roiling, slow-motion growth of the mushroom cloud, till it filled the entire sky, all taking place in awful silence, until that crack shook us awake again. I could still picture us helicoptering in to pick up our detectors, half expecting to find the tall palms and lush vegetation we had left just days before on a beautiful tropical island, only to find half of it vaporized, the other half nothing but rubble. What would Manhattan look like after a hydrogen bomb? Could I ever order that to be done to other human beings, killing them by the millions in one second?

These images had flitted by in an instant, but I knew what to say. I looked at Johnson, and said "No, I couldn't order anyone to drop a nuclear bomb."

Johnson straightened himself in his seat. "Well," he said, with obvious pride, taking time to dwell on the first word to follow, "I could!" He looked triumphantly around the room, as though he had just proved Fermat's last theorem and was carefully penciling in "Q.E.D." to make his point. He had produced incontrovertible proof: I was unfit for my job.

Congressman Tom Lewis was also impressed. As soon as the meeting was over he rushed back to his office and though it was late in the afternoon, issued the following news item "for immediate release":

> LEWIS DISAPPOINTED WITH SMITHSONIAN DIRECTOR'S EXPLANATION OF JAPANESE BIAS IN UPCOMING EXHIBIT OF ATOMIC BOMB
>
> Director Admits He is Opposed to Using Nuclear Weapons in Wartime
>
> ". . . In my opinion, this exhibit goes beyond being 'politically correct,' and is outright offensive to veterans such as myself," said Lewis. "Not only does it go out of its way to treat Japanese as victims, but it actually glorifies kamikaze pilots. I never thought I would see the day when American servicemen would take a backseat to kamikaze pilots in the Smithsonian Institution." . . . Dr. Harwit admitted that if a decision was left to him, he would not use atomic or nuclear weapons in wartime. . . .
>
> The Smithsonian has drastically varied from their chartered obligation to portray service men and women as inspirational. . . . In the next week, Lewis will be meeting with other Members of Congress to explore remedies, including Congressional action. Concluded Lewis: "Taxpayers fund this museum, and I will not allow their money to be spent revising history."[30]

Lisa Metheny, on Lewis's staff, sent a copy of this to John Correll, with a handwritten note: "Thank you for your extensive background information. We'll be in touch!"[31]

The AFA had scored another point!

Sam Johnson also returned to his office to issue an immediate news release:

PURPOSE OF *ENOLA GAY* EXHIBIT DISPUTED

... Dr. Harwit has been very public about his personal feelings on atomic weapons. I have to wonder if his personal feelings and motives, instead of those of America, are being served through this exhibit ...[32]

WHO IS PAYING FOR THE EXHIBITION?

I got back to my office. Around 5:30 P.M., Connie Newman called to say that Congressman Lewis had one more question: "Was any Japanese money supporting the gallery?"

I thanked my good fortune for having had the foresight a year earlier not to ask for any. I told her "Definitely no! We made the decision long ago not to accept money from anyone, just in expectation of such an inquiry."

But Lewis wanted it in writing. Hence a letter penned to me that same day:

> Thank you for meeting with me and other members of Congress....
>
> As you may be aware, there have been comments made in the press regarding Japanese funding of this exhibit. Is this correct? In addition, there have been accusations that the Japanese exhibits were provided with certain conditions as to their exhibition. Were there any conditions? Finally, does the Air and Space Museum receive any funding for its activities from the Japanese government or other Japanese sources?
>
> Given the abrupt end to our meeting, I was unable to raise these questions. . . . I would request that these questions be answered by the close of business on Friday....[33]

I answered this two days later:

> ... I can tell you unequivocally that funding for the exhibition is derived entirely from within the Institution. No funding whatsoever has been solicited, accepted or even contemplated from any outside source, including any Japanese source.
>
> ... [As] is the case with all of our exhibitions where we include source material from outside the Museum, the donating entities or individuals are provided the opportunity to review the exhibition concept or script so they may be aware of how those artifacts and other materials will be represented. This is a courtesy we extend to all individuals or organizations when we request to utilize their materials. While there is a tentative agreement with several Japanese organizations to utilize selected artifacts and materials, they may ultimately decide not to lend the Museum these materials, which is their prerogative. Let me assure you, however, that if this is the case, the Museum will accept that decision and make no accommodations nor accept any conditions in order to secure a loan.
>
> Finally, the National Air and Space Museum does not receive any funding from the Japanese government or other sources. What may have been referenced in some recent press accounts is the large-scale exhibi-

> tion entitled "American Festival Japan 1994," which is currently underway in Japan. . . . The costs of staging this exhibition have been covered entirely by NHK and other Japanese sponsoring organizations. Since opening on July 9th, more than half a million Japanese have visited the Smithsonian component of the exhibition. The Smithsonian exhibition included artifacts from the National Museum of American History and the National Air and Space Museum. I have attached a fact sheet describing the Festival for your information.[34]

A LETTER FROM CONGRESS

The day was not yet over. Congressman Peter Blute, Republican of Massachusetts, had not participated in the afternoon's session. But on the way up to Congress, we had heard that he had sent Bob Adams a letter signed by himself and twenty-three other members of the House, eighteen Republicans and six Democrats. From this group, only Duncan Hunter had been at the afternoon session with us:

> We write to express our concern and dismay about the National Air and Space Museum's intended exhibit on the fiftieth anniversary of the bombings of Hiroshima and Nagasaki. . . .
>
> As you are aware, there has been much controversy about this exhibit based on the fact that many respected military historians and veterans' organizations found the original exhibit and accompanying script to be lacking in balance and context. It portrayed Japan more as an innocent victim than a ruthless aggressor, and cast Americans as being driven to drop the bomb out of revenge and for political reasons rather than out of concern for the hundreds of thousands of American lives that would have been lost during an invasion of Japan.
>
> Air and Space tried to quell criticism by revising the exhibit and script after consultation with veterans, historians, and representatives of the armed forces. However, after review we have found that the revised script is still biased, lacking in context, and therefore unacceptable. . . .
>
> There are 32 photographs of Japanese casualties during the war for the Pacific, but only 7 photographs of American casualties. Attention to the fact that the atomic bombs prevented an invasion of Japan and an estimated one million American casualties is limited to one small wall label at the end of the exhibit. There are 84 pages of text and 97 photographs relating to Japanese suffering, but less than one page and 8 photographs relating to the suffering caused by the fierce Japanese aggression between 1930 and 1945.[35]

The hand of the Air Force Association could not have been clearer if this letter had been written on AFA stationery. Correll's statistics had not yet been published. The September issue of *Air Force* magazine, which contained them, had not yet been released.[36] These figures had come directly from the AFA.

To maintain the momentum gained through Congressman Blute's letter, Aubin again wrote to Blute's office on August 22: "As museum officials work frantically to 'satisfy' members of Congress and veterans, it is our view that their overall approach and response to criticism thus far has been flawed, and very well may continue to be flawed...."[37] His single-spaced memorandum, over two pages long, itemized a detailed restructuring of the exhibition that would have totally changed it. All five sections were to be renamed; "Section *200*, 'The Decision to Drop the Bomb' should be renamed 'The Decision that Ended the War' and revised to reflect widely accepted scholarship." ... the section dealing with Hiroshima and Nagasaki on the ground should be "dramatically restructured and cut down in size"; the section on " 'The Legacy of Hiroshima and Nagasaki,' is so out-of-place and out-of-context that it should be entirely eliminated. The speculative and sophomoric treatment of nuclear deterrence has no place in this exhibit...." Much of this wording may have been taken from a nearly identical proposal Hatch sent me two days later.[38]

THE AFA LEGISLATES

The AFA's biggest success, however, was the passage of a Senate resolution sponsored by Senator Nancy Kassebaum. Stephen Aubin had been in close touch with Ed Bolen, on the senator's staff, since the previous March. On September 15, Aubin sent Bolen a telefax: "Here are some thoughts on the *Enola Gay* per our discussion. I also include a letter from Secretary of Veterans Affairs Brown. I borrowed some words from him."

Jesse Brown had written to me on September 6. Aubin evidently had received a copy, which he now paraphrased in his second paragraph. His first paragraph, once again, was a misapplication, to the National Air and Space Museum, of legislation for the nonexistent Armed Services Museum:

> U.S. Code states that "the Smithsonian Institution shall commemorate and display the contribution made by the military forces of the Nation toward creating, developing and maintaining a free, peaceful and independent society and culture in the United States." It also states that "the valor and sacrificial service of the men and women of the Armed Forces shall be portrayed as an inspiration to the present and future generations of America."
>
> In memorializing this nation's role in armed conflict, the National Air and Space Museum has an obligation to portray history in the proper context of the times. The role of the *Enola Gay* was momentous in helping bring World War II to a merciful close, saving both American and Japanese lives. The museum should mount this exhibit with appropriate sensitivity toward the men and women who served this country so faithfully and selflessly during World War II, and should avoid impugning the memory of their comrades who gave their lives for freedom.[39]

Four days later, on September 19, 1994, Senator Kassebaum introduced Senate Resolution 257, which duly cited much of Aubin's submission verbatim:

> To express the sense of the Senate regarding the appropriate portrayal of men and women of the Armed Forces in the upcoming National Air and Space Museum's exhibit on the *Enola Gay.*
>
> Whereas the role of the *Enola Gay* during World War II was momentous in helping to bring World War II to a merciful end, which resulted in saving the lives of Americans and Japanese;
>
> Whereas the current script for the National Air and Space Museum's exhibit on the *Enola Gay* is revisionist and offensive to many World War II veterans;
>
> Whereas the Federal law states that "the Smithsonian Institution shall commemorate and display the contributions made by the military forces of the Nation toward creating, developing, and maintaining a free, peaceful, and independent society and culture in the United States";
>
> Whereas the Federal law also states that "the valor and sacrificial service of the men and women of the Armed Forces shall be portrayed as an inspiration to the present and future generations of America"; and
>
> Whereas, in memorializing the role of the United States in armed conflict, the National Air and Space Museum has an obligation under the Federal law to portray history in the proper context of the times: Now, therefore, be it
>
> Resolved, That it is the sense of the Senate that any exhibit displayed by the National Air and Space Museum with respect to the *Enola Gay* should reflect appropriate sensitivity toward the men and women who faithfully and selflessly served the United States during World War II and should avoid impugning the memory of those who gave their lives for freedom.[40]

On the day of its introduction, Ed Bolen telefaxed Steve Aubin, "Thanks for your help. Please call if you have any comments," and attached both the printed version of the resolution and the text of the speech Senator Kassebaum had given in introducing it.[41]

Senator Kay Bailey Hutchison, Republican of Texas, was tempted to alter, substitute, or amend this legislation. That same day, her aide David W. Davis telefaxed new draft language to John Correll, with the note "Attached is the Senator's Sense of the Senate Resolution on *Enola Gay.* I would appreciate your review and any suggestions."[42] It appeared to be the senator's request to Correll to accept a friendly amendment.

But no substitute language was accepted. Three days later, the Senate agreed to Senator Kassebaum's resolution by unanimous consent. Not only had the senators accepted Aubin's proposed wording with only minor modification, they had failed to realize that the thirty-three-year-old federal law they cited referred to a nonexistent Armed Forces Museum.

A grateful Monroe Hatch wrote Senator Kassebaum, "I am writing to commend you for bringing the *Enola Gay* exhibition issue to the attention of your colleagues on the floor yesterday.... We also believe, as your resolution points out, that the Smithsonian Institution has an obligation to 'commemorate and display the contributions made by the military forces of the Nation toward creating, developing, and maintaining, a free peaceful, and independent society and culture in the United States.' "[43]

The AFA had also worked with Congressman Pat Roberts, Republican of Kansas, on identical wording in House Resolution 531. While that did not pass, a conference report accompanying the House appropriations bill for Interior and related agencies for 1995 did include the amendment, "The managers expect the Smithsonian's exhibit surrounding the *Enola Gay* to recognize properly and respectfully the significant contribution to the early termination of World War II, and the saving of lives of both Americans and Japanese, by its crew, the Army Air Services Command and President Truman."[44]

To appreciate the magnitude of the Air Force Association's influence, one needs to note that they had first used John Correll's articles in *Air Force* magazine, as well as their appearances on radio talk shows and on television, to alarm veterans' organizations and the public. They had then used the "Special Report" Correll had produced to provide an "analysis" of the museum's script, as in Aubin's letter to congressional staffers Stroman and Morgan on April 8.[45] Having gained credibility in this way, they had been able to write the text that, with minor editing, became Senator Kassebaum's resolution. For senators like Kay Bailey Hutchison, they had then been called on for a further opinion on wording of and potential amendment to the resolution. And finally, in a show of deference, they had congratulated Senator Kassebaum for her wisdom in authoring a resolution that they themselves had drafted for her.

20

An Intricate Military Web

THE MILITARY SERVICE HISTORIANS

When they had first met with us at the museum, I had assumed that the congressionally mandated 50th Anniversary of World War II Commemoration Committee would be largely autonomous. Later, I was somewhat surprised to learn that they were headquartered at the Pentagon. At the time, I thought this largely a cost-saving measure, to run the program with a minimum of overhead in manpower and office expenses. As we got to know the committee better, however, their apparent independence turned out to be an illusion.

On January 12, 1994, just before the first draft of the exhibition script was to be issued, General Kicklighter and some of his staff visited us at the museum. In this meeting, Kicklighter was particularly concerned with obtaining the Air Force's backing for the exhibition.

I mentioned that General Hatch had been angry in our November meeting but that I did not understand his reasons. In response, Kicklighter proposed that if we could obtain the approval of historians attached to the various branches of the military services, in addition to that of the advisory committee we had already established, those opinions would count heavily with the veterans and veterans' organizations.

Following up on his proposal, Kicklighter called a meeting of service historians and museum curators for April 13, 1994. Navy commander Luanne Smith, on his staff, made the arrangements and took notes on the day's discussions.[1] In her report to Kicklighter a few days later, she summarized,

> A meeting between Service Historians and Smithsonian Air and Space representatives was held on April 13 at the Air and Space Museum. Originally scheduled for 0930–1200, the meeting was extended until 1600 to permit a thorough, page-by-page review. . . . There is considerable distrust that the Air and Space Museum representatives will not make key changes to the script that will allow the Service Historians to recommend Department of Defense (DoD) support. Air and Space Museum representatives, however, thought the discussions were the most productive they have had with outside historians, and have indicated that many changes will be made as a result of the meeting. Copies of future script revisions will be provided to the Service Historians for their review.
>
> Present were historians from the Army, Navy, Marine Corps and Joint Chiefs of Staff historical offices, as well as key members of the Museum's exhibit team, including Dr. Crouch and Dr. Neufeld. Dr. Harwit came in to welcome participants, and later to listen to concluding remarks. The Air Force Historian's office did not participate because they have already made comments on that version of the script. Given the hard line they have taken, however, including a recent critical memo to the Chief of Staff of the Air Force voicing opposition, it is unlikely that the Air Force [Historian's office] will ever support the exhibit, whatever changes are made. . . . Fortunately, the DoD Historian, Alfred Goldberg, has offered to participate in future discussions to moderate the debate.[2]

HALLION AND THE AIR FORCE CHIEF OF STAFF

Unbeknownst to us, Air Force historian Richard Hallion had written to Air Force Chief of Staff General Merrill A. "Tony" McPeak, on April 8, 1994. Exulting over the radio talk show blitz John Correll had just initiated after publishing his two articles in *Air Force* magazine, Hallion wrote,

> Herewith a brief update on the National Air and Space Museum's *Enola Gay* exhibit and script: the discussion on this issue has gone public and has reached cacophonous proportions. . . .
>
> Correll's presentation on WRC was a *tour de force*; moreover, as is evident on the tape, the talk show host, Jeff Kamen, became highly energized, deploring the NASM's script and describing it as "garbage." . . .
>
> Talks with [Army, Navy, Marines, and JCS] historians indicate that their view, like ours, is highly critical of the script for the exhibit.
>
> Although NASM has made some progress with the script, they presently appear unwilling to repair it, in the following areas:

> The script gives the impression that Truman was more concerned with the atomic bomb as a diplomatic weapon against the USSR than as a route to shorten the war and avoid heavy American casualties.
>
> President Truman's deep concern with potential American casualties, should an invasion be necessary, does not come out clearly in the script.
>
> The script never mentions that President Roosevelt was the leading American official advocating the bombing of Japanese cities "heavily and relentlessly." Outraged by Japanese brutality in China, FDR indicated even prior to the Pearl Harbor attack that he wanted to see Japanese cities bombed.

This was the third time Roosevelt's insistence on bombing the Japanese had specifically been mentioned by the Air Force historian or the Air Force Association.

Hallion's letter to McPeak now concluded,

> In the script, four or five sentences take us from 1931 to 1941. Missing in proportion to other events, are Japanese aggression and their brutality against subject peoples.
>
> "For most Japanese, it was a war to defend their unique culture against Western Imperialism." Who was the aggressor here? The U.S.?
>
> There are additional problems that Dr. Harwit and his curators appear unwilling to address. These include a vastly unbalanced visual presentation, that emphasizes Japanese civilian casualties from bombing as opposed to Japanese brutality and atrocities.
>
> Incredibly, Dr. Harwit has apparently asked Lt. Gen. Kicklighter, heading the DOD World War II Commemoration Committee, to help fund this exhibit with $250,000. We are hopeful that Gen. Kicklighter will turn this request down.
>
> I shall keep you informed.[3]

Since I never saw a copy of this letter until thirteen months later, when the AFA published it, I was still under the impression, at this time that Hallion was supporting us and that Correll had misquoted him in *Air Force* magazine as being opposed.

BEHIND THE SCENES

Other correspondence we never saw till much later also worked against us. Besides writing to the Air Force chief of staff on April 8, Hallion also had written to General Kicklighter on April 19:

> Given the curator's unwillingness to consider DOD's viewpoints, I feel deeply that it would be entirely inappropriate for the Department of Defense to, in any way, help fund this exhibit, unless extraordinary changes to the exhibit script are undertaken.[4]

He sent copies of this letter to the heads of the Army, Navy, Marines, and Joint Chiefs of Staff history offices—but not to anyone at the Smithsonian. Chief of military history at the Department of the Army was Brigadier General Harold W. Nelson, who also expressed his opposition to the exhibition in a letter to Kicklighter bearing the same date.[5] Copies of none of these letters were sent to the museum or the Smithsonian.

THE SERVICE HISTORIANS ANALYZE THE SCRIPT

We were now receiving detailed critiques from service historians. Typical, perhaps, was the response to our script that Navy historian Kathleen Lloyd sent us at the end of April. She sent a single-spaced, nine-page analysis, with myriad detailed comments appended to an opening statement, in which she argued,

> In general the script is repetitions of certain unproven views that the atomic bombs were dropped to impress the Soviets and because the U.S. had racist attitudes against the Japanese. It is unbalanced in its approach and represents views not held by the majority of historians or supported by modern scholarship in U.S. and Japanese sources, and in using Ultra. ["Ultra" was the name for the overall decrypting scheme that had broken the various German and Japanese secret codes.] It ignores the fact that Germany was required to surrender unconditionally.
>
> For balance, more details should be included on the Japanese treatment of the Asian countries that they conquered and of the Allied prisoners of war. Along with the coverage of the atomic bomb's civilian victims, information about the Japanese soldiers and military installations destroyed by the bombs should be included.
>
> Also for balance it should be made clear that no one knew the effect that the bomb would have or the lasting effect that radiation would have on individuals.... Even in the 1950's, the U.S. conducted above-ground atomic tests with U.S. troops nearby in trenches.
>
> In addition, the script reflects a lack of understanding of Japanese culture and their beliefs about surrendering.[6]

Kathleen Lloyd was not alone in holding such views. I was now beginning to see the deeper problem we were facing in our attempt to produce a balanced exhibition. One major source of contention, once again, was the existence of two communities—historians attached to the universities and historians attached to the military services. The selection of diary entries, minutes of meetings, memoranda, letters, and memoirs presented by these two communities to establish a coherent history of events tended to differ dramatically. Documents and speeches considered crucial by one group were dismissed as misleading, unrepresentative, or merely irrelevant by the other. The museum was going to have to navigate through these differences and make equitable choices on behalf of the public.

This we began to do. But it was slow work. Michael Neufeld had to go through Kathleen Lloyd's nine pages of detailed recommendations, all of which had to be checked out for accuracy and then checked again to decide on genuine relevance for public presentation. The same had to be done also for the equally detailed recommendations arriving from the other service historians. This took a great deal of time, and in the meanwhile, the Air Force Association was vilifying the museum in the press, on radio talk shows, and on television. All this put great pressure on the curators.

LOBBYING THE COMMEMORATION COMMITTEE

While nobody from the Air Force historian's office had attended the April 13 meeting of service historians, Hallion had been busy lobbying the heads of military history to voice no confidence. On April 18 General David A. Armstrong, head of the Joint Chiefs of Staff history office, met with General Kicklighter and Commander Smith also to urge Kicklighter to withhold any contemplated funding for the museum's exhibition. Kicklighter assured him that he had no funds to hand out, but that he thought it important for the military historians to remain involved as advisors. He mentioned that the museum was taking the service historians' advice seriously and was making changes to the exhibition.

On April 25, 1994, Commander Smith alerted Kicklighter to one of the major difficulties she discerned:

> Behind the scenes, Dr. Hallion has been orchestrating the various Service historical offices to not wait to see what changes the Air and Space Museum is prepared to make and to go on the record that they cannot support the exhibit. He knew about the April 13 meeting and was invited to attend, but chose instead to write a memo prior to the meeting to the Air Force Chief of Staff, attacking the Smithsonian and declaring that the Museum would not make any significant changes to its exhibit.... Given Dr. Hallion's efforts to date in opposition to the exhibit and his personal predilections, it is unlikely that he will ever support the Smithsonian's exhibit, whatever major changes are made.[7]

Kicklighter was now also coming under pressure from other quarters. It began with a letter from W. Burr Bennett, Jr.

> This is a letter to express my concern and opposition to the rumor I heard this noon that your office is planning to fund the exhibit at our National Air & Space Museum, entitled, "Crossroads: The End of World War II, The Atomic Bomb, and the Origins of the Cold War." ...

He then listed his many complaints about the exhibition and finally pleaded,

> Sir, please tell me that this rumor is unfounded. If it is true, please reconsider the plans of your committee to financially support this attempt to rewrite the history of our war with Japan.
>
> With all due respect, General, the 50 year anniversary of the end of the war with Japan is not the place for a psychologically twisted exhibit at our National Air & Space Museum that is an affront to the Air Force veterans of the war.
>
> I hope and pray for your favorable response. Please don't let us down.[8]

To impress Kicklighter, he indicated that he had also sent copies of the letter to his two senators and the congressman for his district. Five days later, William A. Rooney followed suit and also complained to Kicklighter about the exhibition.[9]

Bennett and Rooney wrote again the following month, and Bennett wrote twice more after that in June and August 1994. By this time, he noted, he had gathered 11,500 signatures on the petitions he and others were circulating.

KICKLIGHTER IS SUBJECTED TO A LETTER WRITING CAMPAIGN

Letters from others were coming in as well. To Kicklighter and Smith it became obvious that a deliberate letter campaign had been unleashed on them. They were pretty sure they knew who was orchestrating it, and decided to end it by writing to General McInerney, assistant vice chief of staff of the Air Force, to whom Hallion reported. Commander Smith's draft for this purpose went straight to the point:

> We do not know why we are getting letters from Air Force veterans claiming that we are funding the Smithsonian Air and Space Museum's *Enola Gay* exhibit.... This appears to be the beginning of a letter writing campaign, apparently orchestrated from within the Air Force or Air Force Association.... We would appreciate if you would look into this matter and, if possible, put an end to it.[10]

This was a little too direct for Kicklighter. The version he signed read,

> We have recently been receiving letters from veterans and historians (attached) with regard to their displeasure with the Smithsonian ... exhibition. Somehow they are under the impression that we are providing funds for the exhibit.... We are worried that this may be the beginning of a letter writing campaign.... Could someone on your staff look into this for us. That would certainly be appreciated.[11]

McInerney's reply came on June 30 and was brusque. He returned the letters Kicklighter had forwarded and attached "... a draft response for you to send to veterans and historians." It recommended that Kicklighter respond to inquiries with these words:

> Thank you for your recent letter. Please rest assured that the 50th Anniversary of World War II Commemoration Committee will not under any circumstance provide funding for the National Air and Space Museum's *Enola Gay* exhibit.
>
> The Committee's efforts will continue to focus on helping the NASM to present a factual and properly balanced exhibit that honors the valiant service of America's World War II veterans.
>
> We are now working with the NASM to make certain that its revised exhibit script reflects the comments and suggestions recently offered by historians of all the military services.[12]

This was not much to the liking of Kicklighter's staff, but by the end of the summer, their outgoing mail regarding the *Enola Gay* exhibition was regularly funneled through the Air Force History Office, to avoid offending the Air Force. Thus, when veterans began writing the committee to ask whether Air Force historian Richard Hallion had indeed written the curators a congratulatory note on the first draft of the exhibition script, in February 1994, Kicklighter's staff sent the Air Force History Office their proposed reply to one of the veterans for review. It consisted of a confirming letter, to which was attached a copy of Hallion's review.[13] A few days later, the Air Force History Office's answer came back:

> The Air Force History Office at Bolling asks that the letter not be sent to the veterans as proposed....
>
> The History Office proposes [that you write instead] "In response to your question about whether Dr. Hallion had commented positively about the script, the Office of Air Force History has provided us the attached comments by Dr. Herman S. Wolk, senior historian, about the revised script for your review," ... [and] that Dr. Wolk's comments of 8 September 94 be attached to the letter as a substitute for the proposed attachment, which is out of date.[14]

Commander Smith notes,

> Ultimately, the letter was never sent, as it did not truthfully answer the veteran's question. It was inappropriate for the Air Force Historian's Office to ask the Committee to cover up Dr. Hallion's earlier statements that were supportive of the exhibit.[15]

By this time, the Air Force History Office was increasingly strident, demanding far more changes to the script than they had ever requested in their original critique. They were also insisting that they had never liked the script and had only written their positive comments to be collegial.

THE AIR FORCE

By midsummer 1994, I had for months been seeking a way to have Tony McPeak provide his good offices in arbitrating a discussion between Monroe Hatch and myself. I knew that the AFA's ideas about an exhibition would be totally unacceptable to Japan and would precipitate an international incident if followed through. In the meantime a way had to be found to mount a realistic display of the bomb's consequences for the populations of the two cities.

With these hopes for arbitration I wrote to McPeak on July 15, particularly emphasizing the differences in Hallion's report to the museum in February and to others later.

> I feel, at this point, that if we could reconcile the views that the Air Force Historian has expressed to us with those I understand he has conveyed to Monroe Hatch, the essence of our differences might be resolved.
>
> If you, in your dual role as Air Force Chief of Staff and ex-officio member of the Museum's Advisory Board, wished to call a meeting of Dick Hallion, Monroe Hatch and myself—and, I would add, possibly Tom Kilcline [President of The Retired Officers Association]—to clarify and talk out the remaining issues, I would be more than happy to participate. . . .
>
> If you had the time and inclination to arrange for a meeting along the lines I described, it could well constitute the catalyst needed to bring this unfortunate disagreement to a suitable resolution.[16]

Before I could get an answer to this, I was told that Under Secretary of the Air Force Rudy DeLeon had invited Kicklighter's staff to a meeting on the exhibition on July 19, to which none of us from the museum had been invited. I didn't know DeLeon at the time, but I knew Dr. Sheila Widnall, the secretary of the Air Force, quite well, and wrote her a message to be hand-delivered at once.

> . . . some weeks ago, you informally spoke to me about the exhibition *The Last Act: The Atomic Bomb and the End of World War II*, which the Museum is planning for 1995. . . .
>
> Yesterday I wrote . . . a response to Tony McPeak, Air Force Chief of Staff, suggesting a meeting with Gen. Monroe Hatch, executive director of the Air Force Association and Dr. Richard Hallion, the Air Force Historian, to see if matters could be resolved:
>
> The Air Force Association has been raising fears about our exhibition among veterans, and have claimed that they have the support of the Air Force Historian. This is puzzling because . . . as you will see from Dick Hallion's letter, which I also attach, he was strongly supportive, calling the script "a most impressive piece of work, comprehensive and dramatic, obviously based upon a great deal of sound research. . . ."
>
> I should mention that we have also tried to make sure that the exhibition of the *Enola Gay* will not further strain U.S. relations with Japan. For sixteen months we have been working first with Ambassador Armacost and more recently with Ambassador Walter Mondale and his staff in Tokyo, with Ambassador Kuriyama at the Japanese embassy in Washington, and with the State Department, to make certain that the exhibition we mount will truthfully report the events at Hiroshima and Nagasaki . . .

> I have briefly spoken about the current dilemma with Ambassador Winston Lord, Assistant Secretary of State for Pacific and East Asian Affairs, and also had occasion, some weeks ago, to informally talk about the exhibition with Secretary of State Christopher. I am most seriously concerned that the changes in the exhibition demanded by the Air Force Association would, if accepted, cause an uproar in Japan when the exhibition opens.
>
> My reason for writing you now is that I understand the Under Secretary of the Air Force, Mr. Rudy DeLeon, has arranged for a meeting on the exhibition tomorrow, to which we at the Museum have not been invited, but members of the 50th Anniversary Commemorations have.
>
> If this is a meeting to exchange information, I think it would be useful for the Museum to be invited to make a presentation and participate. I wonder whether I could ask you to invite us. Tomorrow may be a difficult day, because I have to be officiating at a number of [25th Anniversary of Apollo 11] commemorative events, but I am sure a mutually acceptable date could be set.
>
> We would very much like to see this difficult issue resolved. Your help to that end would be greatly appreciated.[17]

The encounter I mentioned with Secretary of State Christopher had been a chance meeting at a reception. He had listened courteously, though without comment. I was glad that he at least was aware of the situation, but I knew that the real work would be done at a lower level and I stayed in touch with Winston Lord.[18]

While none of us from the museum did get invited to attend the meeting DeLeon had scheduled, he and I met sometime later and had a cordial conversation. Meanwhile, the staff of the 50th Anniversary Commemoration Committee of World War II met with him. Commander Smith recorded,

> The [Air Force] Under Secretary was informed that our Committee is charged with the responsibility of commemorating the major events of World War II, and that one of our most important roles is to teach the history and legacy of World War II. He understood that it would be wrong to ignore one of the major events of this century, and that shutting down the exhibit would be an embarrassment to our country as it would indicate that our country feels guilt about our actions. Dr. Goldberg briefed the Under Secretary about the efforts of the Service Historians.... Dr. Goldberg's comments were extremely supportive of the exhibit. Mr. DeLeon asked Dr. Goldberg what the Air Force Historian's views were, and Dr. Goldberg indicated that they were initially extremely positive, but that they had inexplicably changed....[19]

MEETING HATCH BY HIMSELF

The Air Force appeared unable to help us, and I decided to call Hatch and arrange to have lunch together without anyone else present to see whether we could sort matters out. We met on August 8 and talked for quite a long time; but I seemed unable to elicit any kind of agreement or support. In hindsight, now that I am aware of the scale of his efforts, I am not surprised. Hatch probably could not have stopped his own campaign to have Congress censure the exhibition. Massive lobbying builds up its own momentum.

THE MILITARY

I was rather disappointed with our colleagues in the military. McPeak and Kicklighter were always sympathetic, but seemed unable to help us. To an outsider it seemed as though, if there was even one dissenting voice among the military or the veterans, everyone else fell into step. One could have all the friendly supporters one would wish, but in the military and among veterans any one individual or organization appeared to hold a veto. With that kind of a system there was little chance for reaching an understanding on reasonable grounds.

21

Internal Dissent and Regrouping

DISSENTING DOCENTS

In mid-February 1994, just after the first draft of the *Crossroads* script had been circulated within the museum and to the Air Force Association, Sandy Murdock got in touch with me. An attorney specializing in matters of aviation and an enthusiastic museum docent who volunteered his services as a guide on weekends, he had become a good friend. Now he sent me an unsigned, four-page, single-spaced diatribe he had found pinned to a bulletin board in the docents' meeting room. The distribution list showed it had been sent to Secretary Adams, several regents, a dozen members of the House and Senate, editors of aviation magazines, former directors of the museum, Washington mayor Sharon Pratt Kelly, Monroe Hatch, General Paul Tibbets—but unfortunately not to me. Secretary Adams's office had not forwarded a copy either, though normally such documents were. Perhaps his staff considered it a crank letter since it was unsigned.[1]

The note echoed many of the concerns that had already been voiced the previous summer by the Collings Foundation, and by *Aviation* magazine, namely that the museum was not restoring enough airplanes, was too concerned with aviation and spaceflight history and science; was displaying airplanes and spacecraft in a context that was not uniformly celebratory, was emphasizing the international development of aviation and spaceflight, and was teaching children about future possibilities rather than concentrating on aerospace history and in particular on American achievements alone. They also resented the museum's efforts to augment dwindling federal resources by means of gift shops and other enterprises.

I understood these feelings and the nostalgia they represented. But the museum could not stand still. We needed to provide the public with a better understanding of how the wonderful technology we displayed actually worked, and how it was being used in the service of society—for better or worse. I also wanted to have children understand what the future offered them in this exciting field. Both those aims required an expert staff. We certainly needed to preserve and restore airplanes and spacecraft, but we also had to display them in meaningful ways. If we showed the visiting public only American achievements, they would not realize the stiff competition we were already facing from abroad. In the launch of commercial satellites, Europe had drawn ahead of us and now controlled sixty percent of the international market. How many Americans knew that? And should they not know if they had any wish to see their country recapture its lead?

In March, a new problem arose. For some weeks, Arthur Hirsch, a correspondent of the *Baltimore Sun*, had been working on an investigative article on the *Enola Gay* and had spent a great deal of time at the museum, including the Garber restoration facility.

While Hirsch was working on the story, I heard that Frank Rabbitt, a docent who gave guided tours at the Garber facility, had interfered with one of Hirsch's planned interviews, as detailed below. I asked Mr. Rabbitt to stop by so I could see whether this report was true.

Rabbitt came in and verified the story, saying he felt he had acted in the best interests of all. For him it was a personal mission to see that the exhibit would properly honor American servicemen of World War II who had lost their lives for their country. In January 1945, his older brother had been killed flying with the American forces in China. Rabbitt didn't think the exhibit was doing enough to honor the fallen, and didn't think that Arthur Hirsch should be interviewing General Tibbets. He felt he owed it to his brother to block that interview.

I explained that a public institution is precluded from interfering with investigations by the press, and wrote to him,

> At your request, I am putting in writing your suspension as a docent for the National Air and Space Museum for three months, beginning today.
>
> In our meeting with Tom Crouch, last Friday afternoon, you verified for me that you had been at the Garber facility, wearing your badge, which clearly identified you as a docent, when you were addressed by . . . Arthur Hirsch, reporter for the *Baltimore Sun*.
>
> Mr. Hirsch talked with you about the *Crossroads* exhibition planned by the Museum for 1995 and told you that he had obtained an appointment from Gen. Paul Tibbets to interview him in Ohio the following day.

> You told me that you felt Gen. Tibbets needed protection from reporters, and decided to use the information Mr. Hirsch had given you to place a call to an associate of Gen. Tibbets, who had informed you he was trying to protect the General from interviews by the press. The associate then placed a call to the General to inform him. . . .
>
> The end result was that Mr. Hirsch traveled to Ohio and to Gen. Tibbets's home, only to be met by the General at the door and told that he would not allow him the promised interview.
>
> You explained that you acted out of personal concern for Gen. Tibbets, and because you don't think the Museum is mounting the right kind of exhibition. Your not liking the exhibition is a personal matter. This is a country in which free speech is every citizen's right. But working as a docent for the National Air and Space Museum is a privilege for which many compete. The assumption on both sides is that the Docent will work with the Museum, not against it.
>
> One important obligation of the Museum, as a nationally supported facility, is to leave itself open to public scrutiny. That includes scrutiny by the press. Interference with the legitimate activities of a member of that press, by any member of the Museum's staff, whether paid or unpaid, is unacceptable and against the obligations of the Museum and the best interests of the public we serve. Such behavior cannot be tolerated, and that is why I am regretfully suspending you for the indicated period.[2]

Unfortunately, the story did not end there. Shortly after his reinstatement in June, I was shown a couple of clippings that showed Rabbitt soliciting signed petitions opposing the museum's *Enola Gay* exhibition.[3] I felt that volunteers joined the museum to help, not to oppose us in our work, and wrote Rabbitt again on June 17, this time to dismiss him permanently.

> On March 14 I sent you a memorandum suspending you for three months from your activities as a docent for the National Air and Space Museum.
>
> Since then our staff has received numerous indications that you have been actively engaged in undermining the exhibition the Museum is planning for 1995. . . .
>
> I attach two newspaper clippings naming you a disseminator of petition blanks protesting the exhibition and a collector of those petitions. That you oppose the exhibition is your right, and I do not contest it. However, since you are so strongly and actively opposed to one of the Museum's key activities we cannot keep you on as a member of the docent corps, whose primary function is to help the Museum carry out its activities, not to oppose them.
>
> Given the evidence I attach, I have no alternative: I am hereby permanently dismissing you from the Museum's docent corps.[4]

In all my years at the museum, this was only the second time I had dismissed one of our four hundred volunteers. They were a wonderful group of people, and a terrific help.

The docent corps at the Garber facility was very upset even with Rabbitt's first suspension. Later, his permanent suspension hit them even harder. Frank Rabbitt had worked for the museum loyally for seven years, and they all felt for him. Tom Crouch, who met with the docents at the end of March 1994, wrote me the next morning,

> It did not go well with the docents last night. Many of them have now read the script, and the majority of those in attendance were very angry about the exhibition. I have always regarded the Garber Docents as a distillation of our traditional constituency. Their reaction is probably an indication of what is to come....
>
> I am sure that what we have seen thus far is only the tip of the iceberg.
>
> All of this has convinced me that your recognition of the importance of this subject was well-founded. This is an exhibition that Americans need to see. The strength of the growing opposition leads me to wonder, however, if the importance of the show is worth the price we are probably going to have to pay. This project is becoming very visible. My guess is that while our exhibition is solid, the controversy over this one will make "West As America" look like a tempest in a teapot.[5]

I was troubled that the docents were so upset. As a rule they were fabulously loyal to the museum. But many of the docents at Garber were long retired and had served in World War II. I could not expect their reactions to be very different from that of other veterans. Frank Rabbitt was not alone in the strength of his feelings.

I did not agree with Tom Crouch's implication that we were at this point anywhere near ready with "an exhibition Americans needed to see." My position was that given the outcry from the docents, from the Air Force Association, and recently also from the press, we needed to take a much more careful look at just what upset them so much. There had to be a way of reaching an accommodation that made sense to all well-intentioned individuals and groups.

I believed the historic facts brought out in the exhibition script to be largely correct. We had, after all, been working with some of the nation's leading historians to make sure we had not overlooked any major factors, or erred in other ways. The exhibition's opponents instead appeared upset by the way these facts were presented, the choice of laudatory or pejorative wording, the juxtaposition of evidence implying certain conclusions, and the choice and balance of artifacts, images, and factual material.

TIME

In late May 1994, *Time* magazine carried an article by Hugh Sidey:

> Air Force veterans have seen the 559-page proposal for the [National Air and Space Museum] show. And they are feeling nuked.

> The display, say the vets, is tilted against the U.S., portraying it as an unfeeling aggressor, while paying an inordinate amount of attention to Japanese suffering. Too little is made of Tokyo's sneak attack on Pearl Harbor or the recalcitrance of Japan's military leaders in the late stages of the war—the catalyst for the deployment of atomic weapons. John T. Correll, editor in chief of *Air Force* Magazine, noted that in the first draft there were 49 photos of Japanese casualties, against only three photos of American casualties. By his count there were four pages of text on Japanese atrocities, while there were 79 pages devoted to Japanese casualties and the civilian suffering, from not only the atom bombs on Hiroshima and Nagasaki but also conventional B-29 bombing. The Committee for the Restoration and Display of the *Enola Gay* now has 9,000 signatures of protest. The Air Force Association claims the proposed exhibition is a "slap in the face to all Americans who fought in World War II" and "treats Japan and the U.S. as if their participation in the war were morally equivalent."
>
> Politicians are getting in on the action. A few weeks ago, Kansas Senator Nancy Kassebaum fired off a letter to Robert McCormick Adams, secretary of the Smithsonian Institution. She called the proposal a "travesty" and suggested that "the famed B-29 be displayed with understanding and pride in another museum. Any one of three Kansas museums." . . .
>
> Meanwhile curators Tom Crouch and Michael Neufeld, who are responsible for the content of the display, deny accusations of political correctness. Crouch claims that the critics have a "reluctance to really tell the whole story. They want to stop the story when the bomb leaves the bomb bay." Crouch and Neufeld's proposed display includes a "Ground Zero" section, described as the emotional center of the gallery. Among the sights: charred bodies in the rubble, the ruins of a Shinto shrine, a heat-fused rosary, items belonging to dead school children. The curators have proposed a PARENTAL DISCRETION sign for the show. . . .[6]

While this was rather one-sided, it was not possible to call it factually wrong. But it was clearly influenced by the Air Force Association and others strongly opposed to the exhibition. Long after the event, someone on the staff at the Garber facility showed me a copy of a handwritten note from Sidey to Rabbitt that had been posted at the facility for some time. It was handwritten on *Time* magazine stationery, and dated May 24, 1994:

> Dear Frank,
> Thank you for the kind words. But you were the source, so any good done must be credited to you for your sense of honor and determination. I had a much longer piece written which I, of course, thought better. But space is always the tyrant in my business and so I got chopped.
>
> But we got in print—thank God for every inch.
>
> Cheers,
> Hugh[7]

CORRESPONDENCE AND CONGRESS

In response to the *Air Force* magazine article and the newspaper coverage that was quickly developing, the number of letters the Smithsonian and the National Air and Space Museum were receiving was suddenly surging. The secretary's office, to which many of them were addressed, was swamped, and on April 5, 1994, Kathy Boi, who worked directly for Secretary Adams, wrote to me, "The Secretary has asked that we forward all future letters such as the enclosed to you to respond directly."[8] The secretary wanted to detach himself from the routine letters that were now streaming in, but wished to keep involved on more important policy issues.

In turn, I began to pass these letters on to the Department of Aeronautics to answer. But some were violent and their effect on the curators depressing. They were now working at breakneck speed to try to follow up on suggestions arriving from military historians and other sources. I decided they needed to be isolated from this added pressure. At the beginning of summer 1994, I asked Mike Fetters and his staff in the museum's public affairs office to write responses. Eventually, when the number of incoming letters reached a hundred or more a week, we had to give Fetters added help.

Every letter had to be sifted, to see whether it required any special attention. If so, it had to be sent to the appropriate department. If not, it had to receive a polite answer in a detailed form letter we had generated.

Not atypical was a letter from William A. Hulbert, of Akron, Ohio, who wrote in early April 1994, shortly after reading John Correll's *Air Force* articles, "I am a fairly long subscriber to the *Smithsonian* magazine and also a WW II veteran." He had been a B-24 pilot, flying out of Saipan, in the Pacific, in World War II. He had read the *Air Force* magazine article and wrote, "I am very disturbed about your exhibits and the message you seem to be sending. I will no longer subscribe to your magazine. It is the least I can do."[9]

I answered on April 14. "I am not surprised that you were shocked by the article that appeared in *Air Force* magazine. So was I." I described for him what we were actually doing and attached my response to John Correll's article that was to be published in the May issue of *Air Force*. I also wrote I hoped he would change his mind about canceling his subscription.[10]

Over the next few months, we received many letters like that; but on checking, I was told that only a tiny, one-third of one percent of all subscribers actually canceled their membership in the Institution's *Smithsonian* or *Air & Space* magazines.

Pressure, however, also came from the Congress. On April 12, Bob Adams wrote to Congressman G.V. "Sonny" Montgomery, chairman of the Committee on Veterans Affairs.

> In my almost ten years as Secretary of the Smithsonian Institution, this is the first time I have ever written to Members of Congress in advance of an exhibition. However, "Crossroads" will be one of the most important exhibitions ever presented by the National Air and Space Museum. . . . Dr. Martin Harwit, Director of the Museum, and the curatorial staff responsible for "Crossroads" would welcome the opportunity to brief you on the exhibition. I am confident that once all the facts are known, there will be consensus that the Museum has created a comprehensive and objective exhibition which chronicles a decisive period in history. . . .[11]

REASSESSING THE SCRIPT

Given the many apprehensions about our exhibition, I could see that it would have to state clearly that the United States had not started World War II, had not invented strategic bombing, and had indeed sustained terrible losses in the war. My assumption all along had been that everyone knew this. John Correll, Monroe Hatch, and the docents were saying in effect we could not take that for granted. Unfortunately, instead of telling us just that, the AFA was accusing us of willful bias through the press and stirring up public protest.

All this concerned me a great deal. On April 16, I decided we needed to reexamine the script. Identifying specific sections and page numbers, I wrote to the exhibition team:

> All of us associated with the exhibition have always known that the most difficult task before us would be to achieve accuracy and balance.
>
> Though I carefully read the exhibition script a month ago, I evidently paid greater attention to accuracy than to balance. Accuracy is somewhat easier to check, at least for the aspects of the exhibition that are familiar. Balance is more difficult to assess, since it requires an overview that allows one to see the script as a whole. One reading apparently was not enough to afford me that overview.
>
> A second reading shows that we do have a lack of balance and that much of the criticism that has been leveled against us is understandable. Most strikingly:
>
> We talk of Hitler's vow not to bomb civilians (100-29) but dwell on the corpses in Dresden (100-29, 200-13) due to Allied bombing without showing in similar detail the prior bombings of Nanking, Warsaw, Antwerp, Nottingham, and other cities that had earlier been heavily bombed by the Axis powers. We talk of the heavy bombing of Tokyo (100-32, 33), show great empathy for Japanese mothers (100-34), but are strangely quiet about similar losses to Americans and among our own allies in Europe and Asia.
>
> We show terrible pictures of human suffering in Hiroshima and Nagasaki in section 400, without earlier, in section 100, showing pictures of the suffering the Japanese had inflicted in China, in the camps they set up for Dutch

and British civilians and military, and U.S. prisoners of war. We mention internment camps for U.S. citizens of Japanese extraction (100-41) but go into nowhere near as much detail into the internment of Koreans and other non-Japanese in Japan (100-49) providing statistics alone, but no pictures. Nor do we show pictures of Japanese racism against Americans. We do not note that conditions in the American internment camps were far more favorable than in Japanese internment camps, where slave labor conditions prevailed.

We show virtually no pictures of Allied dead or wounded either in sections 100 or 300. Section 300 is almost clinically military in its tone, when contrasted to section 400 which speaks about the action on the ground entirely in human terms. Section 400 has any number of heart-rending, tragic stories of suffering on the ground. Where are the corresponding tragedies in section 100 in China, in the Philippines, in Singapore, in the former Dutch possessions? We go into American racism against Japanese (100-43) but show nothing equivalent on the Japanese side.

The alternatives to the atomic bomb are stated more as 'probabilities' than as 'speculations,' and are dwelled on more than they should be.

Section 400 has far too many explicit, horrible pictures.

I suggest the following cures:

1. Take out all but about one third of the explicit pictures of death and suffering in section 400. Add to section 400 pictures of prisoners just released from Japanese internment camps. (Lin Ezell has a wonderful letter from a woman who had been released as a young girl and who might have pictures too.)
2. Put in an equal number of pictures of death and suffering in section 200 for soldiers on both sides. This will document the enormous casualties that preceded the atomic bombing.
3. Put in more pictures of allied cities that were destroyed before we ever reached Japan or started bombing Germany with effect. Show Japanese bombing in their Asian campaigns. (See e.g. the film in the World War I gallery.)
4. Contrast the hardships of war in Japan with hardships the Allies in Europe and in the Pacific were suffering. America, with its great wealth was entirely exceptional. Not doing that makes it look as though the Allies had no reason to complain. In the U.S., show a copy of a telegram received by a family announcing their son's death. Show gold star mothers, yellow ribbons, etc.
5. Reduce much of the speculative material about what might have been possible without the atomic bomb. I have made some specific suggestions on the relevant pages, but further deletions might be useful as well.

If we make these changes, I think we will have a better exhibition. I do not think that these changes should be difficult to implement, since most of them require deletion of material rather than addition, except

> where pictures of Allied suffering are involved, and those should be readily available.[12]

Here I might call attention to my first recommendation, which proposed taking out all but one third of the "explicit pictures of death and suffering" in section *400*. The Air Force Association immediately gave this a new spin, by tacitly claiming that *all* of the pictures in that section fitted that description, and alleging that the curators were refusing to cut the *entire* section *400* down to one third its previous size. In fact, what I was asking the curators to do was to cut out two-thirds of the pictures of the dead and dying, but not aerial photographs, pictures of the mushroom clouds taken from the ground, or of rubble, damaged buildings, destroyed machinery, or the artifacts we wished to borrow. Monroe Hatch would later admonish me to enforce my "original direction to reduce this section by two-thirds," recommending that,

> **Twenty pictures and a half dozen artifacts would be ample.** These photos and artifacts should not unduly emphasize women, children, and religious objects.[13]

Hatch used bold print to emphasize his intent.

To put this in perspective, the exhibition was devoting five times this number of photographs just to B-29 aircrews, their aircraft, and their activities. To reduce this part of the exhibition to just six objects, and to studiously avoid reference to women, children, or religious artifacts, when over a hundred thousand civilians had died on the ground, seemed bizarre. Hatch's request to foist such a truncated exhibition on the public seemed so out of proportion that I was astonished to see him put it in writing.

A VISIT TO CONGRESS

As the AFA's assault on the museum began to spill into the daily press and radio talk shows, at the beginning of April 1994 I began to work closely with Under Secretary Newman. She, in turn, made available to us a broad range of central Smithsonian resources. For the next ten months, we worked together as a small but well-matched team. Newman took an active role and was always available when needed. Only as November dawned, and a new secretary of the Smithsonian Institution was taking the reins, did cracks begin to break the alliance.

On April 18, worried about the potential fallout in Congress, Mark Rodgers, Tom Crouch, and I went up to Capitol Hill, where we met with majority chief counsel/staff director Mack G. Fleming of the House Veterans Affairs Committee, and with minority chief counsel/staff director Carl D. Commenator and several members of their staffs. We explained our predicament with the Air Force Association and the press and asked them for their support. As the meeting ended, Fleming cheerfully told us that they would be glad to work with us but if his office got too many opposing letters, they'd come down on us like a ton of bricks. We thanked him: I felt, at least, he'd been honest.

THE TIGER TEAM

The next day I spoke with retired Air Force brigadier general Bill Constantine, one of the docents at the museum, and asked him whether he would be willing to chair a "Tiger Team"—an independent team with broad access to any information needed—to conduct their own study on the exhibition and give me advice. Serving with Constantine would be Don Lopez, a World War II fighter pilot ace and a longtime curator at the museum who had served as deputy director for many years; Gregg Herken, the chairman of the museum's Department of Space History and an expert on the politics of the atomic bomb; Ken Robert, a docent at Garber, who worked at the National Security Agency as a researcher on aviation; and Thomas Alison, who had only been at the museum for a year and had retired as an Air Force colonel. Steven Soter, special assistant to the director, would act as secretary to the committee.

None of these individuals had been directly involved with the exhibition script. Given the advice and recommendations we had received from the original advisory committee in early February, and more recent, conflicting advice we were receiving from historians working for the military, I felt I needed an independent body, familiar with both the military and the museum, to examine these differences and report back to me.

Constantine agreed to serve and got started at once since I had asked for a quick reaction from the team.

That same afternoon, Under Secretary Newman and Assistant Secretary for Museums Tom Freudenheim came over for a presentation with drawings, a scale model of the gallery as planned, and a talk on the contents. Two days later Secretary Adams and his executive officer Jim Hobbins were given the same briefing. All had been sent the exhibition script two and a half months earlier. We had freewheeling discussions, and both groups seemed satisfied.

I sent an official letter of appointment to the Tiger Team the next week on April 26:

> Ten days ago, the National Air and Space Museum invited a team of historians from different branches of the Military to comment on the script for the exhibition *Crossroads: The End of World War II, The Atomic Bomb and the Origins of the Cold War.* In addition to questioning individual points where corrections might be in order, they expressed dissatisfaction with the script's overall balance. In their opinion it was flawed in its portrayal of Japanese and American history, activities and customs.

This response runs counter to the critique of the script the Museum received just two months ago from a distinguished panel of scholars, including historians from the military, the government and academic institutions. While giving advice on individual points, their overall assessment had been highly satisfactory.

Whatever the origin of these differences, the Museum must be certain that the exhibition we mount is indeed balanced. I am therefore, asking you to serve on an independent Tiger Team, once more specifically to look for any signs of imbalance, and to report back to me by Friday May 13. I regret the short deadline, but the exhibition schedule is tight and corrective action if needed will have to be taken at once.[14]

At Constantine's request, Mike Neufeld attended the Tiger Team's first meeting. Clearly worried, a few days later he wrote the Tiger Team, the exhibit team, and me a thoughtful response to document his concerns:

During the preliminary meeting with the "Tiger Team" last Friday, we had a passing discussion about the purpose of the exhibit and the so-called decision to drop the bomb which, in hindsight, has bothered me. The clear implication was that the atomic bombings of Japan were "debatable," as I put it, only because some of the pacifists think that they were immoral. Attempts to reevaluate the decision were largely dismissed as "speculation." In fact, thirty years of scholarly research (ably summarized in J. Samuel Walker's article, "The Decision to Use the Bomb: An Historiographic Update,"[15] which I have distributed to many of you), has shown that the decision is debatable on its political and military merits—a conclusion which goes to the heart of the exhibit concept. This research obviously has not resolved the debate; indeed it cannot resolve the debate, since too many factors that went into the decision remain subject to dispute. But this research has produced much new knowledge that must be communicated in some form to the public.

One of the most important conclusions one can draw from this research is that, although it is certainly still possible to argue for the correctness of Truman's decision to use the atomic bomb without warning, the traditional justification used in this country is no longer tenable. That justification, which is endlessly repeated with almost religious fervor, asserts that Truman was faced with only two options: a) drop the bomb without warning, or b) invade Japan at the cost of a quarter of a million, half a million, a million or many millions of American and/or Japanese lives, depending on what version is being told. This account is untenable for at least four reasons:

1. *Casualties.* We know for a fact that the military chiefs never presented such numbers to Truman. [Here, Neufeld gave numbers offered by different military figures, which John Correll had also mentioned in his articles.]

   In any case, a fundamental assumption behind half a million American dead figures is wrong: the U.S. military never planned to conquer every square inch of Japan by force, but only to occupy two key areas: southern Kyushu and the Kanto Plain around Tokyo. All indications are that this would have sufficed to force the Japanese to surrender, if "Olympic" [the Kyushu operation] alone did not.

2. *Options.* Truman was never presented with a clear set of options to dropping the atomic bomb without warning, such as can be constructed in hindsight. Options were, however, discussed in his presence or by his advisers. [Neufeld here gives details on those that were discussed—an Emperor guarantee, a bomb demonstration, the shock effect of Soviet entry into the Pacific, or waiting for the naval blockade or conventional bombing to take their toll.]
3. *An Early End to the War*... With the advantage of hindsight and knowledge of the workings of the Japanese government, it now seems clear that the odds of Japan quitting in the summer and fall of 1945 were considerable. Soviet entry or an Emperor guarantee or both might well have sufficed. The Japanese elite, especially the militarist fanatics, have a great deal of responsibility for what happened at Hiroshima and Nagasaki, since they could not come up with a realistic surrender offer, but they did tentatively communicate their interest in quitting through [secret messages transmitted to their ambassador in Moscow and intercepted by the United States. Japan hoped that the Soviets might broker a peace treaty].
4. *Momentum and the Soviet Factor.* The traditional justification also grossly oversimplifies by leaving out contributing factors to Truman's decision (such as it was—in the context of 1945, Truman was unlikely to have done anything else): the momentum of the Manhattan Project and the "bonus" that American use of the bomb might have in Soviet-American relations. Although most historians agree that the key factors in Truman's decision were casualties and an early end to the war, evidence clearly suggests that these other factors added weight to the decision.

All of this is not to say that the exhibit should take any position on these controversies—indeed it does not. But it is clearly in the Smithsonian's charge, "The increase and diffusion of knowledge," to make this scholarly research accessible to the public. It is also important to realize that, if one of the central concepts of the exhibit is that the use of the bomb is debatable, this research clearly supports that assumption.[16]

The Tiger Team worked as carefully and speedily as they could, and provided me with a detailed twenty-two-page report by May 25, 1994.[17] It reviewed the exhibit section by section, commenting both on strengths and weaknesses, and recommended a significant number of changes. Ultimately, we adopted a large majority of these recommendations. But they had to be carefully checked out, often requiring substantial research, and could not be implemented overnight.

In the meantime, the Air Force Association accused us of refusing to accept the advice of a team we ourselves had set up. On July 19, Monroe Hatch wrote to me,

> You have a long list of those you have consulted with, including your advisory committee, military historians, your "Tiger Team" and even the Air Force Association and other veterans groups. Yet, the results of these efforts include relatively few script changes, and the structural exhibition plans have not changed very much at all. I don't believe the problems can be corrected without starting over again on several sections.[18]

A month later, on August 22, John Correll went a step further in a summary on "Developments in the *Enola Gay* Controversy." He wrote,

> AFA does not speak for anyone else on this issue. We can confirm, however, that there is considerable opinion that the Smithsonian's fix-up plan is *too little, too late,* and that it's time to *shut down this exhibit and start over with different curators.*[19]

THE SMITHSONIAN REGENTS

On May 9, I gave a presentation on the exhibition to the board of regents, with Chief Justice William Rehnquist presiding. I told the board about the procedures followed, the advisory committee that was set up, the discussions with Gen. Kicklighter's organization and with the Japanese, and the opposition from the Air Force Association. To my relief they were supportive. Hanna H. Gray, former president of the University of Chicago, felt that if the process we had followed was proper, the regents should not become involved with content. Senator John Warner, a congressional regent, suggested countering the AFA attack, with its selective quotations from the script, by publishing the entire script in the *Congressional Record.* But nothing ever came of this suggestion, and to my knowledge, the senator never mentioned it again. Congressman Norman Mineta also commented supportively.

I was pleased that the museum had this high-level internal support, given the pressures from the media and the huge numbers of letters that were beginning to come in from veterans stirred into action by the AFA and the other veterans' organizations.

A FINAL EXHIBITION TITLE

On May 31, I had one more meeting with Adams and Newman concerning the exhibition title. I mentioned that "Crossroads" was problematic, because it was also the name of the atomic bomb test series run at Bikini in 1946. The subtitle also was a problem; because many believed that the onset of the Cold War was an event logically separate from the start of a massive nuclear arms buildup. Finally, translated into Japanese, the title had no coherent meaning. Adams appreciated all this and we finally agreed on "The Last Act: The Atomic Bomb and the End of World War II." I was pleased that the sequence he agreed to in the subtitle was chronological. The curators had asked for that; it made more sense.

22

A Search for New Allies

THE INSTITUTION CONTACTS THE AMERICAN LEGION

Even before the regents' meeting on May 9, 1994, Connie Newman had come up with a new proposal to deal with the worsening situation. Before joining the Institution, she had been head of the Federal Office of Personnel Management. In that role she had frequently dealt with the major veterans' organizations, and had established good rapport with their leadership. Given the unyielding attitudes of the AFA, she thought we might be better off seeking the support of these larger organizations. If that worked, the AFA, whose membership was only about 180,000 would have to defer to such giants as the American Legion, with its 3.1 million members.

I agreed that this was worth a try, and Newman sent letters to the leadership of the American Legion, the Disabled American Veterans and the Veterans of Foreign Wars. After describing the exhibition and the way in which it had been shaped, she invited them to meet with us.[1]

To this request the American Legion alone responded. On May 20, Newman, Mark Rodgers, Tom Crouch, and I met in Newman's office with Herman Harrington and Hubert "Hugh" Dagley II, from the Legion's Internal Affairs Division. Harrington, white haired, blue-eyed, avuncular and about seventy, and Dagley, with light brown hair and in his forties, sounded sympathetic, but also told us that the Legion already had a draft resolution on its books to condemn the exhibition. It would be taken up at the Legion's Labor Day weekend convention in Minneapolis. They suggested that we might wish to attend the convention to make a presentation and see whether we could change the minds of the Legion's membership.

Newman and I readily accepted that offer. The meeting had been quite friendly, and we thought we might be able to gain the support of the Legion's leadership by convincing them of the strengths of our exhibition.

Newman's letter to National Commander Thiesen had reached him just days after the Legion's national executive committee, in its May 4–5, 1994 session, had passed Resolution No. 22, "Smithsonian Exhibit of the *Enola Gay*." This was the resolution Dagley and Harrington had mentioned. Not surprisingly, perhaps, it had got its start on Ben Nicks's typewriter the previous November—two months before the first script had been completed.[2] Nicks, an inveterate letter writer and a member of the American Legion, the Veterans of Foreign Wars, the 20th Air Force Association, and the 9th Bomb Group Association, had convinced the Legion to act. In part, the resolution read,

> visitors . . . will need to go through an area, as described in the Smithsonian Exhibit plan as, portraying the desperate actions of the Japanese to defend their homeland from the brutal, vindictive, and racially motivated conduct of the War by the United States, . . . [and] will then have to pass through the "emotional center" of the exhibit where the horrors and destruction of the use of the atomic bomb is depicted in such shocking realism that the plan notes parents might find some parts unsuitable for viewing by their children and warns "parental discretion is advised" and includes a collection of burnt watches, broken wall clocks and photos of victims in a variety of conditions, from death to the remains of maimed and diseased bodies, . . . whereas . . . the valor and sacrificial service of men and women of the Armed Forces shall be portrayed as an inspiration to the present and future generations of America. . . .
>
> Resolved; By the National Executive Committee of The American Legion . . . in Indianapolis, Indiana, on May 4–5, 1994, That The American Legion is concerned the planned exhibit . . . is politically biased and is in violation of the prescribed direction and function of the Smithsonian Institution and the Air and Space Museum. . . .
>
> Resolved; The American Legion strongly objects to the use of the *Enola Gay,* and the heroic men who flew her, in an exhibit which questions the moral and political wisdom involved in the dropping of the atomic bomb, and which infers that America was somehow in the wrong and her loyal airmen somehow criminal in carrying out this last act of the war which, in fact, hastened the war's end and preserved the lives of countless American and Japanese alike. . . .
>
> Resolved; The American Legion expresses its objection to the planned exhibit at the Smithsonian by directing the National Commander to forward a copy of this resolution to the President, Congress and officials of the Smithsonian Institution.[3]

As background material for this resolution, the Legion cited a request that had been submitted as early as January 21, 1994, and signed by the department adjutant, Charles G. Norton, in Nashville, Tennessee, asking that "the display of the *Enola Gay* shall not be used as a springboard to decry the morality of the Allied strategic policy of nuclear bombing of hostile cities during World War II. We are proud of *Enola Gay*'s historic role in history."

This request had been made even before the museum had released the first exhibition script.

A further piece of background material dated back to the Legion's seventieth annual national convention held in Louisville, Kentucky, September 6, 7, 8, 1988, resolving that "... the *Enola Gay* ... B-29 Bomber that dropped the first atomic bomb on Japan, ending World War II, shall be restored to insure that this historic aircraft will never fade into obscurity, but shall be displayed as a symbol for present and future generations that the price of freedom and peace has always been an awesome sacrifice of human lives." That resolution had been submitted by the adjutant in Indiana, and coincided with the period in 1988 when we had been working with Frank Stewart and General Paul Tibbets to seek support for more rapid restoration. Its wording also bears strong resemblance to the resolution Ben Nicks submitted to the 9th Bomb Group in October 1988.

At this point, the American Legion was apparently reacting primarily to forces set in motion by the "five old men." W. Burr Bennett, Jr., had motivated the Air Force Association into taking up the group's cause. Now Ben Nicks, who belonged to several veterans' organizations, including the Legion, had persuaded the Legion to draft its condemning resolution. Don Rehl, who had also written to the Legion recently, received a reply from Hugh Dagley soon after Dagley had met with us in Newman's office. In addressing Rehl, Dagley struck a considerably tougher tone than he had taken with the Smithsonian. He criticized the exhibit and assured Rehl,

> ... I am prepared to keep at it until the National Air and Space Museum's ill-conceived display conforms to the facts of history.
>
> I want you to know that I deeply respect what you and your comrades did during World War II and I am aware of and impressed with your decade-long fight to see the *Enola Gay* properly exhibited and honored. With the support of the National Executive Committee, the National Commander and the Chairman of the National Internal Affairs Commission [Herman G. Harrington]—himself a World War II combat veteran—I am committed to seeing the aircraft justly presented....
>
> Mr. Harrington and I agree that major revisions in the philosophy, content and language of the exhibit will be necessary to satisfy our members' deep concern over this exhibit, and that officials have so far not indicated such sweeping changes will be made....
>
> We have also contacted other veterans' groups and some general circulation publications about our efforts, and have received quite a bit of

> correspondence and phone calls from interested—and outraged—veterans. . . .
>
> The American Legion is challenging this exhibit in the most strenuous way. . . . Please know that we will not consider our job done when the letters fulfilling Res. #22 are written; rather we will continue in appropriate ways to see that only the full truth is included in any *Enola Gay* exhibit.
>
> Within the next couple of weeks we will begin to have a feel for our next step, and I would be pleased to keep you informed as we progress toward the convention.[4]

PAUL TIBBETS

Dagley's letter also quoted a strongly worded speech General Tibbets had just made on June 8, 1994, on receiving the Air Force Sergeants Association's Freedom Award.[5] I was badly shaken by those words, because I had talked with Tibbets only two weeks earlier.

Joanne Gernstein, one of the two young curators working on the *Enola Gay* exhibition, had talked with Tibbets on the phone, late in the summer of 1993. During that conversation, he had invited us to meet with him when he and his former crew were coming to Washington in the winter of 1994. That trip had, however, been postponed to the week following Armed Services Day, in late May. For two days, the Greenwich Workshop, a Connecticut-based company producing a television program, had arranged to film some of the 509th's key members at the museum's Garber restoration facility, with *Enola Gay* in the background. Joanne and Patti Woodside, introduced me to many of the men whom they knew so well by now through long-established contacts. Gernstein had been corresponding with them for many months to discuss loans of their memorabilia for the exhibition. Woodside also wanted to film them for the gallery.

When the group broke for lunch, I had an opportunity to talk with General Tibbets, whom I had not seen since 1988 when he and Frank Stewart had come by to initially offer support for the restoration of the *Enola Gay*. In the six years since, he had aged considerably. He was not as erect, and his hearing now required aids in both ears. While I was speaking with him, he seemed preoccupied with autographing artwork that had been laid out for him and his crew to sign. He was noncommittal when I asked whether we could work with him and obtain his support for the exhibition. As the group sat down to eat, I welcomed them to our facility and asked them to help us with the exhibition. Their knowledge, memorabilia, and memories would greatly strengthen the exhibit we would be able to offer the public.

I felt that if we could have General Tibbets agree that our display was fair in presenting the mission of *Enola Gay*, the veterans' associations would probably be satisfied and would give us their support as well. That was why the speech he gave two weeks later was such a blow. Tibbets had declared,

> . . . Today, there is a debate on how to present the *Enola Gay* and the use of the atom bombs to the American public and the world at large. There are questions as to how best to present the events of the summer of 1945. . . .
>
> I suggest that the *Enola Gay* be preserved and displayed properly—and alone, for all the world to see. She should be presented as a peace keeper and as the harbinger of a cold war kept from going "hot." The *Enola Gay* and her sister ship *Bockscar* should be remembered in honor of the scientists who harnessed the power of the atom for the *good* of mankind. The talents and skills of those men and women who gave us the means to use, regulate and control atomic energy. Such notable positive contributions are worthy of Smithsonian recognition.
>
> The *Enola Gay* has become a symbol to different groups for one reason or another. I suggest that she be preserved and given her place *in the context of the times* in which she flew.
>
> Her place in history has been dealt with unfairly by those who decry the inhumanity of her August 6th mission. Ladies and gentlemen, *there is no humanity in warfare.* The job of the combatants, the families, the diplomats, and factory workers is to win. All had a role in that "all out" fight.
>
> I am not a museum director, curator or politician. I am a pilot. I am a military man trained to carry out the orders of a duly elected commander-in-chief. . . .
>
> Today, on the eve of the 50th Anniversary of the end of World War II, many are second-guessing the decision to use the atomic weapons. To them, I would say "STOP!" It happened. In the wisdom of the President of the United States and his advisors at the time, there was no acceptable alternative but to proceed with what history now knows as Special Bombing Mission No. 13.
>
> To those who consider its proper presentation to the public, I say: "FULL SPEED AHEAD!" We have waited too long for all the wrong reasons to exhibit this aircraft. Too many have labeled the atomic missions as war crimes in an effort to force their politics and their opinions on the American public and to damn military history. Ironically, it is this same segment of society who sent us off to war that now wish to recant the flight of the *Enola Gay.*
>
> Thus far the proposed display of the *Enola Gay* is a package of insults. . . . If nothing else, it will engender the aura of evil in which the airplane is being cast.
>
> I am unaware of any positive achievements being credited to the men and women who built the B-29 bombers that carried the war to the Japanese homeland, or the soldiers, sailors, marines and Seabees who fought, lived and died fighting to take Pacific Islands that were needed for airplane bases within striking distances of the mainland. What about the airmen who flew those strikes and lost their lives? And, those who survived. Are they to be denied recognition for their efforts? Something is wrong with this scenario.

> In closing, let me urge reconsideration and let the exhibition of the *Enola Gay* accurately reflect the American spirit and victory of August 1945. Those of us who gained that victory have nothing to be ashamed of; neither do we offer an apology. Some suffered, some died. The millions of us remaining will die believing that we made the world a better place as a result of our efforts to secure peace that has held for almost 50 years. Many of us believe peace will prevail through the strength and resolve of the United States of America.[6]

I was convinced that General Tibbets had never read our script and knew about it only through newspapers or the warnings of close associates, perhaps the same ones that had warned him not to speak with Arthur Hirsch from the *Baltimore Sun*. I was particularly taken aback by the general's statement because two of his closest friends for more than fifty years, Dutch Van Kirk and Tom Ferebee, were working so closely with Patricia Woodside and Joanne Gernstein. Why would they work with us on an exhibition that the general was opposing, if they did not think they could convince him to back it?

I could understand General Tibbets's wish to "STOP" those who wanted to inquire into the ways President Truman's decision had come about. In a democratic society, the military obeys the commands of the duly elected civilian leadership. They are not to inquire or go off on their own. Orders are to be obeyed on the assumption that they were issued, as Tibbets had put it, "in the wisdom of the president of the United States and his advisors." That works two ways. It places responsibility in the hands of the people and their elected representatives; it also absolves the military of responsibility for carrying out legitimate orders. The more debatable a decision, the more important it becomes for the military to emphasize this separation of responsibilities.

General Tibbets's "STOP" may still be our best policy in time of war. But it is not a useful long-term approach. A country needs to learn from its history, and particularly from the most important and difficult decisions of its times—not to cast blame, but to understand how best to deal with an ever-changing world. There, the experiences of past leaders need to be tapped. We cannot afford to disregard either their wisdom or their errors. Both are important and must be understood.

A curious sentence is, "Ironically, it is this same segment of society who sent us off to war that now wish to recant the flight of the *Enola Gay*." Which segment is Tibbets accusing?

A FORMAL INVITATION TO TIBBETS

On June 17, 1994, I wrote to Tibbets to persuade him to examine our exhibition himself:

I am writing to follow up on our conversation a few weeks ago, when we briefly spoke at the National Air and Space Museum's Garber restoration facility.

You may recall that I asked whether you would be willing to provide your recollections and express your informed views on the bombings of Hiroshima and Nagasaki, in the exhibition which the Museum is planning to open in May 1995, *The Last Act: The Atomic Bomb and the End of World War II*. In the many interviews we have conducted in preparing for the exhibition, we have found that the public—particularly the younger generation—carries so many misconceptions about the raids, what happened to the crews after the war, and what your personal role might have been, that any video-recorded statement you could make would be of the greatest value in setting the record straight.

The main aim of the exhibition is to tell the story of the last days of World War II for the benefit of the generations of Americans who have grown up since 1945 and don't understand either why the United States is the only country to have ever used the atomic bomb in war, or how powerful a weapon the bomb really was. It is of paramount importance that this history is not forgotten or trivialized and thereby misinterpreted or lost.

For the future of our nation it is essential that people understand the rational arguments leading to President Truman's decision to drop the bomb. It is also important for the public to realize just how powerful and destructive a weapon the atomic bomb was and is. That is why one section of our exhibition shows the awesome extent of destruction suffered at Hiroshima and Nagasaki. Misinterpreting our display of these effects of the bombing, critics have taken us to task for displaying undue sympathy for the Japanese.

The Museum has also been criticized for inadequately caring for the *Enola Gay* in decades past. The evidence, however, denies that allegation. Aircraft of the *Enola Gay*'s size cannot be transported off the grounds of an airfield. Never having had an airport of its own, the Smithsonian was always in need of a host Air Force base for housing the aircraft. At Andrews AFB, the airplane was denied hangar space even though my predecessor Paul Garber must have pleaded that the aircraft be brought indoors to stop its deterioration. Ultimately in 1960, when vandalism and the weather threatened total deterioration, the Smithsonian took the draconic step of disassembling the airplane so it could at least be transported to and brought under roof at our restoration facility.

I am told that, in the early 1970s, the Museum offered the *Enola Gay* to two Air Force bases, Offut AFB in Nebraska and Kirtland AFB at Albuquerque. Both had expressed interest in displaying her, but neither could guarantee to keep the aircraft under roof, and so the *Enola Gay* stayed at the Garber restoration facility, where work began ten years ago to fully restore the aircraft for history. Since coming to the Museum in 1987, I have directed that this restoration be maintained as our highest priority, and it will continue to hold that preeminence until fully restored next year.

I know it pains you that we will not yet have a building adequate for displaying the fully assembled aircraft in 1995. Congress only last summer authorized us to plan a facility that will be permanently able to house the *Enola Gay* indoors, at Washington's Dulles airport; but that action came too late for a display there in 1995. Nevertheless, the Museum believes that the front 60 feet of the aircraft's forward fuselage resting on its landing gear, and with a *Little Boy* casing on a dolly beneath the open bomb bay doors, will make an impressive sight in the gallery we are assembling.

These are the circumstances surrounding our planned exhibition. They are not ideal, but I believe they provide the Museum a never-to-be-repeated opportunity to bring to the public a thoughtful exhibition on the most pivotal historic event of this century, the dropping of the atomic bomb on Hiroshima. In my mind, it is essential that we make every effort, for the sake of our country, to make you and your views a part of the exhibition.

I am writing you now for two reasons. First, I would like to again invite you, as we have invited so many other members of the 509th, to present your personal views to our visitors in a video-recorded statement. The lessons of the historic ending of World War II and the role that you and the members of the 509th played must be properly remembered, so future generations of Americans might learn from them and apply them to problems our nation might face in a future decade. With eight to ten million visitors coming to the National Air and Space Museum each year, the opportunity to present your personal views to such a large fraction of our public should not be overlooked.

Secondly, I have read the statement you released last week and am concerned that, while rumors about the exhibition have reached you, you have not had an opportunity to see for yourself the full script, which comprises the exact text of every label to be displayed, a compendium of every picture to appear, and the list of artifacts to be exhibited. A new draft of the script was completed last week.

I invite you to come to Washington, as a guest of the Museum, so that we might walk you through our plans for the exhibition, give you a copy of the script that you can study at your leisure, and listen to your comments or answer any questions that you might have. If it is inconvenient for you to travel to Washington at this time, members of the exhibition team and I are ready to personally visit you in Columbus, at any time convenient to you, to provide you with the same information.

I know that you will wish to think about this request, and you may also want to avail yourself of further information before you make a decision. Should you wish to speak with [any of us at the Museum who could be helpful] we would be glad to provide you with any information you might like to have.

I plan to call you in the next few days, when you have had an opportunity to consider our request.

General Tibbets, you have always served your country when it needed you. Could we not ask you to once more take up that challenge and provide the American people your views on the events that so dramatically brought World War II to a conclusion and forever changed people's lives? [7]

I called Tibbets at 8:30 A.M. on June 22. My notes on the conversation show that Tibbets answered the phone himself, and said he had received my letter.[8] After I had recapitulated our intent, he said he would be willing to provide us with a video-recording, but he could not do it briefly. There were too many matters to be mentioned about the complicated state of affairs toward the end of World War II.

I said that the typical attention span of a visitor on his feet was four minutes. He replied that he had stopped giving Kiwanis and similar talks, because he had found he could not do the topic justice in the typical eight minutes they wanted to allow. Wouldn't we have some place where visitors could sit down. I told him that there would be limited seating and probably far too many people coming through the gallery, but that we would somehow arrange it. I asked what he would need, and when he thought that fifteen minutes would probably do it, I said that we would be glad to work around fifteen minutes or somewhat more if that were needed.

As to format, he thought he might feel more comfortable just speaking into a camera than in a question and answer pattern. He would prefer recording it in Washington rather than in Columbus, but would want to have one or two months of warning, especially because it was summer now, and he got around a lot better in the summer. He and his wife had plans to travel.

I told him we'd work out a date with him that would give him a couple of months' notice. That would also give him time to sort out what he'd wish to say. I also added that he might be called by Tom Crouch or others on our staff working on the exhibition to work out a format for the video and arrangements for the trip. We would pay all his travel expenses. He said he'd be pleased to work with our team. The best time to reach him on the phone was early morning or early evening.

And finally, I told him I would write him a letter recapitulating the essence of the discussion.

On June 24, I followed up:

> I am very pleased that I had a chance to reach you on the phone on Wednesday, and that you are willing to consider having your views video-recorded in connection with the National Air and Space Museum's exhibition *The Last Act: The Atomic Bomb and the End of World War II.*
>
> As I had mentioned in my earlier letter, the central role that you and the men you commanded played in bringing World War II to an abrupt end is so important to our exhibition that it would be difficult to conceive of its succeeding without your personal input. Your views of the situation faced in the last months of the War will be of the greatest interest to our visitors.
>
> You mentioned that the message you wish to emphasize cannot be conveyed in less than about 15 minutes, and the Museum will be pleased to accommodate that requirement. The format of the recording you would find most fitting can be worked out with our staff. Dr. Tom Crouch, Chair-

man of the Museum's Aeronautics Department, should be in touch with you soon to discuss it.

Over the telephone I also assured you that we will seek to arrange a convenient date for the video-recording, making sure that you have several weeks, perhaps as much as two months' notice, so that you might prepare your remarks at your leisure.

Finally, as I had written you, you will be our guest when you come to Washington. We can help to arrange travel plans and make hotel reservations if that is convenient. And if you would like to bring Mrs. Tibbets when you come, we would be able to cover the cost of her travel and stay as well.

I look forward to seeing you again when you come, and will be greatly interested in the statements you will be contributing to the gallery.[9]

But on June 27, Tibbets responded disappointingly:

Thank you for your letter of June 17th and your telephone call of June 22, 1994.

I am not at this time ready to travel to Washington, D.C. or to receive your interview team here in Columbus for the purpose of making a "video-recorded statement." In my mind, such an act would be a *de facto* approval of the exhibit plans which, I understand, are incomplete and under review. I am not willing to participate in such a presentation until I know all of the facts. I may consider participation at that time.

As for the exhibit script, you may, if you like, send me the most recent version of text, concepts, etc.. At that time, I will give some thought as to whether or not it is appropriate for me to participate.

You should know, too, that I am not interested in entering into any debate about the *Enola Gay*. I have made my position quite clear. For this aircraft and this subject to have been of the "highest priority" for so long is indeed a surprise to me. My only contact with the Air and Space Museum has been relatively recent and only for the purpose of answering the questions of Curatorial Assistant Joanne Gernstein. No-one "in charge of the exhibition" as you say has been in contact with me. And, at this point, we will leave it that way.

To sum it all up, let me add this. I am not ready to participate in an interview. I would not do so in any event until after I have read the exhibit script or final plan. Should I ultimately elect to make a statement, I would want to review questions in advance and be comfortable with what we are trying to accomplish with the final product.

Again, it is not my purpose to enter into any debate. Neither do I want to receive a flurry of telephone calls or visitors. I am confident that you can understand these matters.[10]

I was puzzled by the shift in tone. We had begun a dialogue with Tibbets back in 1988, when he had asked us to work with his former 509th Composite Group men. We had done that, and more recently had worked extensively with his crew on the *Enola Gay*. He had been friendly in a telephone call his men had suggested Joanne Gernstein make to him. I had talked with him just a few weeks earlier, when he was at the Garber restoration facility. On the one hand he was private and wished to be approached only through his former men in the 509th; on the other, he seemed upset at our having not approached him directly.

Two weeks later, I decided to write to him again:

> I am responding to your letter dated June 27, in which you raise a number of reservations about participating in the National Air and Space Museum's exhibition of the *Enola Gay,* planned to open in late May 1995.
>
> The points you raise are certainly appropriate:
>
> We would not expect you to make a decision on contributing a statement to our exhibition, without first having seen the script it will follow. I am, therefore, conveying to you, with this letter, the revised draft for the script *The Last Act: The Atomic Bomb and the End of World War II*. It is the blueprint we will follow in preparing the gallery.
>
> I know you are fully aware that the Museum has been accused of doing the veterans and the 509th an injustice with this exhibition. After reading the script I hope you will see, instead, that we have tried to deal as fairly as possible with a difficult subject.
>
> The Museum has been keenly aware that the public has many questions about the use of the atomic bomb. These should not be ignored or swept under the rug. That would only give the appearance that the United States is unable to face the issues squarely. As a national museum I think we have both an opportunity and an obligation to show that the questions under debate can be discussed openly and without apology; but we also realize that this is only possible if we are willing to touch on *all* the issues that have been raised from time to time, and not just those that are unanimously accepted as uncontroversial. In that respect the *Enola Gay* differs from the *Wright Flyer* or the *Spirit of Saint Louis*, which can be exhibited with just a single label. That format would not work for the *Enola Gay* without exposing us to accusations that the museum into whose care the aircraft had been placed by our nation was playing it coy.
>
> This is the context in which I view the exhibition. I think you will understand that it is a defensible and honorable point of view, though you may disagree with it.
>
> As to the format of the statement we would like you to consider providing for the exhibition, our only concern will be to make it as true to your views as possible. We have no particular questions in mind, though we could let you have some that would typify queries to which the public might like to have responses. But we had felt that the best way to present your thoughts would be with a statement you yourself had prepared. We would undertake to present it in its entirety, as long as it did not exceed something of the order of 15 minutes in length. The statement would be projected on a video monitor placed outside the *Enola Gay,* near the right side of the cockpit. Its approximate location is shown by the square marked with your name on the attached gallery layout.
>
> I would appreciate your considering this offer and responding to it with full candor.[11]

Attached to the letter were drawings showing the gallery layout, and the most recent version of the exhibition script.

To this I received no answer. After waiting another month I wrote again, in the same vein, on August 23, to say that I would try to call the following week. That too proved impossible. Tibbets now had a new, unlisted telephone number, which I could not obtain from the operator. I felt that my pursuing the matter beyond this would probably only irk the general. I hoped that ultimately his close friends, his former bombardier, Tom Ferebee, and navigator Dutch Van Kirk, who by now had helped us make a fine film for the gallery, would speak to him on our behalf. They would be more likely to persuade him than I.

MEETING THE VETERANS' ORGANIZATIONS

While Herman Harrington and Hugh Dagley had said they would be willing to work with us, they also ruled out sitting down at the same table with other veterans' organizations. Since the Legion is by a wide margin the largest veterans' organization, they felt their influence would be diluted if they had to work out compromises with the many smaller organizations.

Another large consortium, however, is the Military Coalition, which coordinates the activities of twenty-six military associations. The largest of these is the Retired Officers Association (TROA), which has members from all the uniformed services and provides leadership on issues of sweeping concern. On June 15, 1994, I wrote Admiral Tom Kilcline, President of TROA, describing the exhibition we were planning, assuring him that it would "treat our servicemen and women as it should: as skilled, brave, loyal and dedicated members of the United States' Armed Forces who honorably served their country," and inviting him to "meet with the exhibition curators and me to discuss how we might best achieve a positive outcome. I am confident that once all the facts are known, there will be a consensus that the museum has created a comprehensive and objective exhibition which chronicles a decisive period in history."[12]

My aim here, again, was to secure the backing of a large and sensible veterans' organization, so that the AFA's strident stance would not be the sole perspective mirrored in the media.

Kilcline responded the following day, and proposed that the Museum staff meet not just with him, but with representatives of several other veterans' organizations as well. I agreed to that. We decided to schedule a meeting for July 13, 1994, and I began to invite the leaders of the half dozen organizations that Kilcline had proposed we ask.

23

The Military Coalition and the Service Historians

Admiral Kilcline, President of the Retired Officers Association (TROA) had suggested that we invite the Veterans of Foreign Wars (VFW), the Disabled American Veterans (DAV), the Military Order of the World Wars, the American Legion, General Kicklighter's 50th Anniversary Commemoration Committee, and the Air Force Association (AFA) to the July 13, 1994, meeting at which the Museum would present our plans for the exhibition to them.

At the AFA I wrote to both Monroe Hatch and John Correll to invite them.[1] To make sure that their views had the best chance of prevailing at such a meeting, they had Ken Goss, the AFA's director of national defense issues, write to the other members of the Military Coalition the following week. I only learned of this memorandum, "The *Enola Gay* Controversy or—Who Started World War II?" a year later, when the AFA released it to the press:

> ... Because many of you also have received information from the NASM staff, the Association wanted to share our position and supporting information with you.
>
> The Association believes, after hundreds of hours of research and analysis by historical and aerospace experts, that the proposed exhibit is politically biased and reinterprets history in a manner that distorts the role of the United States in World War II.... According to John Correll, Editor in

> Chief of Air Force Magazine, unless major, substantive revisions are made to the exhibit, Americans will be viewed as waging a war of vengeance against the Japanese; the Japanese will be viewed largely as victims of American aggression.[2]

On July 12, the day before the meeting, the Air Force History Office telefaxed their critique of the latest version of the script.

> ... the Air Force does not in any way endorse this exhibit. However, this revised script contains additions, deletions, and changes that make it an improvement over the previous script.
>
> Nonetheless, this latest script falls considerably short as far as balance and context are concerned....
>
> There are some curious omissions. For example, although President Franklin D. Roosevelt is mentioned on several occasions, his role as Commander-in-Chief of America's armed forces never comes into play. The fact is that although FDR in pre-1941 criticized the bombings of civilians, during 1942–1945 he was a strong and impatient advocate of strategic bombing. He put enormous pressure on Marshall and Arnold to bomb Japanese cities "heavily and relentlessly." ... There is no question that strategic bombing policy flowed clearly and directly from the Commander-in-Chief. This is fundamental and most important—nowhere does FDR's intense desire to bomb appear in "The Last Act." ...
>
> To summarize: The overall impression gained from "The Last Act" is that the Japanese, despite years of aggression and wanton atrocities and brutality, remain the victims. The culprits in this version of history are the American strategic bombing campaign (against civilians) and those who directed and implemented it.

The Air Force felt the script was threatening its good name! They had been ordered to do the bombing and therefore were blameless. That is of course true.

In our form of government, all actions of war ultimately are the responsibilities of the president as Commander in Chief. Because this is so natural, the museum had never considered including such a statement in our script. But the Air Force apparently wanted to be seen as "good guys"—not that one could blame them—and wanted the responsibility to be shouldered by President Roosevelt.

This now appeared to be the real reason for wishing to bring in President Roosevelt's dedication to bombing. Unfortunately, I did not receive a copy of this letter until the AFA published the correspondence a year later. Nor had I received a copy of the letter Hallion had sent General McPeak on April 8, 1994, in which Roosevelt's fervor for bombing the Japanese had also been emphasized. Otherwise, I would probably have understood these now explicitly stated fears and done something to reassure the Air Force. Their letter now, went on to list about half a dozen points on which they disagreed with the updated exhibition script. Many of these seemed rather minor. Given the endorsement Richard Hallion and Herman Wolk had written us on February 7 on reading our first draft, and given that they acknowledged we had made "an improvement over the previous script," it was curious that they now so explicitly disavowed it. The main explanation seems to come from the first paragraph in their critique, which stated,

> [This] is not the kind of exhibit that the Air Force would have done for the *Enola Gay*. The Air Force would have chosen a context emphasizing B-29 organization, development, production, training, command, and the evolution of bombing policy and strategy.[3]

In short, Hallion and Wolk would have preferred to treat the aircraft as a typical B-29 bomber, and in the context of military preparedness, logistics, effectiveness, and policy. The trouble was that *Enola Gay* was not a typical B-29, and that her mission had changed world history.

The next morning, July 13, 1994, the historians attached to the military held another meeting to compare notes on the most recent revision of the script the Museum had sent out to them. As usual, Commander Luanne Smith, from General Kicklighter's staff, attended as coordinator on behalf of the 50th Anniversary of World War II Commemoration Committee. Afterwards she called Tom Crouch, who wrote me a note reporting what he had heard from her, so I would be forewarned for the afternoon's meeting.

The gist of his message reflected Commander Smith's record of this meeting, which had been hosted by Dr. Alfred Goldberg, who had, earlier in his career, been the Air Force historian and now was historian for the secretary of defense. Parts of the record read,

> The Air Force, represented by Herman Wolk, indicated that they would not be able to support the script whatever changes are made to it. The other historians, however, viewed the script more favorably and have only a handful of critical issues that require attention. . . .
>
> The Air Force would like to eliminate the entire final segment of the script concerning the nuclear legacy, but would still not support the exhibit if this is done. General Armstrong initially was also in favor of removing the nuclear legacy portion of the exhibit, but changed his mind when it became apparent that most of the historians disagreed, and when he understood that one of our Committee's purposes is to teach the legacy of World War II. Both the Joint Chiefs of Staff and Office of Secretary of Defense historians were basically favorably impressed with the script, but indicated that improvements can be made, including changes to factual errors.
>
> Both Dr. Goldberg and General Armstrong indicated that it is important that the historians keep an objective view and not allow service parochialism to cloud their judgement. . . .[4]

By virtue of their positions, both David A. Armstrong, who headed the history office for the Joint Chiefs of Staff, and Alfred Goldberg, who worked for the secretary of defense, were largely aloof from interservice rivalries. As evidenced by Commander Smith's record, they were, by now, concerned that the Museum's exhibition was becoming an ideological battleground.

In analyzing these concerns a year later, Tony Capaccio, editor of *Defense Week,* would look back on the fall of 1994 and note a set of comments Goldberg had written lead curator Michael Neufeld, in which he had

> praised two sections that the AFA and other veterans' groups worked hard to eliminate—sections on Hiroshima's "ground zero" and the bomb's post-war legacy.[5]

The Air Force Association and the Air Force historian persisted in their opposition to those sections, while other military historians and organizations conceded they were a significant part of the history the museum was attempting to portray.

A MEETING WITH VETERANS' ORGANIZATIONS

At three o'clock, the representatives of the veterans' associations, two members of the staff of the House of Representatives Veterans Affairs Committee, Commander Luanne Smith and Colonel E. Vincent, from General Kicklighter's staff, and museum staff met in the museum's main conference room. I asked Admiral Kilcline whether he would cochair the meeting with me to assure balance, and he and I both welcomed the attendees.

General Hatch, with John Correll at his side and conferring with him, did most of the talking, and heavily attacked the planned exhibition. One point of contention was a note curator Michael Neufeld had circulated to the exhibition's advisory committee and to the service historians three weeks earlier, on June 21. He had sent all of them a revised copy of the exhibition script, which included changes he had made in response to their comments, and added a cover letter, which I reproduce in its entirety, because one sentence from it was later often quoted to indicate an unwillingness on the part of the curators to listen to advice. Neufeld wrote:

> Dear Advisory Board Member or Military Historian:
>
> Enclosed you will find the revised script of our exhibit, now titled, "The Last Act: The Atomic Bomb and the End of World War II." This script revision was based on your helpful and extensive comments plus a review by an internal committee at the National Air and Space Museum. Because many of the issues in this exhibit are controversial by their very nature, this exhibit cannot hope to satisfy everyone. We hope, however, that you find the revised script to be balanced and consistent with the latest scholarship. If you find any factual errors or if you object strongly to certain formulations in the revised script, I would be happy to hear them. But, if the exhibit is to be opened in late May 1995, as planned, we must now move on to the production and construction phase. This script therefore must be

> considered a finished product, minor wording changes aside. Once again I would like to thank you for your willingness to invest considerable time and energy in advising us on this very difficult, but very important exhibition.[6]

Hatch now quoted that one sentence, which the AFA would subsequently cite over and over again: "This script must be considered a finished product, minor wording aside." Later, the sentence was also quoted to suggest that the curators had been "out of control," that I had been unable to keep them under rein. Never did the AFA cite the sentence with the explanation that the script had already undergone five months of revisions, in response to comments from nine members of the advisory committee, six additional service historians, the Air Force Association, and many critics internal to the museum, and that it was now urgent to adopt a schedule that would permit the exhibit to open on time, in May 1995. Nor would the AFA acknowledge Commander Smith's assessment that aside from the Air Force, the other historians had "only a handful of critical issues that require[d] attention."[7] As Neufeld's letter clearly indicated, he was still ready to incorporate such a small number of changes.

In response to Hatch's strong comments, Commander Smith spoke up. Her notes on the meeting read,

> . . . I mentioned that we had met with the Service Historians that morning. I also said that there had been major changes to the script and that the NASM had been very accommodating to making changes. . . . Mr. Correll responded very angrily to my comments saying that that is not the Air Force Historian's position, and thrust a copy of Dr. Hallion's script-review before me. . . .[8]

The representatives of the other veterans organizations, who had not been apprised of the dispute, expressed some bewilderment at this confrontation. The museum staff distributed copies of the most recent version of the script to them, and we agreed to meet again after they had had an opportunity to inform themselves better.

The meeting had lasted a little more than an hour.

After the others had left, I thanked Commander Smith for her efforts on our behalf, and then called Gen. Kicklighter, who was out of town but had been anxious to know how the meeting had gone. Subsequently I also wrote the veterans' organization representatives who had attended to ask them for their comments on the script by August 5, noting that we were "on a tight schedule with the exhibition and would need to have input by that time to give [their comments] adequate thought and make any urgent revisions."[9]

Commander Smith later recorded,

> Subsequent to the meeting, I briefly talked to Mr. Correll, indicating that only the Air Force had real problems with the script. He appeared shocked, and asked whether General Armstrong thought that the script was okay. I

> indicated yes, based on his support for his historian at the meeting earlier in the day. Again, Mr. Correll sounded shocked. . . .
>
> After the meeting, both the AFA and Dr. Hallion made a number of calls to shore up support for their anti-exhibit stance. Mr. Frank of the Marine Corps Historian's office contacted me and indicated that Dr. Hallion had asked him to serve on the script review panel. . . . Dr. Hallion also contacted General Armstrong, to ask him if he thought the exhibit was 'fine,' as I had mentioned to Mr. Correll. The AFA also talked to the *Washington Post*, with Dr. Hallion echoing their views. . . .[10]

Two weeks later, apparently worried that their stance might be undermined if all historians attached to the military services declared themselves satisfied, the AFA called on their director of communications, Stephen Aubin. On July 26, he wrote directly to those historians' supervisors. Addressing himself to Brigadier General Ed Simmons, USMC, retired, director of Marine Corps history and museums, at the Marine Corps Historical Center; Brigadier General Harold Nelson, chief of military history, U.S. Army Center of Military History; and Dr. Dean Allard, director of naval history programs, Naval History Center in essentially identical letters, he wrote,

> I recently learned of your contacts with the National Air and Space Museum on the subject of its planned 1995 exhibit. . . . I have been talking regularly with Dr. Richard Hallion on the same matter. As you may know, the Air Force Association has expressed its strong opposition to the museum's current plans. We believe that the current script . . . still lacks balance and context. . . . I am enclosing some recent correspondence and our reports on this subject. If we can be of any help, please do not hesitate to contact me.[11]

These tactics did not prevent the service historians from gradually expressing satisfaction with the script. But the AFA was not overly worried. They, as well as we, knew that the real battleground would be in Congress and the media.

A PHOTOGRAPHIC INTRODUCTION

The most significant outcome of the afternoon's meeting, however, only emerged after all our visitors had left. Several members of the staff and I had congregated just outside the conference room and were having an informal postmortem, when Nadya Makovenyi, head of the Museum's exhibits department, came up with a brilliant suggestion: Why not have an introductory photographic exhibition on "The War in the Pacific" outside the gallery in which we were going to exhibit the *Enola Gay*. It could be arranged so that visitors would first have to pass through this photo-exhibit to get into the gallery.

A year earlier, Secretary Adams had suggested that we might do more on the history of the war. At the time I did not see how we could accommodate that wish. We already had a full gallery just dealing with *Enola Gay*, and the space immediately outside that gallery was being considered for other exhibitions.

Also a year earlier, I had expected that anyone coming to the exhibit would know enough about the war in the Pacific to remember that America had been attacked at Pearl Harbor and had fought an uphill battle for three years before decisively turning the tide. I now realized that I had been wrong. Our filmmaker, Patricia Woodside, had recently conducted a number of man-in-the-street on-camera interviews in the Washington, D.C. area, which showed that many young people had only the vaguest ideas about World War II. Some did not even know who had dropped an atomic bomb on whom. One person apparently thought it was the Russians who had dropped it on the Chinese, while another conjectured it might have been the Japanese on Pearl Harbor.

I also had never understood what Monroe Hatch had in mind when he criticized the exhibition as having "structural problems." But now, in a letter recapitulating his stand at the meeting, he finally made himself clear on this one issue:

> . . . The exhibit still lacks balance and context. . . . You can't give visitors to the museum and students of history a balanced perspective of World War II if you only show the "last act."[12]

This probably was not the only point of contention for Hatch, but it was one on which I now agreed, and for which Makovenyi was suddenly offering an opportunity to act: at relatively low cost, we could mount an exhibition of perhaps fifty large-format photographs in the space outside the gallery, which she could vacate for that purpose.

We would need a separate curatorial team to handle the planning of this photographic offering at such short notice. I decided to ask Tom Alison, a former Air Force pilot and now our new curator for military aircraft; Don Lopez, who, though now officially retired from the museum's staff, had flown for the Army Air Corps in World War II China; and Tim Wooldridge, who had made his career as a naval aviator. All three agreed to take on this task. The existing curatorial team had its hands full already, and could not be asked to do more than work with the others to assure a smooth transition from the photographic exhibition to "The Last Act" gallery.

I asked Makovenyi to work out the concept to see what resources she would need to go forward. In the meantime, I also began to discuss the idea with the curators who would have to make themselves responsible for the actual work. By mid-August, we were beginning to go forward; at the end of August, we made a public announcement that we would add an entire section on "The War in the Pacific."

VACATION

I had intended to take off for a couple of weeks of vacation starting in mid-August 1994. Instead, I decided I would spend the time entirely on matters related to the exhibition. Normally I did not have that much uninterrupted time to devote to it. Now, I felt, I urgently had to.

The first week, the exhibition team and I sat down with Mark Rodgers and Steven Soter for hours at a time and went through the script word by word, looking at ways we could improve it. Tom Crouch was away that week, partly at a meeting and partly on vacation. We also began planning the photographic exhibit on the war in the Pacific, which Tom Alison, Don Lopez, and Tim Wooldridge were prepared to curate.

I also attended an August 16, 1994, meeting called by General McPeak in response to my letters to him and Secretary Widnall—unfortunately not a meeting with him or with Monroe Hatch. Instead, he asked his new vice chief, General Thomas S. Moorman, Jr., to host the meeting, and Moorman invited Correll, Hallion, and historians from the military services Hallion asked to attend. Unfortunately, Hallion had not invited the historian most positive about our exhibition, Alfred Goldberg.

The only useful result of the meeting was my agreeing to a request by the military historians to see for myself how little the museum's curators had changed in the script in response to their suggestions. This I did over the following weekend, and found, to my dismay, that I could understand their disappointment.

The problem was that many of the suggestions the service historians had made appeared to be useful, but more in the nature of detailed commentary that could not always be incorporated into an exhibition, where every excess word needed to be weeded out to keep labels from getting excessively long and cumbersome for the public. The curators had decided in many instances that there was no pressing reason to adopt the recommended material. My feeling now was the opposite: If it made the service historians happy and didn't excessively hurt the script, we might as well adopt their recommendations. We badly needed their support, and right now, they were clearly angry, thinking that we were unwilling to listen to their advice and heed it.

I called Hallion about this, but he was out of town and his colleague Herman Wolk took the call. Wolk in turn called the *Washington Times*, and their regular writer on the subject, Rowan Scarborough, came out with an article headlined "Museum director rebuked curators of A-bomb exhibit."[13] Wolk was quoted as saying that I had been surprised to find the curators lax in heeding the service historians' advice. I wrote to the paper at once to say that my call to Wolk "should be seen as part of a process of discussion and debate," both inside the museum and with outside advisors, "aimed at producing the very best, most thoughtful exhibition." Such discussions and differences in points of view should not be confused with dissension or lack of discipline.[14]

Privately, I was greatly disturbed. Wolk's action had to have had the approval of his boss, Hallion. If we no longer could openly talk with each other without fear of having our conversation reported and analyzed in the newspapers the next day, how could we continue to discuss the contents of the exhibition or seek advice? We had arrived at an impasse with the Air Force History Office.

That week, the exhibition staff, Mark Rodgers, Steven Soter, Mike Fetters, and I sat down with the exhibition team a second time, and went through every comment we had received from all the different advisors for the past few months, deciding whether to accept or reject them. This was slow work. Michael Neufeld, who had just married, was away that week, and we had to go over some of this work again on his return to make certain that he was in agreement. But we made considerable headway, and by August 31, 1994, had a strongly revised script, with which I was much happier.

Tom Crouch was not. We had eliminated or toned down a great deal of the prose he had put into the section that dealt with the effects on the ground, making the labels much more factual and less emotional. This was the part of the script he had originally written, and I suppose it hurt him to see his work treated that way.

In fairness to Tom, I have to say that he had a difficult task. I don't think anyone else could have done better. The Japanese had worried that their unassuming artifacts would never be noticed next to the huge *Enola Gay*. Tom was trying to give these artifacts prominence by emphasizing the painful memories they symbolized.

Writing labels that project just the right amount of feeling is difficult. The labels have to be repeatedly tested on people to see their reactions. The responses we had obtained showed that our wording was overly dramatic and had to be toned down.

THE SERVICE HISTORIANS SUPPORT THE EXHIBIT

On August 26, 1994, after the museum had announced its addition of a four-thousand-square-foot photographic exhibit on the war in the Pacific, thus virtually doubling the originally planned exhibition's size, Commander Smith informed General Kicklighter,

> The Air and Space Museum has been extremely accommodating to making changes to their script, contrary to what is being claimed by exhibit opponents. The NASM's opponents have gone out of their way to misrepresent the Museum's intention....
>
> The opposition is intent on canceling the exhibit as currently planned, and want to merely display the *Enola Gay* with a statement indicating that

> it was the aircraft that ended World War II. . . . The abbreviated exhibit would . . . not fit our Committee's objective of teaching the legacy of World War II, which includes the legacy of nuclear weapons, like them or not.
>
> The NASM's planned exhibit is a superb opportunity. . . . We may have no better opportunity to teach the American public about World War II than this exhibit, which should be among the most visited in our Nation's history . . . and one that would make our veterans proud because the American people would realize the sacrifices they made for our freedom.[15]

By this time only the Air Force historian was opposing the exhibition. The other service historians had been very helpful with their critiques and were now generally satisfied that the script was greatly improved, though additional changes should still be considered. Dr. Alfred Goldberg had long been a supporter, and over the months he had been joined by Edward Drea, chief of research and analysis at the Army Center for Military History for the U.S. Army; David Armstrong, director of the Joint History Office; and Mark Jacobsen from the Marine Corps Command and Staff College. Responding to Goldberg's request for comments, Drea wrote on July 12, 1994, "General observation: I find the script more balanced than its predecessor." Armstrong followed two days later, on July 14, "Some attempt has been made to address virtually every criticism raised, although in some cases the fixes have been minor. . . . Overall, the script is more concise, coherent, accurate and balanced. . . ." Jacobsen followed suit on July 18, "I'm pleased to see a number of substantive changes, both in content and in tone, that should make the eventual exhibit acceptable to a broad range of Americans."[16] In the same vein, Wayne Dzwonchyk, from the Joint Chiefs of Staff historian's office, on September 22, 1994, indicated to Commander Smith that the script was "very good . . . really pretty balanced."[17]

Historians from both the academic communities and the military services had by now largely satisfied themselves that the exhibition's coverage was historically thorough and balanced. The museum still had nine months before the exhibition was to open to make other adjustments as we saw fit. Only the lobbyists were still dissatisfied and were keeping the media and through them the veterans stirred up. Monroe Hatch wrote me a four-page, single-spaced letter on August 24, commenting,

> While we are pleased that you have received the kind of "line-in, line-out" comments provided by the service historians and others who have undertaken a "technical" review of the script, the issues of context and balance need to be addressed on the "broad" structural and conceptual levels.[18]

He then went on to describe a totally different kind of exhibition that he would favor. Given the campaign against the exhibition he was just then waging in Congress, he probably felt he would ultimately be able to dictate the changes he wanted, and saw no reason to stop just because the historians were now generally approving the exhibition. On the other hand, he probably was well aware that his loss of the service historians' support might weaken his position.

He chose to attack the competence of the museum's curators.

HATCH ATTACKS THE MUSEUM'S CURATORS

Hatch's letter continued by asking "how scholars and high-level curators at one of the finest museums in the world could ever produce concept papers and drafts of scripts so out of tune with historical scholarship, the published memoirs of the leaders who made these awesome decisions, and with the firsthand reports from veterans who fought the war[?]" He advised, "I believe you would be better served if you expanded the charter you have given to Col. Tom Alison, Col. Don Lopez, and Capt. Tim Wooldridge, to include all sections of the exhibit, allowing them to make modifications throughout."[19]

Hatch, in short, was asking for curators on the museum staff who had formerly served in the military, and whom he therefore trusted, to take over from the curators who had been in charge. By then, I had already assigned these former military men curatorial duties with the exhibition. Originally, they had not been members of the team, because Don Lopez was scheduled to retire and Tim Wooldridge's fellowship term would end well before the exhibition was to open. Tom Alison had not joined the museum until years after the exhibition team had been selected. Now all three were helping out, but I saw no reason to replace the original team.

Hatch's allegation that the script was "out of tune with historical scholarship," however, was clearly false. On July 15, 1994, Alfred Goldberg, historian for the secretary of defense, wrote to Michael Neufeld, lead curator on the exhibition:

> My overall impression of the *Enola Gay* script is favorable. It shows evidence of careful research and an effort to realize a balanced presentation—that is, taking account of differing interpretations of events and seeking to present them evenhandedly and emphasizing the most significant parts. This is, of course, extremely difficult given the strong emotions and biases on both sides of the issue. It will not be possible to satisfy everybody—therefore some of the decisions may not satisfy anybody. Still, the canons must be honesty, accuracy and objectivity in attempting to present a picture of a two-sided war.[20]

24

The Media and a National Museum's Defenses

THE MEDIA

The first comprehensive article on the exhibition was Arthur Hirsch's carefully researched article in the *Baltimore Sun*.[1] Hirsch did not shirk mentioning that a controversy was brewing. But he treated the different views presented by the museum staff and by opponents with respect. He had talked with W. Burr Bennett, Jr., and William A. Rooney, who had repeated many of the allegations they had so long made in their correspondence with us and their fellow veterans; but he took no sides in the dispute, allowing readers to formulate their own opinions.

An April 2, 1994, article by David C. Morrison in the *National Journal* quoted solely Correll's "special report." Morrison apparently never spoke with anyone at the museum. In part, his long article stated,

> One might think that air power enthusiasts would be, well, enthusiastic about the pending display. In fact, as John T. Correll, editor in chief of *Air Force* magazine, a monthly put out by the Air Force Association (AFA), notes in a "special report" on the *Enola Gay* display, for the past two years, "museum officials have been put under fire from veterans' groups, who charge the exhibition is politically biased."[2]

A May 7 column by Tom Webb, on the staff of the Knight-Ridder-Tribune News Service, appeared in the *Wilmington Morning Star* under the headline, "Grisly display for famed *Enola Gay* bomber angers vets":

> . . . unlike the proud display accorded . . . other famous aircraft, officials at the Smithsonian's Air and Space Museum have more chilling plans for the *Enola Gay*. And that rankles a growing number of World War II veterans who wish to evoke the pride of their wartime sacrifice—not have it overshadowed by gruesome photos of dead children and radiation victims.
>
> Ben Nicks, 75, was a B-29 pilot who flew his last mission the day the atomic bomb was dropped on Japan. He is delighted the Smithsonian is painstakingly restoring the *Enola Gay* and plans to exhibit it again. "The *Enola Gay*, the aircraft itself, is nothing but a piece of tin," said Mr. Nicks. "But as a symbol, as a reminder to generations who followed World War II and to whom it's only a memory, we hope that the symbol is one that reflects credit on us."
>
> But this is not quite what he expected. . . Along with the proud memorabilia from the 509th bomb group, there will be ghastly photos of Japanese women and children, representing some of the 100,000 who died from the fireball on Aug. 6, 1945, or from the slow poisoning of radiation.
>
> But most troubling to Mr. Nicks and other veterans is their belief that the exhibit slights their sacrifice. Some even felt Americans would be portrayed as the Bad Guys—a charge Smithsonian officials vigorously deny. "What we're really looking at here is a reluctance to really tell the whole story," insisted Tom Crouch. . . . "They want to stop the story when the bomb leaves the bomb bay. This is an exhibit which goes beyond that, including what happens when it hits the ground."
>
> A full year before the exhibit opens, the controversy shows how the shock waves from Hiroshima reverberate still. . . .[3]

A May 5 article by J. Ivan Potts, Jr., took on a more strident tone:

> The intent of this proposal is to use the *Enola Gay* as a front for an exhibit intended to defame the strategic bombing that took place in World War II. The proposed exhibit rewrites history through omission. It makes Japan look like the victim and the United Sates forcing surrender of an enemy, with a superior weapon, committing a "war crime." . . .
>
> Smithsonian academicians, who determine the Institution's policies and priorities, are perverting the museum's original purpose from restoring and displaying aviation and space artifacts to presenting unprovoked and opinionated social commentary on the uses to which artifacts have been put. They should be replaced with experienced professionals who will display these priceless artifacts "as an inspiration to the present and future generations of America."[4]

While Potts's article, though fairly typical, might be shrugged off as representing merely a regional newspaper with a small circulation, during the summer months, the controversy over the exhibition heightened. On July 21, *Washington Post* staff writer Eugene L. Meyer wrote,

> The critics accuse the Smithsonian of choosing political correctness over historical accuracy in the presentation. They charge that the exhibit as planned will portray the Japanese as suffering, even noble victims and the Americans as racist and ruthless fighters hell-bent on revenge for Pearl Harbor.[5]

Tony Capaccio, as editor of *Defense Week* magazine, might have been expected to side with the more strident critics. But in July 1995, by which time the controversy had long disappeared from the front pages, Capaccio and Uday Mohan severely criticized the press in the *American Journalism Review:*

> The controversy was largely fueled by media accounts that uncritically accepted the conventional rationale for the bomb, ignored contrary historical evidence, and reinforced the charge that the planned exhibit was a pro-Japanese, anti-American tract....
>
> The key story ... was a July 21 *Washington Post* Style section piece by Eugene L. Meyer, that elevated the controversy to national status when it caught the eye of Republican Rep. Peter Blute of Massachusetts.
>
> So alerted, Blute issued a letter on August 10, co-signed by 23 other legislators, condemning the museum for proposing an "anti-American" and "biased" exhibit. The lawmakers wanted "an objective account of the *Enola Gay* and her mission rather than the historically narrow, revisionist view contained in the revised script."
>
> After members of Congress intervened, the story, as covered by the media, degenerated into a shouting match, with the veterans' groups doing most of the shouting. But instead of covering the veterans' charges *and* the historical debate, the media focused narrowly on the allegations of imbalance and anti-Americanism.... Reporters, columnists and editorial writers often used criticism by the AFA, the American Legion and other veterans' groups as a club to beat on the museum.
>
> And while the public was continually informed about the veterans' groups' take on the exhibit plans, news organizations failed to report that a number of historians had actually praised the museum for its efforts.[6]

In response to the hostile press coverage we were receiving in the *Washington Post* and other newspapers, I decided that I should write an Op-Ed piece to show the museum's intent to mount a thoughtful historical exhibition that would try to faithfully present a variety of legitimate perspectives. The article appeared in the Sunday, August 7, 1994, *Washington Post:*

> Forty-nine years ago this weekend the United States dropped an atomic bomb on Hiroshima and then another on Nagasaki. A year from now, on

the 50th anniversary, Americans will commemorate these pivotal events. But we lack a national consensus on what to say.

Two divergent but widely held views define the dilemma. One view sprang up as soon as the bombs exploded and the war ended. Its proponents are united on the many details that need to be included in their story. Properly told it appeals to our national self-image. The other point of view, slower in coming to the fore, is more analytical, critical in its acceptance of facts, and concerned with historic context. It is complex and, in the eyes of some, discomfiting.

The first view recalls the morning of August 6, 1945, when three B-29 Superfortresses arrived over Japan's Inland Sea. One of the aircraft, the *Enola Gay*, named for the pilot's mother, approached its Hiroshima target, released its heavy payload, then veered away to distance itself from the bomb. Seconds later, at 8:15 a.m., the atomic bomb exploded over Hiroshima. The crew was stunned by the sight. The blast rocked the aircraft. The 29-year old pilot, Col. Paul W. Tibbets, commander of the 509th Composite Group, which was trained and tasked to deliver the bomb, was awed by the sight of the burning devastated city below. To his copilot he remarked "I think this is the end of the war." Five days and another atomic bomb later, Japan surrendered.

Our troops were ecstatic. They would not have to die by the many tens of thousands in a bloody invasion of Japan. They'd go home instead, settle down with their sweethearts, have children and lead normal lives. They had been asked to save the world for democracy, had accepted the challenge at great personal risk, and had come through victorious.

Approaching the 50th anniversary of Hiroshima next year, these same men, now in their seventies, have asked the National Air and Space Museum, into whose care the *Enola Gay* was entrusted after the war, to put their aircraft on exhibition. They want the Museum to tell their story the way they have always told and retold it—a story of fighting a ruthless enemy, perpetrator of barbaric massacres in China, the infamous attack at Pearl Harbor, the death-march at Bataan, torture and executions in prison camps, kamikaze raids on our warships, and deaths by the thousands for every Pacific island wrested away; a story of the world's top physicists working in secrecy to perfect a mighty weapon; a story of a powerful new aircraft, designed, built and first flown in just 24 months; a story of ordinary citizens, men and women, working together to defeat a ferocious enemy.

These are the themes emphasized by those who fought so hard to secure freedom for their children and grandchildren.

Those children and grandchildren by now are mature citizens. For them the atomic bomb has added associations: ICBMs, megaton warheads, the DEW line, 45 minute warnings, first strike, Mutually Assured Destruction, nuclear winter.... Theirs was not a world of two small atomic bombs, but of 50,000, many of which are 1,000 times more powerful than those that destroyed Hiroshima and Nagasaki. Next year these younger people will not only commemorate a bomb that ended the most terrible war, but also they will have reason to celebrate the restraint that has prevailed for half a century in which no man, woman or child has been killed by an atomic bomb. They want to extend that record to all time.

The *Enola Gay* symbolizes the end of one era and the beginning of another. For an older generation the aircraft meant the end of World War II; for younger people, it ushered in the nuclear age. The postwar generations respect their fathers for the sacrifices they made, but they also realize that the nuclear bombs that saved their fathers' lives continue to threaten their own and their childrens'.

These conflicting views pose the dilemma the National Air and Space Museum faces as we prepare an exhibition of the *Enola Gay* for 1995. We want to honor the veterans who risked their lives and those who made the ultimate sacrifice. They served their country with distinction. But we must also address the broader questions that concern subsequent generations—not with a view to criticizing or apologizing or displaying undue compassion for those on the ground that day, as some may fear, but to deliver an accurate portrayal that conveys the reality of atomic war and its consequences.

To that end, the Museum proposes to tell the full story surrounding the atomic bomb and the end of World War II; to recall the options facing a newly installed President Truman, who had never heard of the bomb until the day he was sworn in; to examine the estimates of the casualties Truman anticipated if U.S. troops had to invade Japan; to consider the extent to which his wish to impress a threatening Soviet Union influenced his decision to drop the bomb; to exhibit the destruction and suffering on the ground at Hiroshima and Nagasaki; and to recall the escalating numbers of weapons in the superpowers' nuclear arsenals during the Cold War, and their current decline.

Faced with a number of alternatives, the Museum has chosen to provide not an opinion piece, but rather the basic information that visitors will need to draw their own conclusions. This is our responsibility, as a national museum in a democracy predicated on an informed citizenry.

We have found no way to exhibit the *Enola Gay* and satisfy everyone. But a comprehensive and thoughtful discussion can help us learn from history. And that is what we aim to offer our visitors.[7]

This did not fare well. The editors of the *Washington Post* either were not aware of or did not believe the polls I have cited earlier that clearly established the generational differences in outlook on the atomic bomb.[8] The following Sunday, the *Post* had a refuting editorial, which attacked the exhibit on quite different grounds:

The two sides are talking past each other, and it is the Smithsonian that needs to do more listening. . . . Mr. Harwit's terms betray the assumption that has rightly made critics so unhappy, namely that the difference between him and his critics is not simply one of political opinion but of

> intellectual sophistication.... What the tenor of the debate suggests instead is a curatorial inability to perceive that political opinions are embedded in the exhibit, or to identify them as such—opinions—rather than universal, objective assumptions that all thinking people must necessarily share. This confusion is increasingly common in academia and owes much to the fashionable and wrong notion that objectivity is unattainable anyway and that all presentations of complex issues must be politically tendentious.[9]

These last two sentences seemed contradictory, but the more important point was the reluctance to discuss the historical facts presented in the exhibition, or the balance projected. Important also was the hostility of the editorial and the matching hostility of the five letters to the editor that accompanied it under the headline "The Mission That Ended the War."[10] One of these was by John Correll, another by Manny Horowitz, a former B-29 navigator who worked closely with Correll on various occasions and had done a detailed analysis of successive versions of the exhibition script.

Altogether, there was twice as much space given to these six negative pieces as to my original Op-Ed piece. In testimony offered a year later before the Senate Committee on Rules and Administration, Kai Bird and Martin Sherwin, both well-known historians, found this to have been a general trend in the *Washington Post*'s coverage of the *Enola Gay* exhibition. They wrote,

> The *Post*'s coverage of the *Enola Gay* was unbalanced: the newspaper reported the controversy as a dispute between thousands of veterans—armed with their irrefutable authentic memories—and a handful of wooly-headed curators. The Smithsonian's curators are described as men of a younger generation who never saw combat, and in some cases were not even American-born or citizens of America. The curators, according to the *Washington Post,* were influenced by left-wing revisionists, the anti–Vietnam war movement and a latent anti-Americanism.... In the *Post*'s coverage, historians were rarely quoted, and the historical evidence was rarely cited.... In stark contrast to the *Post*, the *New York Times* editorialized that the curators should be left alone to do their job, and the *Times* reporters frequently quoted both historians about the controversy and quoted from some of the key archival documents.[11]

A series of cartoons now began to appear. Initially, they just recorded the controversy, as in Garner's *Washington Times* cartoon on August 31, 1994.[12]

But they became increasingly bitter in two of Harry Paine's drawings, on September 2 and October 7, followed by Cullum's cartoon on November 7 and Bruce Tinsley's strip on November 28.[13–16]

"I THINK THE SMITHSONIAN'S GETTING CARRIED AWAY WITH THIS HISTORICAL REVISIONISM..."

A SMITHSONIAN PRODUCTION
NATURAL BORN KILLERS
THE U.S. ATOMIC BOMBING OF JAPAN
COMING SOON
IT BOMBED WITH THE CRITICS, SO THE SMITHSONIAN HAS AGREED TO RE-EDIT.

SMITHSONIAN MUSEUM
GALLERY
SELF-FLAGELLATION. THIS MUST BE ONE OF THE PURITANS...
NO... ACTUALLY, IT'S THE MUSEUM DIRECTOR.

THIS CONTROVERSY OVER THE SMITHSONIAN'S "END OF WORLD WAR TWO" EXHIBIT IS REALLY HEATING UP!
ORIGINALLY, THE MUSEUM PLANNED AN EXHIBIT THAT PORTRAYED THE U.S. AS A BLOODTHIRSTY AGGRESSOR...
...AND JAPAN AS A NATION SIMPLY TRYING "TO PRESERVE ITS UNIQUE CULTURE."
POGO WAS ALMOST RIGHT: "WE HAVE MADE UP THE ENEMY, AND HE IS US...."

Some of the media coverage was sympathetic to the exhibition, as in a *New York Times* editorial on September 5, which correctly noted,

> The Smithsonian is the premier cultural institution in the country; surely it can find a way to incorporate various criticisms without line-by-line supervision from members of Congress who are neither historians nor curators. The problem with endless tampering by Congress is that some critics will not be satisfied with anything short of complete vilification of the Japanese and uncritical glorification of the American war effort. . . .[17]

But this was immediately countered on September 10 by two letters to the editor, one by John Correll, the other by W.E. Cooper, of Dallas, Texas, who had flown B-29s in World War II and had corresponded with us at the museum on occasion. Between them, they obtained more space than had been devoted to the editorial they challenged. Certainly the AFA was quick to organize opposition to any positive coverage, making the net effect a loss for the museum. As the summer of 1994 was drawing to an end, I could see no way of parrying this.[18]

Every time we or our supporters tried to voice our point of view in the press, the opposition was able to obtain twice the column space in response. There is little doubt, judging from the names of respondents we often recognized, that an energetic organization like the AFA has the means to organize a flood of "letters to the editor" from all over the nation with a few well-placed phone calls. To the uninitiated these will look like a true public response reflecting the nation's opinion.

One more article needs to be cited. On August 29, the *Wall Street Journal*, in an unsigned editorial wrote,

> The Smithsonian Institution's mania for revising American history is in evidence again, this time in the National Air and Space Museum's proposed exhibit on the *Enola Gay*, the B-29 that delivered the atom bomb to Hiroshima. In presenting the Japanese as the prime victims of the Pacific War, an American Museum has rewritten history as never before. . . .[19]
>
> What we can learn from the proposed *Enola Gay* exhibit scheduled to open in May is a question that has already evoked furious protest from veterans and military historians. The picture that emerges from the script is of a besieged Japan yearning for peace. This Japan lies at the mercy of an implacably violent enemy—the United States—hell-bent on total victory and the mass destruction of women and children. And why? "For most Americans, this war . . . was a war of vengeance. For most Japanese, it was a war to defend their unique culture against Western imperialism."
>
> Now removed from the script—though only after the Air Force Association and other critics weighed in—the line tells everything about the mindset behind the show. So does the script's not so subtle suggestion of the Nuremberg war crimes defense in its reference to the American pilots who were just "following orders."
>
> . . . it is especially curious to note the oozing romanticism with which the Enola show's writers describe the kamikaze pilots. These are the

Japanese suicide pilots whose noble rituals, rites of purification, letters to their mothers and general spiritual beauty are adoringly detailed in the script. These were, the script elegiacally relates, "youths, their bodies overflowing with life." Of the youth and life of the Americans who fought and bled in the Pacific there is no mention. . . .[19]

In their 1995 article, Capaccio and Mohan, called this last paragraph,

one of the most damaging errors just as congressional and editorial pressure was building against the museum. . . .

The Journal's observation ["it is especially curious to note the oozing romanticism with which the Enola show's writers describe the kamikaze pilots . . ."] was picked up the next day by *Washington Post* reporter Ken Ringle, who wrote that "just yesterday, for example, an editorial in the *Wall Street Journal* found it 'especially curious to note . . .' " He then repeated the statement.

The quote the *Journal* attributed to elegiacal curatorial prose was actually written after the war by a surviving kamikaze pilot, Ensign Yukiteru Sugiyama. This was clearly spelled out in the script. *Journal* spokesman Roger May and Ringle refuse to discuss the mistake. "We don't do postmortems on editorials," May says.

Capaccio and Mohan are correct. In an attempt to show visitors to the exhibition how each side viewed itself in the war, how it saw the opposing side, and how such views were translated into policy and action, we had shown cartoons and quoted from writings of the times. The quotation in question was not a museum view, it was cited to show how the Japanese saw themselves.

Referring to Stanford University historian Barton Bernstein, Capaccio and Mohan concluded their article:

No one expects reporters on deadline to be budding Barton Bernsteins. But the realities of time and space do not mean that the conventional wisdom on the A-bomb has to be uncritically passed along to the public. There was ample opportunity and time as the issue unfolded for reporters to incorporate the latest research into their stories.

In this case the media's shortcomings are all too obvious. Journalists did not do enough research and failed to hold the veterans' version of history to the same exacting standard they used in judging the curators' version.

The [initially proposed] exhibit had flaws of context and historical perspective—but not as serious and certainly not as ill-informed as the media coverage led the public to believe.[20]

Coming from the *American Journalism Review* and a member of the national defense journalism community, this criticism of the coverage of the exhibition was welcome, but of course far too late, except possibly as a postmortem.

A NATIONAL MUSEUM'S DEFENSES

Critics among historians, pacifists, and well-wishing reporters have asked why the museum did not defend itself more vigorously. The answer is simple.

An institution like the Smithsonian, devoted to "the increase and diffusion of knowledge," *should* not, and *does* not, have the means to enforce its views the way that an association created to lobby for the interests of its narrowly based constituency does: It *should* not, because true knowledge is not determined by votes, or by money or power; it is shaped by careful research, respect for facts, and dispassionate discourse. It *does* not, because like other organizations of the federal government, the Smithsonian, by law, is precluded from lobbying the Congress. This extends to soliciting others to lobby on the Smithsonian's behalf.

This basis in law was explained to me in a letter from the Smithsonian's congressional liaison soon after I had arrived at the Institution:

> Title 18 of the United States Code . . . specifically proscribes the use of appropriated funds to influence legislation and provides criminal penalties for those who do so. In addition, the committee reports on our appropriation annually include a prohibition on the use of funds "to promote or oppose legislative proposals on which Congressional action is incomplete."[21]

In contrast, the Air Force Association has no such restrictions. In their world, votes, money, and power do determine the outcome of issues. The AFA routinely lobbies Congress on behalf of the military aerospace community. If their effectiveness in their attempts to reshape or else to scuttle the museum's *Enola Gay* exhibition is any indication, they must be very good at what they were set up to do.

The American Legion, with more than ten times the membership of the AFA, is even more powerful. It has 3.1 million members who pay an average of more than $20 each year in annual dues.[22] Even a small fraction of this money can buy a lot of publicity and influence.

VISIT TO A CONGRESSIONAL REGENT

The Smithsonian and the museum did, of course, try to gather support in Congress. I have already reported occasions on which we met with congressmen and their staffs to explain our position, give them an opportunity to examine our plans for the exhibition, and answer questions. On August 17, the week after we had been invited to meet with the six Republican congressmen who had so strongly disapproved of the exhibition, Mark Rodgers, Mike Fetters, and I visited Congressman Bill Barrett, ranking Republican on the House Administration Committee's Subcommittee on

Libraries and Memorials, and thus in oversight of Smithsonian functions. Barrett was considerably more accommodating than his fellow Republicans had been a week earlier, and we were able to calmly discuss the exhibition with him.

Later, Rodgers, Fetters, and I were joined by Connie Newman and went to see Norman Mineta, Democrat from California and Congressional regent on the Smithsonian board. We met in his office at 4:00 P.M. and talked with him and his staff for an hour and a half, telling him what we were doing and answering his questions. He was highly sympathetic—again an absolute difference from the meeting we had held with Congressmen Sam Johnson, Tom Lewis, Robert Dornan, and their colleagues just one week earlier.

Three months later, however, the Democrats would be out of power, the Republicans would be starting a revolution in the House, Johnson would be taking his seat on the Smithsonian board of regents, and Mineta, though reelected would be contemplating resignation from the House and thereby automatically giving up his own seat on the board. These changes would bode ill for us.

THE IMAGE MAKERS

We did whatever we could do legitimately. But unlike the larger veterans' organizations, which often engage public relations firms, we could not hire such image makers to promote our cause; we could not write articles asking supporters to contact their congressmen on our behalf; and we could not contribute money to political parties or election campaigns.[21] These were powerful tools that those opposed to the exhibition had at their disposal. Even if they did not use all of them, the Congress knew they could be brought to bear.

In such a confrontation, the museum was also handicapped in other ways. A contest between a public and a private institution is uneven. The Air Force Association could and did use the Freedom of Information Act to demand copies of letters, reports, or memoranda written by the museum's staff. It could and did leak many of these to the press. And, as Jack Giese, AFA chief of media relations, pointed out, they could and did add commentary of their own to suit their purposes. In contrast, the museum had absolutely no right to demand access to internal documents or any other materials generated by the AFA or other veterans' organizations.

In an interview with a reporter Jack Giese described his operations this way:

> As AFA got information . . . it was pumped immediately to Congress. . . . It started broad. Then we got Blute and Johnson. The debate is going back and forth. . . . I'm feeding (*Washington Times*) reporter Josh Young. I'm feeding *(Washington Post* reporter) Gene Meyer. . . .

> One of our strengths we have here—it seemed silly to some people—we had a printer on the third floor. We could have this material done at the speed of light. That is a resource government agencies don't have. We could build packages depending on where the story was going. . . . We had Blute, Johnson, all those guys. Kassebaum. As soon as we did something we fed them. We fed all levels. Any Joe Blow off the street. Any American who called got the same treatment as the media or anybody else. We'd send it to vets. We'd send it to high school students, teachers. Get the information out quickly and in a good readable manner. When dealing with the media or Congress 'quickly' meant Fed Ex the next day or courier—speed of light, get that information out to them.
>
> I'd say we did a hundred copies of script one—the only one we could copy because it wasn't copyrighted; ten renditions of clipping books, 50 at a time. . . . August [1994] was four to five couriers a day, four to five Fed Exes a day. . . .
>
> August was the heat month. . . . An unbelievable feeding frenzy. . . . We broke down our messages into a single page. "It was an act of vengeance. We were portrayed as the bad guys."—that was one of our sound bites. "Tell history the way history happened,"—stolen from Blute, who took it from a vet. Another one of our messages was, "We are worried that American youth will get a distorted view of what America did in the War." . . . Various sound bites like that.
>
> I come on the *Today* show. I've got my sound bites, boom they go to him. He starts building a clock, talking about "Well, you have to understand,". . . He's getting no points across. He's doing a rational discussion. He does not know the media he is in.
>
> If a reporter said, "I was looking in here and I don't like the way it sounds, especially on the history," we'd say, "Call Dick Hallion. See if he knows any historians." If something would break, we would stop everything we were doing at AFA Communications until we got an approved message.[23]

In the early summer of 1995, the AFA wanted to recapitulate the entire history of the *Enola Gay* confrontation. They issued a press kit consisting of four volumes. One contained the five-hundred page first script of the exhibition. Two others, roughly as thick, reproduced much of the press coverage, respectively, from 1994 and 1995. The fourth and thickest volume comprised correspondence, memoranda, resolution, reports, and other relevant documents. The entire set runs some two thousand pages. Jack Giese estimates the AFA may have run off as many as a thousand copies for distribution.[24] The cost must have been staggering.

With the AFA, the American Legion, the VFW, and any number of other, smaller veterans organizations willing and able to spend unmatchable time, manpower, and money, the museum did not have the ability to compete adequately. This does not mean we gave up. On many days I gave interviews to two or three different media representatives; appeared on early morning TV shows; talked with reporters from all over the world;

wrote articles; gave speeches; and worked with our public affairs officer, Mike Fetters, who took on the added task of manning the many radio talk shows. Together, we tried to present our view as coherently and disseminate it as widely as possible.

Defeat of a museum with a total staff of 280, by veterans' organizations whose summed membership stands six million strong is not shameful. I like to believe we fought valiantly but were badly outgunned.

25

Negotiating the Script

In late June 1994, Smithsonian under secretary Newman wrote to Herman Harrington and Hugh Dagley of the American Legion's National Internal Affairs Commission to thank them for having met with us in May. She forwarded the latest version of the exhibition script we were just sending out to our advisory committee that day, and offered that we meet with the commission, anywhere, to discuss any aspect of the exhibition.[1]

Hugh Dagley responded on August 2 with an invitation to meet with the commission on the morning of September 3 at the Minneapolis Convention Center, where the American Legion would be convening over the Labor Day weekend. In my letter of acceptance I indicated that Mark Rodgers, in charge of the Smithsonian's government affairs, Mike Fetters, the museum's public affairs officer, Tom Crouch, and I would attend.[2]

In the interim, the Legion proceeded as directed by Resolution No. 22, passed on May 4–5, 1994, by the National Executive Committee: On August 12, National Commander Bruce Thiesen wrote to President Clinton:

> Smithsonian officials have argued, not disingenuously, that nowhere does the exhibit explicitly state an anti-American bias. That is true. But the preponderance of visual material, personal perspectives and descriptive narrative, taken as a whole, prompts the unmistakable conclusion that America's enemy in the latter days of World War II was defeated and

demoralized, ultimately the victim of racism and revenge, rather than a ruthless aggressor whose expansionist aims and war fervor yielded more than a decade of horror and death for millions of the world's people.

As a visual experience, the potential of the exhibit to engender a revised view of history will be even more powerful, and render a serious disservice to the post–World War II generations—of both sides. . . .[3]

A few days later, Jerry May, in the legislative division of the Legion's national headquarters, telefaxed a copy of this letter to Ken Goss at the Air Force Association, with a cover note: "Attached is a copy of the letter we sent to the President. We have also sent copies to all members of Congress. We have informed all State Commanders and asked them to pressure Congress . . ."[4]

By the last week in August, Patti Woodside had completed a rough-cut video-film. It gave a retrospective on the Hiroshima mission as recalled by three of the *Enola Gay*'s crew, navigator Dutch Van Kirk, bombardier Tom Ferebee, and radio operator Dick Nelson, and included the thoughts of Ray Gallagher, *Bockscar* assistant flight engineer on the Nagasaki mission. I have recorded their remembrances in the first chapter of this book. Patti Woodside had sent tapes to each of the men for their approval, and they liked what she had done. But on August 24, when I asked curator Joanne Gernstein to check with them to see whether they would permit me to show the film at the American Legion convention, they were reluctant.

Joanne wrote me a note saying Ferebee had called the Legion:

In the past . . . [Ferebee] was willing to wait for the final outcome of the script to make judgements. . . . I think he was comfortable with waiting, even though he had read various articles about the script and had listened to Paul Tibbets' "people." . . . he is not as comfortable with the controversy, now that he has communicated with the American Legion.

He didn't realize how mad they were. He hadn't heard before his phone conversation that the American Legion was going to approach the President.

He kept on saying that unless the Director changes his platform he can't release his interview.

He describes the platform of the exhibit in this way: The exhibit, in Ferebee's opinion says that 1) the bomb should not have been dropped and that 2) Americans were vindictive, cruel, and racist.

He feels that the exhibit does not provide information that will help visitors decide, but instead, the exhibit decides for them.

He also feels that from what he hears the exhibit is still slanted toward the Japanese.

Tom Ferebee said he would be willing to talk to you. Van Kirk feels the same way.[5]

I talked to Joanne to decide on the best strategy and then called Ferebee to make my request, saying that we were being accused of not showing the veterans sympathetically. Here was a film that was doing just that, and if I could not obtain his and his colleagues' permission to show it, I would not be able to convince the Legion of our intentions. This was the first completed item in the exhibition we could display. I'd have a Catch-22 on my hands if I could not show it. I volunteered to screen the film at the convention, with a disclaimer that would protect all four of them. Ferebee understood, and said he'd consult with the others.

By the next day they had agreed and I wrote each of them a letter of thanks and formal reassurance:

> I am writing to thank you for your willingness to permit the National Air and Space Museum to show the videotape, recently filmed by Patti Woodside on our staff, in which you spoke of your experiences with the 509th Composite Group and its missions to Hiroshima and Nagasaki. I understand that this permission has been granted for a one-time showing only at a meeting of the American Legion on September 3, 1994. I undertake in making this presentation to precede the showing of the tape by the following statement:
>
> "Permission to show the videotape you are about to see was generously granted to the Smithsonian Institution by all four of the participants who appear in it. None of them has had an opportunity to see the script for the exhibition in which we hope to show this footage. And, knowing the nature of today's presentation to the Internal Affairs Commission of the American Legion, they asked me to tell you that their permission to show the tape should not be taken to imply that they are either endorsing or opposing the exhibition."
>
> Once more, I thank you for helping the Smithsonian in this way. This is the first tangible product that we are able to show to represent the spirit of the exhibition we will open in May 1995. The videotape you have helped us produce speaks far more eloquently than any words. It should persuade those who see it that the gallery we are planning will honor our veterans and speak well of America and our nation's role in fighting for freedom in a war we didn't start.[6]

Although I did receive special permission to show this footage on two more occasions, the museum never was able to obtain a full release from the four men. They worried that it would cause resentment among other veterans, and be seen as divisive.

This would become a perpetual problem for us in trying to mount the exhibition. Individual veterans or veterans' groups were satisfied with portions of the exhibition that involved them, but would not give us permission to use the material for fear of breaking rank. We could not get the four crew members to sign a release; General Tibbets would not consent to be videotaped; and eventually, as we will see, The Retired Officers Association would be satisfied with what we had done, but would not say so unless all the other veterans' organizations agreed. It seemed as though any of the groups with whom we were dealing held veto power. Eventually, it would take only one organization to bring down the whole exhibition—because not one of the others would step out of line to come to our defense.

For now, I wrote Dagley that we would be showing the video footage at the American Legion convention, enclosed a copy of the letter I had sent to the four veterans, and made some other final arrangements with him on August 26.[7]

Dagley had invited us to make our presentation on Saturday morning, September 3, 1994. The previous Monday, Mark Rodgers, Tom Crouch, Mike Fetters, and I met with Connie Newman in her office to work out our strategy. I was pleased that Newman had decided she would go with us, since she knew the Legion leadership best.

THE TOLL

Crouch had been deeply affected by changes we had been making to the script the previous two weeks and seemed to be under great strain. Much of his effort to add a fervent human touch had been rewritten, at my behest, and was more in line with my July 2 memorandum the previous year. There, I had written him, "drama and striking examples are part and parcel of good exhibits, but only so long as they do not distort the message, so that only one point of view comes across. And this particular exhibition, more than any one can think of, does not need any added drama. If anything, the labels must be dispassionate, perhaps even bland...."[8]

Newman and I decided there was no need to put Crouch through the extra pressure of attending a potentially hostile meeting, and I told him that he did not have to come to Minneapolis.

To Father John Dear, a Jesuit and peace activist, who came to the museum around this time to read the script, Crouch said, "You have no idea of the forces opposing this exhibit, not in your wildest dreams—jobs are at stake, the Smithsonian is at stake."[9]

Father Dear organized a meeting later in September at which representatives from a number of peace groups presented me with their concerns about changes I had made in the exhibition script.[10]

In contrast to Mike Neufeld, who a year earlier had started out somewhat cavalier about the exhibition, but now was steady and thoughtful in his responses as the pressures were mounting, Crouch was clearly apprehensive. He had come to my office just a few days earlier and asked whether I could not do something about the attacks leveled at Neufeld and himself by the veterans and the media. I had no way of controlling that, but I did add a paragraph to a letter I was writing to Monroe Hatch:

> Let me end with a request: The responsibility for this exhibition rests with the Director of the Museum. If anyone is to be singled out for blame, it

> must be I. I would appreciate, as a personal favor, your asking John Correll to direct any criticism of the exhibition at me rather than our curators.[11]

Whether this influenced the Air Force Association, I do not know, but I had the sense, thereafter, that their tactics had changed. Instead of attempting to drive a wedge between the curators and director, the emphasis now seemed to shift to driving a wedge between the director and the Institution's new Secretary, I. Michael Heyman.

I felt concerned that I had asked Crouch to help with this exhibition, which all of us had known would be controversial and difficult. Ten years earlier, he had worked at the National Museum of American History for several years and had gained enormous respect for his exhibit on the plight of Americans of Japanese descent interned during World War II in clear violation of the United States Constitution. He had taken a great deal of unfair criticism for that, and had admirably stood his ground. I thought perhaps I should not have asked him and his family to make such sacrifices a second time. Unlike men and women entering the military, people do not become museum curators to engage in combat. Crouch had already shown his mettle and his ability to stand up for his principles. I felt I should not have asked him for a repeat performance.

MINNEAPOLIS

Late Friday afternoon, Connie Newman and I took a flight out of National Airport. Mark Rodgers had gone ahead to see about last minute preparations. Mike Fetters was attending an annual meeting of the 509th Composite Group in Chicago, and was arriving separately.

In the evening the four of us went to supper. Newman had been in law school in Minneapolis and knew her way around. We ended up in a family-style restaurant and all ordered "Minneapolis meat loaf," which was pretty good. Over a bottle of wine we discussed strategy for the next day's presentation. Newman would give an introduction on the Smithsonian and its function and then introduce me. I would then make the presentation on the planned exhibition. Rodgers and Fetters would concentrate on relations with the Legion and the media.

The next morning, Newman had an early morning meeting with the Legion's leadership, some of whose members she had known several years. Around 9:00 A.M., we all gathered in the convention center where we had been assigned an hour and a half to make our presentation. The Legion was set up to video-record the entire proceedings.

Mike Fetters suddenly spotted Jack Giese, the AFA public affairs officer, and worried that he might wish to disrupt the meeting. But Dagley assured Rodgers that he would not allow that to happen.

After a number of procedural matters on the Legion's agenda, Newman was introduced, and began by talking about the Smithsonian and the purposes for which the Institution had been established. She then introduced me, making sure to emphasize that I had served in the U.S. Army and was a distinguished scientist. While she spoke, someone handed me a list of questions from the floor that we could expect after the presentation. I deemed that particularly thoughtful and quickly scanned it while also trying to listen to her.

For my presentation, I spent about twenty minutes showing slides of images that would be used in the exhibition. These demonstrated the spirit of what we were planning. I then showed the video of the four crew members of the *Enola Gay* and *Bockscar*. The video included footage of Tinian Island and the *Enola Gay* in flight, intercut with recollections of the four men, and of the mushroom clouds they had seen from the air on their missions. It was a powerful and moving fifteen-minute film, which the audience of about a hundred veterans, many of them World War II servicemen, clearly appreciated. At the end of the film, I showed one more slide from the exhibition—the first page of the first draft of Roosevelt's "Day of Infamy" speech. The text is full of penned-in revisions, and the word "infamy" does not even appear. I let that sink in and then said Roosevelt would not have wanted to be judged on that first draft. I asked them not to judge us on our first draft, either, or on the quotations from it that had been circulated against our wishes in the newspapers. We had been working on revisions ever since and felt we had a good exhibition on our hands now.

The questions then started and I was impressed by the Legionnaires' politeness. Nobody raised a voice. Newman and I answered the questions as best we could. A Japanese television crew was filming much of the proceedings.

After the session, which seemed to have gone well, one of the Legionnaires, Earl Gleason, from New York, asked me when I had served in the Army, looked up the dates and said that I would be eligible to join the Legion if I wished. I said OK and joked that it would be better for me to sit on both sides of the table when we began discussing the script in earnest, which we had told the Legion we would be glad to do later in the month.

The four of us then had lunch with Herman Harrington, who joined us rather late after attending another meeting. He thought our presentation had gone over well, and he was prepared to ask the full convention to stay the condemnation of our exhibition, pending discussions that they and we would hold to see whether a script acceptable to all sides could be worked out. It was a happy gathering. I got teased for ordering a banana split as lunch instead of something more sensible.

Afterwards we went back to talk with Hugh Dagley and make further arrangements before heading for the airport, where Newman called both Secretary Adams and Secretary-Elect Heyman at their homes to tell them that things had gone well. Heyman was to assume office in less than three weeks.

We got back to National Airport fairly late that Saturday night, having stayed longer than originally anticipated in order to make sure we had settled all essentials.

The following week we learned that Harrington's appeal to the Convention had been successful. He and Dagley would be visiting us in the next few weeks to further discuss our plans for the exhibition.

NEGOTIATIONS

Three weeks later, on September 21, 1994, we sat down at the museum for our first serious discussion on the script. Newman, Rodgers, all of the exhibition team, Steven Soter, and I were there, and the Legion was represented by the new national commander, William M. Detweiler, who had just been installed at the Legion's Labor Day convention for a normal one-year term of service. Hugh Dagley and Herman Harrington were there with him.

We began at 9:00 A.M. Detweiler led off in a highly accusatory tone with sweeping statements. When he had finished, I said that we did not see matters at all that way. Could they please show us where in the script they had found the kinds of features he had mentioned. I said "the devil is in the details. Unless you tell us where you see these things, we can't do anything about them."

We agreed to go through the script line by line to see whether we could come to a resolution. Detweiler then left to take care of other business, and Harrington and Dagley started working with us. Dagley did most of the negotiating, but Harrington, as his boss, had the final word.

At 11:15, I had to leave for a previously arranged lunch talk I had promised to give at the National Aviation Club, but the others continued to work while I was gone. At the luncheon were Admiral Tom Kilcline, of the Retired Officers Association, and AFA's Jack Giese. Giese and I had never met before, though we had been debating the merits of the exhibition on television. As the club members sat down to eat, I went over and introduced myself. We acknowledged that we were on different sides of the issue.

I gave the same slide presentation I had given at the American Legion convention and felt I had convinced the members that we were doing a reasonable exhibition.

When I got back to the museum and rejoined the deliberations with the American Legion, I found they were going slowly. While I was away, one label change that had been especially problematic had been resolved. We had previously spoken of the Japanese "defenders" on Okinawa. That rankled the Legion. After debating for about twenty minutes our team had found that if we agreed to substitute the word "troops" for "defenders" they would be satisfied. "Defenders," the Legion representatives felt, suggested sympathy for the Japanese.

It went like that all afternoon, but we were making steady progress.

Shortly after I had returned Connie Newman had pulled me aside. She had talked with the Legion representatives over lunch and had arranged to have a press conference the next morning where we would declare that we were working together on the exhibition. I was taken aback by that, but the decision had been made. I would try to make the best of it, though I knew that this would raise a howl of anguish from the academic community and threaten any support we might expect from them.

At the end of that day, I did not think we had made any changes that violated history. Most of the alterations involved words and phrases that might be judged pejorative or flattering. The language was becoming more bland and dispassionate, in line with my memorandum to our curators fifteen months earlier.[12] At the end of the afternoon, Dagley made a rather nice emotional statement, saying he felt that our historians had dealt with the Legion professionally and courteously and that he appreciated it. In turn, I responded that all of us were genuinely struck by the evident, careful homework their side had done. Dagley's performance had particularly impressed me.

A PUBLIC DEAL

The next morning, a correspondent for the *American Legion Dispatch* interviewed me. Then all of us gathered in the museum's West End where the staff had set up a podium. Detweiler, Newman, and I made presentations to say that the American Legion and the museum would work together to see whether a mutually acceptable exhibition could be worked out. A sizable media turnout included several foreign television and press services, among them Japanese representatives.

In my welcoming statement I said,

> As most of you know, the museum has for many months been planning an exhibition *The Last Act: The Atomic Bomb and the End of World War II*. Last January, we assembled a first draft of the entire label script. We knew we would have to go through many further drafts as we sought advice and suggestions from the widest possible range of interested and knowledgeable scholars, veterans who had actually been there and experienced the war, and military historians. Right from the beginning we sought to make this review broad and inclusive.
>
> With advice from all these quarters, the exhibition has gone through several reviews, as all major exhibitions produced by this museum do—a process for which we always allot many months. That's why we started

with a first fully assembled draft fifteen months before the expected exhibition opening.

This particular exhibition, however, has had even more scrutiny than most others, because of the sheer immediacy of the events we describe, particularly for those who were there in the war in the Pacific. For them the mission of the *Enola Gay* and the dropping of atomic bombs on Hiroshima and Nagasaki were highly personal; they led to the rapid conclusion of the war, obviated the need for invading Japan, and thus saved untold numbers of lives.

Given the sensitivity of the subject, the Museum has always considered the greatest difficulty in preparing this exhibition to be a correct conveyance of perceptions: Does the script for the exhibition convey the same meanings to the reader—and ultimately the museum visitor—that the curators and exhibition staff of the museum intended to convey?

We have found, over the past months, that the exhibition's intended thrust was often misunderstood and that veterans were alarmed by what they perceived the museum to be doing. Fortunately, in discussions with us, The American Legion volunteered to work with us to help weed out points of contention. We agreed to sit down to clarify points of view with the aim of improving the exhibition so that no false perceptions could arise. This is what we now intend to do.

We look forward to working with representatives of the Legion, and hope that we will be able to reach agreement that will lead to an exhibition which will be unambiguous in its text, objects and images. Our aim is to recapitulate the last few months of the war to provide visitors clear insight into the single most pivotal historic event of our century.

We look forward to this partnership and expect it to work to the benefit of a stronger exhibition to open in May 1995.[13]

Detweiler's statement was,

The National Air and Space Museum is truly a national treasure, and is so highly regarded by the people of the United States that the current controversy over the World War II exhibit is causing not only pain to many veterans, but also concern and confusion among our citizens, and active interest among our lawmakers. No responsible organization can—or should—refuse to work with the Institution to rectify this situation.

When we were asked to add our perspective to the barrage of criticism leveled at the exhibit, it was cause for deep consideration and discussion. During that period, we tried to balance our primary charge—that of representing the interests of veterans—against our broader concern for the welfare of the nation. Guided by the expressed conviction of 3,500 delegates at our just-concluded national convention, we have decided to participate in a line-by-line review of the script. That review began yesterday and will continue next week. We have reason to believe this process will help immensely, but the review has not yet touched on the core issues that underlay our deepest concerns.

More than anything else, our disagreements center on the estimate of the number of lives saved by the use of atomic weapons in 1945.

> Was it 30,000 or was it 500,000 potential invasion casualties? Does it matter? To the Museum, it seems to be a matter of some significance. To the historians, it seems to be a matter of great importance to determining the morality of President Truman's.
>
> To the American Legion, it matters less, if at all. The use of the weapon against a brutal and ruthless aggressor to save 30,000 American lives was as morally justifiable as to use it to save a half a million. We are aware of the recent historic discoveries that appear to discount the higher estimates, but we also know that the points of comparison by which the lower estimates were determined are questionable.... They do not take into consideration that Pulitzer Prize–winning historian Richard Rhodes has pegged the 30,000 casualty figure to the first thirty days of an invasion of the southernmost home island—and concluded that an invasion of the main island of Honshu across the plain of Tokyo would be far more violent.
>
> Nevertheless, the number of potential casualties is really not the issue. It is our view that, for any government with the means to end the slaughter on both sides, not to use those means would be morally indefensible....
>
> We want this exhibit to succeed, but we insist that it be accurate in that it present the service and sacrifice of America's veterans as the legislative charge to the Institution mandates ... and that the role of the Japanese as the cause of the conflict be fully detailed. Failing that, because of our broader responsibilities to the people of the United States, to our common heritage, to future generations and, most specifically, to those who have taken up arms in defense of our way of life, we will have no choice but to exercise the options available to us to actively oppose the exhibit.

FURTHER MEETINGS

We met twice more with the Legion—in Washington on September 28, 1994, when we started at 7:15 in the morning and kept at it until late in the afternoon, and again, three weeks later, in Indianapolis on October 20. At the September 28 meeting we finished going through the entire script. One of the toughest, if not *the* toughest, part of the first two sessions was the label dealing with casualty figures. We worked that out after going through several drafts. Harrington finally accepted one form of wording. Dagley would have preferred not to, but deferred to his boss. I will reproduce it in full, because later it led to considerable contention:

> INVASION OF JAPAN—AT WHAT COST?
>
> Estimates of the number of American casualties—dead, wounded, and missing—that the planned invasion of Japan would have cost varied greatly. At a June 18, 1945, meeting, General Marshall told President Truman that the first 30 days of the invasion of Kyushu could result in 31,000 casualties. Admiral Leahy pointed out that the huge invasion force could

> sustain losses proportional to those on Okinawa—about 35 percent—which would imply a quarter of a million casualties, or at least 50,000 dead.
>
> Had the Kyushu invasion failed to force Japan to surrender, an invasion of Honshu, with the goal of capturing Tokyo, would have followed, and losses would have escalated. If even that failed, and Japan continued fighting in the home islands and in the captured territories in Asia, casualties conceivably could have risen to as many as a million (including a quarter of a million deaths). Added to the American losses would have been perhaps five times as many Japanese casualties—military and civilian. The Allies and Asian countries occupied by Japan would also have lost many lives.
>
> For Truman, even the lowest of the estimates was abhorrent. To prevent an invasion he feared would become "an Okinawa from one end of Japan to the other," and to try and save as many American lives as possible, Truman chose to use the atomic bomb.[14]

I was surprised that Dagley and Harrington did not ask us to take out much of the material in section 400, dealing with Hiroshima and Nagasaki on the ground. Since the Air Force Association always considered that a crucial issue, I had expected it to become a major sticking point. But evidently they had other things in mind.

The September 28 session lasted approximately ten hours.

All along, Dagley had said that once they committed themselves, they would not revisit an issue. I therefore felt we had crossed a major hurdle. The next few days were spent in actually making the corrections that we had agreed on in these two sessions. At the end of that time, we ran off another, updated version of the script, and sent it out to Dagley and Harrington to review before we were to meet with them in Indianapolis.

We also sent copies to the Retired Officers Association, the VFW, and the AFA, and then met in Connie Newman's office with Hatch and Correll, from the AFA, Kilcline and Cooper, from TROA, and Bob Manhan and Bob Currieo, from the VFW. The curators and I had gone through their recommended changes ahead of time, and we discussed those suggestions to see whether any remaining differences could be ironed out.

The following morning, October 20, 1994, Newman, Rodgers, Fetters, and I took an early flight to Indianapolis. The talks with Dagley, Detweiler, and Harrington only started around noon, but then lasted till 10:00 P.M. This session was more tedious than earlier ones. The Legion delegation had not gone over the revised script ahead of time, and now wanted to reexamine it and look at each change we had made. On three or four of the points, their notes differed from ours, and I promised to check with the curators who had also been present at the two earlier sessions, and then send Dagley a note about it. I did that in two letters, dated October 25 and November 9.[15]

By the time we left Legion headquarters, we had long missed our flight. Rodgers had left earlier in the afternoon, but Newman, Fetters, and I spent the night at an airport hotel and returned to Washington on the first flight the next morning.

PRESS COMMENTARY

During these weeks, the American Legion's comments to the press were uniformly conciliatory. Speaking with a reporter for the *Indianapolis News* in late September, Dagley commented, "I will tell you, this is one interesting fight.... I don't think the people at the Smithsonian are bad people, I think they have.... I don't know exactly how to say this ... failed to take into consideration the human aspect, the human consideration of this thing.... the curators are most open to our points of view, and that's encouraging. But I think some veterans out there will never be satisfied."[16]

A few days later, Dagley told the *Washington Post*, "We worked with the Smithsonian to use only verifiable facts, to eliminate speculation or unattributed ideas.... This exhibit is taking a more balanced direction. It's not a propaganda piece by any means. I don't think the process is over, but I think the most difficult part of the process is behind us. The debate started fifty years ago, and it has never been resolved. It's not going to be resolved by the Smithsonian or the American Legion; I don't know it if ever will be. The point is to put it in context and let the erudite visitors make some decisions themselves."[17]

But others felt left out and wanted to be sure they were heard. Quoted in the same *Washington Post* article, Stephen P. Aubin, of the Air Force Association, threatened, "If it's not right this time, many of us will call for the cancellation of the exhibit."

W. Burr Bennett, Jr., the ever-vigilant B-29 reconnaissance veteran, speaking to the *Washington Times* in early October 1994, also warned,

> I think the American Legion has done a credible job, but you just don't understand the kind of game these Smithsonian curators are playing.... There are two Air Force historians who have put forth some very competent reviews of a number of the various scripts, and they haven't been invited to attend [the rewrite sessions]. I think it's another game playing kind of thing that's so annoying.

To which Phil Budahn, the Legion's public affairs officer retorted,

> There are a lot of people who are not happy that [the Legion representatives] are talking to the Air and Space Museum. They don't think anyone ought to be talking to the Air and Space Museum. I'm not sure how these people expect the problem to be resolved.[18]

Dagley also felt compelled to reply to Bennett's accusation. In a long letter printed a few days later in the *Washington Times* he asserted,

There is no evidence to suggest, as "some World War II veterans grumble," that Mr. Harwit sought to divide veterans groups by doing business face-to-face with only one group. We informed National Air and Space Museum Director Martin Harwit, Smithsonian Under Secretary Constance Newman and the two principal curators at the outset of our discussions that we would rely heavily on the work already done by other veterans' organizations. Prior to our September 21 meeting with NASM officials, I had frequent and detailed discussions or correspondence with at least two *Enola Gay* crew members, other members of the 509th Composite Group, the Air Force Association, historians from two branches of the armed forces, renowned historians and biographers, former prisoners of war and countless unaffiliated World War II veterans. I informed NASM officials of all these contacts.[19]

But the following day, the *Washington Times* published an article in which VFW executive director Bob Currieo complained,

... we continue to have remaining concerns that there is an obvious, clear and distinct bias and sympathy for the Japanese people and their culture.

Currieo's associate, Bob Manhan, the VFW assistant legislative director, added,

Smithsonian officials agreed to meet with ... a group that includes the VFW and other military organizations except the Legion, on September 23. Mr. Harwit's office canceled the meeting on September 19. The next day, Mr. Currieo sent a letter to Mr. Harwit expressing his disappointment.... "This is particularly vexing," Mr. Currieo wrote, "because I was told you would be meeting only with the American Legion ... I fail to find any logical connection between these events." [20]

I could understand the VFW's annoyance. I had scheduled the meeting with the Military Coalition before the dates for the American Legion discussions had been set, but then canceled it when Connie Newman suggested that we first settle with the Legion—which had insisted on meeting with us alone—before again meeting with the other groups.

These other veterans' organizations now added their voices. In the November 1994 issue of the *Retired Officer*, the magazine's editor, Colonel Charles D. Cooper (USAF, Ret.), wrote about Title 20 §80 of the U.S. Code, and mentioning the AFA but not the American Legion, wrote,

Regretfully, NASM's early stonewalling has produced a new mushroom cloud, this time of resentment and disgust, over the entire Smithsonian system. While the two more recent scripts have incorporated some superficial changes, the basic philosophical problem remained. The curators were still applying Band-Aids when major surgery was needed.[21]

The pressure on the American Legion leadership was mounting. They could not stay entirely aloof from their own membership, which had long been stirred up by the AFA's and even the Legion's own earlier propaganda, and they could not entirely defy the assembled strength of the other veterans' organizations. Only days after we had completed our negotiations in Indianapolis, the November issue of *American Legion* magazine published a long article on "Rewriting *Enola Gay*'s History." It recalled the Legion's national executive committee's Resolution 22 of May 1994, condemning the museum's exhibition, and the national convention's Resolution 391, which stayed the condemnation pending further discussions; and it quoted Ben Nicks as saying,

> We believe the Smithsonian wants to use the *Enola Gay* as a springboard to discuss the morality of atomic bombing and strategic bombing. And we don't think the national museum of the United States, funded by the United States, is the place to discuss this question.[22]

The same issue also reprinted an extensive excerpt from Paul Tibbetts's June 1994 speech, in which he had told those who were "second guessing the decision to use atomic weapons" to "Stop! It happened. In the wisdom of the President of the United States...."

Such coverage undoubtedly also continued to generate the flood of letters to the editor published in the *American Legion* magazine month after month, all hostile to the exhibition. How could it have been otherwise, given everything that had appeared in the media?

All this was reflected in a change in attitude that became apparent as early as December, when Dagley spoke at a videotaped forum organized in Washington by American University. By late December 1994, when the January/February 1995 issue of the American Legion Auxiliary's *National News* was going to press, Dagley was calling for further concessions from the Smithsonian and threatening action if the Institution did not comply. I will return to those remarks later, but the attitude at the Legion was hardening, just as the other veterans' organizations with whom discussions had resumed in October and November were showing themselves increasingly satisfied with the exhibition.

One more remarkable set of comments needs to be mentioned. How much in them reflects the truth and how much was trotted out for the benefit of their membership is not obvious:

In May 1995, four months after the exhibition's cancellation and seven months after the Smithsonian's September 22, 1994, press conference with the Legion, an *American Legion* magazine article appeared with the headline, "How the Legion Held Sway on *Enola Gay*." In interviews with Detweiler, Harrington, and Dagley, Hugh Dagley recalled the first time he and Herman Harrington had met us, when Connie Newman had invited them in May 1994:

> They talked to us like our heads were screwed on. That was the first time they underestimated The American Legion.

A few paragraphs down, Detweiler chimed in with,

> It became clear to them pretty quickly that we intended to confront them on their own turf—no hysterics, no emotionalism, no threats. And then, NASM made another crucial blunder. For some reason, which I don't understand to this day, they agreed to a joint press conference in which the NASM/Legion working relationship would be announced to the public.

Dagley then added,

> That press conference conferred on the American Legion the imprimatur of legitimacy in the process, propelled us to leadership on the issue, and focused intense pressure on NASM from its supporters in the peace movement, the anti-war and anti-nuclear movements in academia. In my opinion, they were iced from that point forward.[23]

If these were the feelings of the Legion's representatives at the time of our negotiations with them, I was not aware of them, and I don't think anyone else at the Smithsonian was either.

STAKEHOLDERS AND CHANGES IN THE EXHIBITION

It was clear that the museum would need to correct certain weaknesses in the script. That invariably happens after a first draft of a script is completed. The museum staff asks for advice, considers it, and decides on alterations.

Over the summer, the Air Force Association had not been content just to offer advice; they insisted on seeing their wishes carried out. Some of their demands were reasonable; others were not. The museum had to decide which changes to make and which to reject. Unfortunately, all this had to be done in the glare of publicity. Each change the museum made evoked a triumphant cry from the AFA and a howl of dismay from academic historians. The historians were taking the museum to task for "caving in" and preparing four successive revisions of the initial draft.

Four or five revisions was not unusually many. Between February 1991 and July 1993, we had gone through at least that many revisions on just the exhibition proposal and later the planning document. We had also changed the title for the exhibition that many times—all before any veterans' organizations had appeared on the scene.

Revisions to the script could not be made all at once. Successive stages addressed different problems. The first two revisions were devoted to historical points raised, respectively by the historians on the exhibition's advisory committee and by the service historians. These alterations had concentrated on the accuracy and balance of the script. By late summer, when the historians were largely satisfied, we turned to two more revisions, which dealt mainly with perceptions the exhibition would raise among visitors. This meant assessing the amount of information the visiting public might be expected to bring to the exhibit, and then working on the script's format and wording to tune it to people less familiar with World War II than the experts among our advisors. On that level, the sessions with the American Legion were useful because repeatedly they pointed out wording in the script that could be misunderstood or might be misinterpreted as anti-military or unappreciative of our veterans' service to the country. We did not always agree with them or take their advice, but it was useful to hear them out.

Historians, pacifists, and other critics voiced their concern over these sessions with the Legion. Their views, as expressed at the time, were later reflected by Prof. Richard H. Kohn, a highly respected professional and the immediate predecessor of Richard Hallion as Air Force historian. For many years Kohn had served as chairman of the museum's Research and Collections Management Advisory Committee, and many of us at the museum still consider him a good friend. Now a professor of history at the University of North Carolina, Kohn is thoroughly at ease in the worlds of both the military and the academic historian. His criticism, for all these reasons, needs to be taken seriously.

In a careful analysis of the first exhibition script, written months after the exhibition had already been canceled, Kohn points to some of its most serious shortcomings; but he also takes the museum to task for beginning

> ... to negotiate content with groups outside the museum with political agendas and no claim to scholarly knowledge, museum expertise, or a balanced perspective. Museums customarily consult, consider, react, and modify their products according to the best advice they can gather, but to negotiate a rendering of the past in exchange for acquiescence poses special dangers. To negotiate an exhibition on the labor movement in American history with the American Federation of Labor–Congress of Industrial Organizations (AFL-CIO) or one on medicine with the American Medical Association risks making the exhibition hostage to a constituent group that then wields a veto over fact or interpretation.[24]

I do not agree with Kohn. All of us at the museum would, of course, have preferred to respond to friendly criticism from our peers on the exhibition script's original advisory committee. But since the committee had not provided us with the substantive critique required, were we to reject legitimate concerns simply because they were voiced by a hostile group? Is it wrong to make a sensible change under pressure, rather than in response to collegial advice? In Kohn's analogy, is advice from the AFL-CIO to be rejected outright when doing an exhibition on the labor movement? Does the AMA have nothing useful to say about American medicine? Is the academic community so elitist that advice from those with "no claim to scholarly knowledge, museum expertise or balanced perspective" has to be rejected out of hand?

I think not!

The museum was often accused of making changes only under pressure. For us, the problem was that we had originally turned to the Air Force

Association for advice on the exhibition from a veterans' perspective. Rather than giving advice, they had decided to launch a campaign to force their views on us. We could not exclude the veterans' recollections of the war if we wished to properly mirror the times. And the only way to obtain the advice we sought was to tolerate the AFA's tactics or turn to another veterans' organization for help.

There is an undeniable danger in such confrontational tactics. A museum often has to turn to communities that have most at stake in an exhibition. Their subject matter knowledge and their sensibilities need to be taken into account. But when such groups turn the museum's search for added insight to their own advantage to force a political agenda, an important trust is abused.

With a topic as sensitive as the display of *Enola Gay*, the museum had to listen carefully to the perceptions visitors were likely to carry away from the exhibition. The difficulty we had encountered in dealing with the AFA was that their critique of the script mostly was couched in broad accusatory terms. The American Legion, in contrast, was willing to be specific. Many hours were devoted to changes in wording that sometimes seemed to be rather trivial to us, but were important to the Legion. That is not unusual; words carry different connotations for different groups and give rise to different perceptions. That is why the script and design for a new exhibition needs to be widely vetted.

Andrea Stone, of *USA Today,* cited two such "subtle changes of language and tone" that had emerged from discussions with the American Legion. One referred to explanatory text:

> *Old:* As relations between the United States and Japan deteriorated during 1941, the United States began to formulate plans for bombing Japan.
>
> *New:* In 1941, as Japanese aggression in Asia brought war with the United States ever closer, the Army Air Corps began to formulate plans for bombing Japan in the event of a Japanese attack.

The other involved the caption of a photograph:

> *Old:* For Aircrews, capture meant imprisonment in horrible conditions and occasionally even execution. This Australian flier was beheaded in August 1945 after the Japanese surrender.
>
> *New:* For aircrews, capture meant imprisonment in horrible conditions and even execution. Like this Australian intelligence officer, Allied fliers were often beheaded.[25]

Not all the changes we made were so simple. Representatives of the Legion felt that our descriptions of the firebombing of Tokyo in effect accused the United States of senseless barbarism that victimized innocent Japanese civilians. They asked us to point out the decline in Japan's industrial capacity resulting from such raids all over Japan. We did that, but we insisted that the number killed be kept. On the March 9–10, 1945, raid, the label still read,

> Flying in three streams 650 kilometers (400 miles) long, 334 B-29s struck Tokyo for nearly three hours. Within 30 minutes of the first bomb, fires were burning out of control. About 100,000 people perished and a million were made homeless.

Another label on the same subject continued the story with,

> The great Tokyo raid marked the beginning of a five-month period during which Japan would suffer widespread devastation. B-29s bombed one city after another, destroying half the total area of 66 urban centers, burning 460 square kilometers (180 square miles) to the ground. Some cities, like the chemical and textile manufacturing center of Toyama, were completely destroyed. The five-month incendiary campaign lowered the overall industrial output of Japan by 60 percent, reduced production of key materials like oil and aluminum by 90 percent, and took several hundred thousand lives.[26]

ATOMIC AND CONVENTIONAL BOMBING

We did reduce the number of pictures and artifacts from the section that portrayed destruction and suffering at Hiroshima and Nagasaki, because I felt that the original script could be mistaken for a diatribe against the horrors of war in general, a subject well beyond the scope of the exhibition. Section *100* of the exhibit had already presented the destruction of Japan by conventional bombing. Section *400* of the final script could accordingly restrict itself to pointing out the differences between conventional and atomic bombing.

Most Americans tend to think of atomic bombs in terms of megatonnage of conventional explosives. That carries the implication that explosive power is all that counts. The sections dealing with Hiroshima and Nagasaki, therefore, talked about the difference between the effects of many small bombs and the destruction wrought by a single atomic bomb of the same total explosive power—the intense searing flash that inflicts characteristic burns on skin facing the epicenter; the keloids that develop later; radiation sickness among those within a mile of the explosions; babies born retarded; leukemia developing in children after a lapse of many years; cataracts that appear late in life, and the still-undefined, long-term genetic effects. All this in the more obvious context of instantaneous eradication of all infrastructure, so that the intense firestorms that arise cannot be fought, and casualties in the devastated areas cannot be reached by firefighters or medical teams.

We had extended debates with the Legion about whether a label summarizing the results of the United States Strategic Bombing Survey should be kept. The Legion argued that the survey's claim that Japan was virtually defeated before the dropping of the atomic bomb was postwar propaganda by those who wished to show the power of conventional bombing and to establish an air force independent of the army. I retorted that the survey had been commissioned by President Roosevelt and had been delivered to President Truman. I saw no reason to accuse men like Paul Nitze, who had played a leading role in overseeing the survey and had later been an advisor to virtually every president of the United States since Roosevelt, of such motives.

The label stayed. Months after the exhibition had been canceled, Hugh Dagley cited this particular label as one major reason for the Legion's insistence that the exhibition be canceled.[27]

If any of this sounds like caving in, as some historians declared, I'd like to know where.

GUIDELINES

In carrying out these negotiations, I set a number of guidelines. First, the exhibition would need to state the major caveats surrounding a broad claim, but then should get on with it and not repeatedly mention those provisos. Otherwise, in an exhibit format, they would appear to gain in importance on each successive mention. That guideline became important in presenting the reasons given for President Truman's decision to use the bomb. While it is clear that postwar diplomacy and suspicion of Joseph Stalin's tactics played a role, it seemed to me, at least, that the evidence was equally clear on Truman's overwhelming desire to end the war quickly to stop the killing and "bring the boys home." As a result, the script mentioned the influence of diplomacy on Truman's decision only once and thereafter concentrated on his desire to end the war quickly and decisively. In a scholarly work, such a presentation might be criticized as overly simplistic; in an exhibition it should not.

Second, the exhibition had to be self-contained. Where a topic was especially charged with emotion and strong objections were raised, I felt we had only two choices. We would either include incontrovertible evidence for our claim, or else we would drop it.

This played a role in the treatment of the mutual hatreds of Japanese and Americans. While MIT historian John Dower's careful study of this subject makes those hatreds quite apparent, I felt we could not make a convincing case for them in the few labels we could devote to the subject.[28] We dropped the topic, because most experts agreed that these hatreds had not played an essential role in the decision to drop an atomic bomb, and concentrated on others that were more important to the mission of the *Enola Gay*.

This same argument applied to the final section on the postwar era. Initially, that section was weak. It needed to be either lengthened or considerably abbreviated. For lack of additional space we eventually decided to use only one long summarizing label displayed against a huge wall showing an array of seventy thousand small images of warheads, two of which would be shown in a different color. Together, the label and backdrop would convey the dilemma posed by nuclear proliferation since 1945.

The third guideline was that images and descriptions should not be unnecessarily grisly. Brutality, death, and injury—all could be shown without becoming overly graphic. Initially, we had a "parental discretion" warning at the exhibition's entrance. But we decided it was important to have children see the exhibit, so we eliminated some images to make that possible.

Guidelines such as these are not extraordinary. Normally, they play a tacit role in many exhibitions a museum undertakes. For this particular exhibition, I felt I had to articulate them with special care, at least for myself, because the selections we made were likely to be attacked.

CHANGING ETHICS

Throughout our discussions, veterans' organizations expressed fears that the exhibition would portray America as the aggressor and Japan as the victim. Those fears were understandable. Today, a dispassionate account of the 1945 bombing campaign, with its huge toll in Japanese civilian lives, can raise serious questions about America's wartime bombing policy, particularly if taken out of context. That policy reflected a gradual change in ethics and an escalation of force, on all sides, as this long and terrible conflict dragged on. World War II claimed some fifty million lives. To some it seemed that the hundred thousand who had lost their lives at Hiroshima and Nagasaki were an insignificant proportion. How does one weigh the loss of lives on such scales of devastation?

The museum's most difficult task was to provide a historically accurate account without evoking conclusions based on moral standards, which, for most people, dramatically shift between war and peace.

TRUTH AND SALESMANSHIP

One major concern still needs to be mentioned. Museums are scholarly communities dedicated to teaching, as much as any university. The freedom of a museum's scholars to teach the truth, as best as they understand it, forms the basis of the entire profession. It provides the distinction between a museum, whose exhibits have to be dispassionate and fair-minded to serve a trusting public, and a trade fair, where every visitor knows and expects vested interests and tough salesmanship.

Having outlined the way the museum operated in attempting to set up the exhibition of the *Enola Gay*, I should turn to the issue of academic freedom—the freedom of a researcher to express the truth as his research reveals it: How was that freedom, that search for a truthful history, affected by the external pressures that were brought to bear? The answer to that question depends on how one responds to a number of others:

Was the curators' freedom to express the truth curtailed because I, as director, took the lead in deciding what was to be kept in the script and what was to be eliminated?

I believe it was not.

I did my best to adhere to the facts that leading historians had established. And where some parts of the exhibition were dropped and others added, I tried to make sure that the balance among those facts remained equitable and representative of the times. But there was no escaping that I needed to be involved in defining the exhibition's contents and, if necessary, overruling the curators' wishes. There was no way in which an exhibition on the *Enola Gay* would be viewed as anything but a museum and even a Smithsonian product. Decisions, therefore, had to be made by those responsible for the conduct of the museum and the Institution, and that meant that the board of regents, the secretary, and I all had to become involved. My task was to do what I considered best and to keep the secretary and the board of regents informed. Through the board, furthermore, the leadership of the country needed to be kept informed because relations between the United States and Japan were easily strained by graphic displays of the bombings.

Was the museum's freedom to present a truthful history violated? I don't think it was violated by our sitting down with the veterans or with members of any other organization willing to serve as a focus group. The intent of the veterans may have been questionable, but a violation would have taken place only if the museum had capitulated on important issues, and I felt that we had not.

HISTORIANS AND PACIFISTS

When the newspapers began to publicize that the museum was meeting with and listening to veterans' organizations, historians and pacifists began to object. But it took them a long time to react. We had come under fire in the media as early as March and April 1994. The first meeting the peace groups asked for only took place on September 20, 1994, the day before the first negotiation session with the American Legion representatives.

The meeting had been organized by Father John Dear, of Pax Christi USA. The exhibition team and I met in the museum's main conference room with approximately eight representatives of peace and environmental organizations.

Dear later recalled,

> We talked about conscience and morality, not just history, and appealed to their integrity. Crouch and the curators did not speak at the meeting, and Harwit seemed exasperated. He said to us: "Where have you been? You are too late. Why haven't you been in before? Why haven't you talked to the media?" Without making any promises to restore or strengthen the script, he thanked us for coming in.[29]

That was fairly accurate.

After our meeting had ended the visiting attendees held a press conference on the steps of the museum to express their concerns. Later, Jo Becker, executive director of Fellowship of Reconciliation, of Nyack, New York, commented in a letter to the *New York Times*,

> By bowing to pressure from the American Legion and other veterans' groups, the Smithsonian Institution compromises its coming exhibition on the atomic bombing of Japan. A high-pressure campaign from veterans' groups had already led the museum to make extensive changes to its script and to double the exhibition's size with an addition depicting Japanese atrocities before and during World War II. These changes present the bombing, a unique event in history, as a justified military response to Japanese aggression. The further changes announced by the Smithsonian only reinforce that view and gloss over the devastation brought on Hiroshima and Nagasaki as a result of United States actions.[30]

On October 19, Robert K. Musil, of Physicians for Social Responsibility, wrote to me on behalf of himself and seventeen others altogether representing sixteen organizations, and stated,

> As representatives of organizations dedicated to public education about the effects of nuclear weapons and to the prevention of nuclear proliferation and nuclear war, we commend the Air & Space Museum for its decision to present an exhibit next year concerning the *Enola Gay* and the atomic bombings of Hiroshima and Nagasaki. . . .
>
> However, we are deeply troubled by your decision, as reported in the press, to delete sections of the exhibit that discuss the impact of the bombings on the people of Hiroshima and Nagasaki and the postwar nuclear arms race. In our view, deliberate omission of the human consequences of the atomic bombings contributes to a dangerous loss of our nation's collective memory about the terrible cost of using nuclear weapons. Such a loss can only weaken our national resolve to prevent such an event from happening again.
>
> Furthermore, any deletion of the exhibit's discussion concerning the postwar nuclear arms race and the impact of nuclear weapons design, pro-

duction, and testing activities on the health of atomic veterans, nuclear weapons plant workers and "downwinders" betrays the memory of the hundreds of thousands of other soldiers and citizens who are in some way casualties of Hiroshima and Nagasaki. . . .[31]

I answered this letter on October 26 to say that I was certainly prepared to meet with Musil and his colleagues: "The Museum has listened with care to all concerned groups and individuals who can bring an informed view to the subject." This was in line with the policy the Institution's new secretary, I. Michael Heyman, had established. The previous week he had announced,

> The process is not finished. We will continue to accept and consider observations from all sources, but we also feel that we are very close to where we want to end up.[32]

Eventually Musil and I arranged for a meeting on the morning of December 15, 1994.

In the meanwhile, however, a second gathering took place, with a group of historians and writers representing forty-eight historians who had composed a letter to Heyman dated November 16, to be delivered to him the following day—the day they were meeting with me. They wrote that they supported the nonbinding Senate resolution that Senator Kassebaum had sponsored and the Senate had passed with unanimous consent:

> We support this sentiment, and indeed, we yield to no one in our desire to honor the American soldiers who risked their lives during World War II to defeat Japanese militarism.
>
> The problem now is that the current (fifth) script of the *Enola Gay* exhibit utterly fails to "portray history in the proper context of the times." Notwithstanding that some additions to the script do add to the historical context, certain irrevocable facts cannot be omitted without so corrupting the exhibit that it is reduced to mere propaganda, thus becoming an affront to "those who gave their lives for freedom."
>
> One of these facts—the observation that there has been a debate from the very beginning over whether the atomic bombings were necessary to bring about an early end to the Pacific war without an invasion of Japan—was accurately reflected in the first few drafts of the exhibit's script. The existence of that debate is an historical fact, and the statement of that fact was removed from the planned exhibit in response to political pressure.
>
> Historical documents essential to an understanding of the historical debate over the atomic bombings likewise have been removed from the exhibit. We refer here to such documents as the June 27th, 1945, memo from Under Secretary of the Navy Ralph Bard reporting his "feeling that before the bomb is actually used against Japan that Japan should have some preliminary warning. . . . The position of the United States as a great humanitarian nation and the fair play attitude of our people generally is responsible in the main for this feeling." In addition, the statements by General Dwight D. Eisenhower and Admiral William D. Leahy which re-

> flected their opinion that the bombings were unnecessary should also be reinserted. One cannot understand the nature of the debate without the inclusion of such statements from prominent U.S. authorities. Nor can visitors understand the context of the debate without seeing at least some of the bombing artifacts—such as the remnants of the personal belongings of civilians killed—which have been removed from the exhibit.
>
> It is most unfortunate that the Smithsonian is becoming associated with a transparent attempt at *historical cleansing*. . . . It is unconscionable, first, that as a result of pressures from outside the museum, the exhibit will no longer attempt to present a balanced range of the historical scholarship on this issue, second, that a large body of important archival evidence on the Hiroshima decision will not even be mentioned; and third, that the exhibit will contain assertions of fact which have long been challenged by careful historical scholarship.[33]

The letter then drew attention to an October 22, 1994, resolution, passed by the executive board of the Organization of American Historians. The OAH is the nation's principal organization dedicated to scholarship in American history. The resolution read,

> The Organization of American Historians condemns threats by members of Congress to penalize the Smithsonian Institution because of the controversial exhibition on World War II and the dropping of the atomic bomb. The Organization of American Historians further deplores the removal of historical documents and revisions of interpretations of history for reasons outside the professional procedures and criteria by which museum exhibitions are created.[34]

On this basis the historians' letter concluded,

> . . . Only by resisting pressures from political sources ill-informed about the relevant historical scholarship can you hope to defend the Smithsonian's credibility as a public institution that faithfully reflects the broad range of debate over our nation's history—and not just what is perceived at the moment as patriotically correct history.

We met with representatives of this group at 9:00 A.M. on November 17, in our conference room. Their primary purpose in meeting with the curators and me was to protest changes made to the exhibition in response to criticism by veterans' organizations. The delegation included Barton Bernstein, of Stanford University, a member of the original exhibition advisory committee.

We talked for two hours. Bernstein was vocal in calling my attention to a grievous misinterpretation in our changed script. He claimed that the casualty estimates for an invasion of Japan given in the text hammered out with the American Legion and presented earlier in this chapter misrepresented Admiral Leahy's June 18, 1945, remarks to President Truman. Bernstein claimed that Leahy's diary entry for that same day stated he had meant that there would be 63,000 casualties—far lower than the "quarter of a million casualties" figure we had attributed to him in our revised casualty label.

Bernstein told a reporter that he had challenged me on these figures, and added that I was unable to refute him.[35]

That is quite true. The reason was simple. I was dumbfounded! The figure of a quarter of a million had come from a paper that Bernstein himself had written in 1986.[36] So, when he suddenly told me, in the middle of a large meeting, that we had made a grave mistake, I did not know how to respond. I was pretty sure we had based our numbers on his paper but on the spur of the moment I could not swear to that. I don't recall Bernstein stating that his article had been in error. I was thoroughly confused and figured I would check it out later with Michael Neufeld—which I did.

As the meeting broke up, I invited Bernstein and the others to please let us have factual corrections to the script if they found any, so we could check up on them. The group then left for a press conference and the release of their letter.

The third and last of these meetings took place in the same room on the morning of December 15, 1994. Robert Musil was there for the Physicians for Social Responsibility, and representatives also came for several other organizations, the Friends Committee on National Legislation, Pax Christi, the American Friends Service Committee, and other peace organizations. One of the most vocal participants was Daniel Ellsberg, director of Manhattan Project II.

The group was not satisfied with the most recent treatment of the script. They said that the atomic bombings had always been controversial, right from the start in 1945; we should be saying that in our exhibition. They also felt we should state that the vast majority of the casualties at Hiroshima and Nagasaki had been civilian, not military.

HISTORIANS AND THE PRESS

Some articles in the press were understanding of the complexities of our exhibition. Notable among these was a letter to the *New York Times* by Robert Jay Lifton and Greg Mitchell, who were completing a book on American post–World War II attitudes toward the bombing of Japan:

> The controversy over the Smithsonian Institution's atomic bomb exhibition will not subside any time soon. . . . One member of Congress said of Smithsonian officials: "Their job is to tell history, not to rewrite it." What history? Whose version of history? . . . Any re-examination leading up to the commemorations next summer—especially one in a museum devoted to national memory—should include all of this: the Pacific war, the decision to use two atomic bombs and the bombs' effects. It needs to draw on all available information and be open to fresh analysis. We must bear witness to what has happened in America as a result of Hiroshima, both the bomb's presence in the world and our having used it.[37]

But others were more concerned with presenting their own historical views than in providing backing for the exhibition. Thus Gar Alperovitz, president of the National Center for Economic Alternatives, wrote an Op-Ed piece in the *Washington Post* that stated,

> The emotional flap over the Smithsonian Institution's plan to exhibit the *Enola Gay*, the B-29 that dropped the first atomic bomb on Hiroshima, reminds us that the nuclear age is now a half-century old and that we have not yet sorted out the basic facts about how it all began.
>
> Unfortunately both sides in the Smithsonian debate mixed and muddled important questions. The initial exhibit plans, for instance, minimized American casualties prior to Hiroshima and paid scant attention to Japanese brutality or to the war before 1945 in general. Martin Harwit, director of the National Air and Space Museum, added insult to injury by gratuitously characterizing the debate as a fight between the new generation and old veterans now in their seventies.[38]

Similarly, historian Kai Bird, in an Op-Ed piece in the *New York Times* wrote,

> It was a humiliating spectacle, scholars being forced to recant the truth. Curators at the Smithsonian's Air and Space Museum in Washington have been compelled by veterans' groups to rewrite the text for an exhibit on the bombing of Hiroshima. . . . During two closed-door sessions with representatives of the American Legion, they agreed to censor their own historical knowledge.[39]

I felt the historians and pacifists had not been helpful. They criticized us for submitting to pressures from the veterans' groups without knowing what we had done or why. And they failed to make their own countervailing views known through the media or in Congress. That combination of criticism and political inactivity acted to add to the criticism advanced by the Air Force Association and other detractors, who often claimed that neither they nor the historians were happy with the exhibition, which therefore ought to be taken out of Smithsonian hands. Dagley was quoted as saying he worried that those visiting the exhibition would leave asking,

> . . . is that something crammed down our throats by the Legion, or is that something crammed down our throats by the historians? What is the value of the exhibit?[40]

JUSTIFYING THE BOMBING

Many of the objections to the exhibition repeatedly brought up by the "five old men" and by the veterans' organizations dealt with the decision to drop the bomb.

While the museum had consistently declared that it was mounting a historical exhibition on the mission of the *Enola Gay*, and that we would not take a stance on the morality of the atomic bombing, the veterans insisted that we do just that—if not in words, then at least subtly through pictures.

General Monroe Hatch's assessment that we were treating the United States and Japan as though "morally equivalent" appeared to be based on our decision not to get involved in moral issues—a decision which, by default, meant treating both sides the same way.[41] Hatch seemingly insisted that the moral issue be raised so Japan could be shown to be at fault. For that, the Japanese atrocities had to be emphasized. An additional way of achieving the same effect was to show large numbers of images of American casualties, to demonstrate how much American troops had sacrificed, and again provide a moral justification for the bombing.

The American Legion, which entered the debate somewhat later, took a different approach to achieve the same end. Their wish was that we first assert the inevitability of an invasion of Japan if the atomic bombs had not been dropped, and then declare that a far larger number of U.S. and Japanese lives would have been lost in such an invasion than had been lost at Hiroshima and Nagasaki.

The Legion objected to suggestions that President Truman had alternatives to an invasion. Mentioning that Japan might have surrendered in response to continued conventional bombing, a continuation of the very effective naval blockade, and the entry of the Soviet Union into the war, undermined this argument and was fiercely resisted by the Legion. Citing the relatively low, though still substantial, casualty rates Truman's advisors had told him to expect in an invasion of Japan also infuriated the membership of the Legion, because those numbers were significantly lower than the more than one hundred thousand who had died on the ground in Hiroshima.

Only if the exhibition agreed that the atomic bombs had actually saved lives would the Legion be satisfied. They insisted that Truman's "courageous" decision had saved far more lives, even far more Japanese lives, than the bombings had cost at Hiroshima and Nagasaki. That is quite possible, but it presupposes that Truman had no alternative to the atomic bomb except to invade. This is difficult to believe. It seems unlikely that Truman would have callously given the order to invade Japan if he had thought that half a million or a million American casualties were at stake. He would, more likely, have considered his other options.

DEFENSE AGAINST ALLEGATIONS OF REVISIONISM

Historians and pacifists took us to task for my deciding to delete two quotations from the script citing Chief of Staff Admiral Leahy's[42] and General Eisenhower's[43] objections to the atomic bombing. I had asked for those to be taken out because there is no archival evidence that either stated those views to Truman. Both men first mentioned them in writing several years after the war.

The historians seemed to want those statements reinstated because they showed that leading figures in 1945 had already had questions about the morality of the bombings. If the historians could satisfactorily demonstrate that, then they would have an effective defense against the veterans' accusations that they were perpetrating "revisionist history."

I was surprised that the historians had overlooked a document that was far more persuasive on that score, had always been part of the exhibition, and had remained in it throughout all the negotiations we had carried out with the veterans. Apparently overlooked by the historians was Secretary of War Henry Stimson's very clear analysis of the moral issues surrounding the atomic bomb. He had presented them in his very first briefing to President Truman, on April 25, 1945, when he informed Truman about the existence of the atomic bomb project. There he called the bomb "the most terrible weapon ever known in human history," and spoke of "a certain moral responsibility upon us which we cannot shirk without very serious responsibility for any disaster to civilization...."[44] This was clear evidence, as shown in Chapter 18, that Truman knew about the issues and that Stimson had discussed them with him. The decision then had been up to the president.

The historians' and pacifists' allegations that the museum had caved in seemed to me both harmful to our efforts and untrue.

26

The New Secretary—Smithsonian Support Wavers

THE SEARCH

In September 1993, Smithsonian secretary Robert McCormick Adams announced to the board of regents his intention to retire the following year. His decision had been anticipated; the Board at once appointed a search committee of six board members chaired by Ira Michael Heyman. Heyman, who for ten years had been chancellor at the Berkeley campus of the University of California, had by then served as a regent for three years.

The search began in earnest in December, and the successful candidate was to be chosen at the Regents' meeting the following May. By mid-March 1994, a list of candidates to be interviewed had been selected and invited for interviews, when Heyman suddenly announced his resignation from the search committee—but not from the board of regents—in order to become a candidate himself. Replacing him as search committee chair was his fellow regent and committee member, Barber B. Conable.

The Washington press reported,

> The quest for a new secretary to lead the Smithsonian Institution accelerated yesterday as the search committee chairman resigned to place his own name under consideration.... The broad responsibilities for the job were described last October in a newspaper want ad, which invited candidates with "a record of superior scholarly accomplishment, a facility for

> written and oral communication, relevant administrative experience and exceptional leadership." . . .[1]
>
> Heyman said, colleagues "had been working on me for a long time within the institution." He said that, as he interviewed people about the strengths the next secretary must have, the response was: "You know it all." . . .[2]
>
> He laughs when asked why he waited until the last minute to resign as head of the search committee. "That is a measure of my confusion over what to do."[3]

That his selection by the regents was a foregone conclusion even while other candidates were being interviewed was obvious. Irwin Shapiro, Director of the Smithsonian Astrophysical Observatory, at once made an appointment with Heyman to brief him on the observatory's current projects and major needs. Some of the other Smithsonian museum and research center directors reacted similarly. By the end of May, his fellow regents had confirmed Heyman's selection.

> Conable said, Mr. Heyman's election does not signal a break with the traditional scientific mission of the institution, but a change in the role of its director. The Smithsonian's previous directors have all been scholars and scientists, but Mr. Conable said the search committee's requirements included more emphasis on fund-raising skills. "We felt he had a record in fund-raising, which is a great need for the Smithsonian, as obviously we are moving into a period of severe fiscal pressure." . . . Fulfilling the search committee's expectations for him, Mr. Heyman said money will constitute his biggest challenge.[4]

THE WARMUP

I had met Heyman briefly on social occasions during the years he served as a regent, but my first substantive meeting with him was on July 22, 1994, when he asked me to meet with him in the Smithsonian Castle to bring him up to speed on major issues confronting the museum. Though he was not to be installed until September, he was beginning to inform himself. We talked for an hour. I told him of our two major activities, planning for the Dulles extension and the *Enola Gay* exhibition. He spoke of his wish to deal primarily with the outside world, to raise money for the Smithsonian, and put the Institution on a more secure financial footing. To concentrate on this priority and relieve himself of dealing with internal problems, he would establish a new "provost" to look after academic concerns. The provost would work closely and as an equal with the under secretary, who would be responsible for day-to-day operations.

At the beginning of September, Heyman joined the Institution, essentially full time, overlapping with Adams for about three weeks until September 19, when he was to be officially installed.

In the afternoon of September 7, 1994, Adams, Heyman, Connie Newman, Mark Rodgers, Mike Fetters, and I met in the secretary's office to discuss strategy for a meeting we had arranged with Meg Greenfield, editor of the *Washington Post*'s editorial page, and her editorial board. The *Enola Gay* exhibition had been getting so much bad press in the *Post* that we wanted to know if there was not some way we could make peace.

Two days later, promptly at 9:30 A.M., the six of us arrived at the *Washington Post* offices. The *Post* was represented by publisher Donald Graham, Meg Greenfield, deputy editorial page editor Stephen Rosenfeld, and two or three others. They lit into us right away for having an atrociously slanted script. At one point I asked Graham whether he had read it. He ventured that he had read all the newspaper reports, and that seemed enough. The others had examined at most an early version, and it was not clear how much of that they had actually read.

I gave a presentation on what we were trying to do and what had happened. They shot a lot of questions at us, most of which I answered with occasional interjections by Adams, Heyman, and Newman. The discussions remained confrontational throughout the full ninety minutes. The *Post*'s subsequent coverage appeared unaffected.

As we left, Heyman mentioned that he had given an early version of the script to a good friend at Berkeley, Emeritus Professor of Law Preble Stolz, who had now written him a long letter with his reactions. I said I'd be glad to see Stolz's comments, and Heyman sent them over the same day. Many of Stolz's suggestions had already been independently incorporated into the script, some we preferred not to accept, but others were appropriate. I wrote to thank Stolz:

> Four of your suggestions do need to be considered. . . . I really appreciate your taking the time to examine the script so thoroughly.[5]

Ira Michael "Mike" Heyman was to be inaugurated on Monday, September 19, 1994. The night before, the regents held a black tie dinner at the National Air and Space Museum for retiring secretary Adams. At the start of the evening, my wife and I briefly talked with Heyman. He spoke of controversial times he had faced at Berkeley, and then, to my astonishment, reached into the pocket of his tuxedo to fish out and hand me a large, orange button inscribed "Don't Panic."

I still have that button. In later months I often wished Heyman had kept it.

One of the Institution's former regents, CIA director Jim Woolsey, was also at the dinner. In the middle of appetizers, his staff informed him that President Clinton was going on national television in the next half hour. Within minutes, the sole member of the museum staff on duty that night had strung cables and transported in a large-screen television projector. By the time the president announced he was ordering troops into Haiti, the regents and their two hundred distinguished guests were able to hear and see him speak, without interrupting their meal.

This quick, seemingly effortless performance by one member of the staff was characteristic of the dedication, skill, and pride in their work of virtually all the men and women at the museum. Throughout my eight years there, I never forgot how fortunate I was to be their director.

IRA MICHAEL HEYMAN ASSUMES OFFICE

The next morning was disheartening. Heyman's inauguration was set for 11:15 A.M. But the *Washington Times* reached us first with an article headlined, "Big man on Mall campus" which touched on the exhibition:

> Military and veterans' organizations have stunned Smithsonian officialdom with a barrage of criticism, arguing that the display, scheduled to open next May would depict the carnage inflicted by the blast in a measure disproportionate to the attention paid to the Japanese atrocities that prompted it. And those organizations are right, Mr. Heyman says.
>
> "We should have avoided it," he says. He uses the term "we" in the institutional sense, since he had nothing to do with organizing the exhibit and only found out about the controversy when it became public.
>
> "We weren't wise in how we were thinking about this show . . . I feel pretty good about the changes we've made. I really think in a funny sort of way it's working out in the best way. We had an ill-conceived show. We'll have a better show because of the accuracy of the debate.
>
> "The exhibit was to be about the dropping of the bomb and its consequences. They took for granted that everyone would know the context. There was a carelessness in not recognizing that, to a lot of people, not displaying the context would make it be viewed in a very one-sided way."[6]

The reporter was probably right in concluding that Heyman had used "we" in the institutional sense. By the fourth paragraph, Heyman had switched that to "They."

The speech Heyman delivered at his inauguration was in the same blunt vein, "Our first script for the exhibition was deficient."[7] Apparently forgotten in all this was Heyman's role as a Smithsonian regent for many years, during which time I repeatedly reported to the regents on progress on the exhibition, and repeatedly received their backing. He had never voiced any objections then.

A "TERRIFIC" SCRIPT

Friday, September 23, 1994, ended Heyman's first week in office. For that evening, he invited Smithsonian museum directors and senior staff to a farewell reception for Bob Adams. As most of the guests began to depart, a few of us still felt like staying.

Connie Newman was also still there, so my wife and I sat down with her. We had just had a busy week with Heyman's installation, the first of the American Legion negotiations, and the press event the following morning, and we chatted about that. But I also wanted to bring up Heyman's inaugural speech and his statements to the press, which worried me greatly. To date, I had always enjoyed the secretary's and the regents' backing on the *Enola Gay* exhibition. Heyman's statements in his first days as secretary had thrown all that in doubt. I had no trouble dealing with the veterans and coming to reasonable arrangements with them. But if the museum's thrust was to be undermined by the secretary, the veterans' organizations would quickly sense that and the museum's position would become untenable.

The caterers were already beginning to clean up, and we were still sitting there talking. So we asked Newman whether we could continue the discussion at our home, just a few blocks away. Over a glass of wine, we talked till midnight, with Newman convincing me that Heyman would stand up for what we were doing with the exhibition, and that this was not an issue over which I should think of "walking," as she put it. I was somewhat relieved, but knew that only time would tell.

The following Tuesday, Heyman came over for an informal "town meeting" with the entire museum staff. I had invited him to come as soon as he was free after assuming office. He spoke briefly, but vivaciously, was open in answering questions, and made an excellent impression.

This also was the second of the two weeks in which we met with Hugh Dagley and Herman Harrington, of the American Legion, and agreed on changes to the exhibition script. A copy of this new version went to Heyman.

On Friday, October 7, I got a call from Jim Hobbins, executive assistant to the secretary, asking whether I would be free to come over to see Heyman around noon. He told me the secretary had read the revised script and really liked it. Since I was free, I went to see Heyman, and while he munched on a sandwich we talked about the exhibition. Heyman had risen at 3:30 that morning to go through the entire draft. He was in an upbeat mood, and called the script "terrific." He asked for two or three small changes. I took his recommendations with me, and the museum's staff made the alterations the same afternoon.

Each week at the museum began with a staff meeting of about thirty people. Generally it lasted not much longer than half an hour, and served primarily to update everyone on the week's planned activities. The minutes of that meeting were also posted so the entire staff would know of ongoing activities. The minutes for October 11, 1994, record my saying,

I met with Secretary Heyman. He read the script for the *Last Act* . . . and he thought it was terrific. We're going forward with the script. He had minor changes. I am going to meet with the American Legion at the end of next week. The changes that were made were primarily of perceptions and one or two substantive changes. I think it's a strong script.[8]

THE NATIONAL BOARD

The Smithsonian National Board is a group of wealthy individuals dedicated to helping the Institution raise money. In late August, then secretary Adams had called me to ask that I get in touch with Bill Anderson, former CEO of National Cash Register. Anderson, who had been previous chair of the national board, was upset about the exhibition. I called him at once, and had a long conversation, which was rather one-sided, because he seemed interested only in telling me he had a two-foot-long shelf of letters and papers that veterans and veterans' organizations had sent him. He himself had been a British prisoner of war of the Japanese and had almost died in captivity under the horrible starvation and punishment the Japanese inflicted on their captives. What we were doing in our exhibition was a scandal. I should listen to Dick Hallion and do what he advised. Anderson had no patience for my telling him that Dick had given us a great deal of praise at first and then changed his mind. He just wanted me to take Dick's current advice. Period.

The national board was to meet on the weekend of October 14–16, 1994, starting Friday morning and going till Sunday. Anderson arrived a day early and had asked to see me Thursday at 10:00 A.M. At the start of that hour, I was scheduled to introduce astronaut Franklin Chang-Diaz in the Museum's theater, where he was to speak to a children's school group. I met Anderson as soon as he arrived and said we could go up to my office, once I had introduced Chang-Diaz. Anderson accompanied me into the theater, but could not contain himself; on our way in, he lectured me, once again, to listen to Dick Hallion. With my mind on what I needed to say in my introduction, I was caught off guard, and told him frankly what I thought of Hallion's reversing himself. Anderson wasn't listening.

Once in my office, Anderson again told me about his treatment as a prisoner of war of the Japanese. But, he assured me, he bore no grudge. In setting up National Cash Register in Japan, he had made many good friends. He also told me he never needed to carry his own briefcase in Japan. They always had a person doing that for him.

Anderson had brought in photographs he had taken in a Chinese museum dedicated to exhibiting the atrocities the Japanese had committed during their occupation. I told him I would pass them on to Don Lopez, who was looking for such images for the introductory section of the exhibition. Throughout, Anderson was incessantly criticizing, despite my efforts to tell him what we were actually doing. I had the impression that he had formed a mental image and wanted to hear nothing that might change it.

After leaving me, Anderson went to see Heyman and complained. I may have told him that we expected the mayors of Hiroshima and Nagasaki to contribute two-minute video-interviews for the gallery, or he may have read that in correspondence circulated by the Air Force Association. I got a call from the Castle a few hours later. Heyman and Newman were not aware of these interviews, although it was something about which I had informed her and Adams many months earlier, in one of my periodic written reports.

I wrote a note to both Heyman and Newman and handed it to them the following morning before the national board meeting's start. It explained how the mayors had been invited and added that Monroe Hatch had also agreed to be videotaped for this same purpose. I quoted from the planning document dated July 1993, and wrote to them that it had been reprinted and widely circulated by the AFA in a July 14, 1994, mailing to Congress, veterans' organizations and media. Included was the following paragraph:

> The conclusion will include a video giving a range of perspectives on the bombing—from 509th or other veterans (or Paul Fussell) talking about how they believe the bombings saved lives and prevented an invasion, to the mayors of Hiroshima and Nagasaki giving their views and peace messages, if they accept our invitation to do so. This video will encapsulate the many perspectives of the bombings, but it (and the exhibit) will embody one common wish: that nuclear weapons never be used in anger again.[9]

Heyman, apparently still pleased with the script he had read the previous week, had asked me to give the kickoff presentation to the national board. The session started at 9:00 A.M. I gave the same slide and video presentation I had given at the American Legion national convention.

I had hardly finished before a barrage erupted. Husbands were there with wives. Some wives were even more incensed than their husbands. "How could the Smithsonian mount such a stupid exhibition?" someone wanted to know. How could we expect them to raise money when everyone they approached embarrassed them with questions about this ill-conceived exhibit.

Barber Conable rose and talked about how he had been slated to land on Kyushu, expecting almost certain death; his life had been saved by the bomb. Then Bill Anderson rose to lambaste me. It went on for a long time. The board had to take a coffee break to reenergize, and then continued with full fury. Only one person, a woman, came up during the break to say some nice things, and she did get up later to speak in the exhibition's favor. Eventually the chair called a halt. They had other business to handle, and I left. Both Newman and I were amazed that the Smithsonian's own friends should have given my presentation so much nastier a reception than the American Legion had just six weeks earlier.

I believe that this board meeting contributed appreciably to Heyman's outlook. He never again sounded as enthusiastic about the exhibition as he had just one week earlier. He had been picked to raise money for the Institution. Anything that affected potential donors, or the congressional budget, greatly concerned him.

Heyman often referred to the September meeting with the *Washington Post* editorial board as having convinced him that something was wrong with *The Last Act.* He also liked to recall the New York taxicab driver who, on seeing that Heyman wanted to be dropped off at the Smithsonian's Cooper Hewitt Museum in Manhattan, asked, "Have you heard what those bastards are doing to the *Enola Gay*?" This anecdote was subsequently repeated in the national magazine *U.S. News & World Report.*[10]

INDIANAPOLIS

As previously mentioned, a final meeting to go over the script with the American Legion took place on October 20 at national headquarters in Indianapolis. They had asked to see a clean copy of the script containing all the changes we had made in response to our two days of meetings. Before we went, Newman, Rodgers, Fetters, and I met with Heyman to discuss strategy. Fetters, who always thought ahead, sent me a list of questions we should consider:

> 1. Are we [willing to tell all the veterans' organizations after the Indianapolis meeting on Thursday, October 20, that we are still willing] to make more changes?
>
>   If the answer is "no," then we should be prepared to strongly defend the current script, both to the veterans and to the Congress, media and general public.
>
>   If the answer is "yes," then it will be increasingly difficult to defend ourselves against the growing number of critics who feel that the Institution has "caved" to a special interest....
>
>   Once the first question has been answered ... we could, for instance, offer to keep Coalition members informed on the remaining steps/issues in the process: the catalogue, public programs, special events ... the production of the three videos....
>
> 2. If we are at all interested in salvaging our relationships with the Hiroshima and Nagasaki museums, then we must be extremely careful in how we characterize ... our relationship with veterans' organizations, the changes that have been made to date, and the "purpose" of the exhibition....
>
>   Issue one and issue two are quite linked....

3. Media Relations: We need to establish a time line, spokespeople and talking points that everyone can agree upon.
4. Public Information: Once we've declared the script "as final as any script gets," we'll need to work with the Secretary's office on the development of letters to the following target audiences: General Public, Congress, Contributing Members, Magazine. . . . Subscribers, Museum Shops. In addition, we should consider an "update" letter to the many exhibition advisory groups and individuals as well as the Regents and other NASM and SI advisory boards and committees. . . .[11]

Unfortunately, when we met in late afternoon the following day, these alternatives were not even considered and no such decisions were forthcoming. Heyman preferred to wait and see. In effect, we were again waiting to react to the veterans rather than deciding what we wanted to do and then sticking to that decision as long as it remained reasonable. This approach was to become permanent. The wavering was evident to all, including the veterans' organizations, which saw a way of capitalizing on this weakness.

GOING IT ALONE

After the Indianapolis meeting with the American Legion, the media wanted to know all about our agreements or disagreements. Heyman and I spoke jointly with correspondents for ABC's *Nightline* and PBS's *MacNeil/Lehrer* shows. I could tell that the new secretary was not at ease with this approach. Where he was eager to say mistakes had been made, I staunchly stood for the approach we had taken.

I had been handling the national media all year, but now Heyman decided he would take on the responsibility. There were also other signs that he preferred to trust only himself. I did not know about it until three months later, at the end of January 1995, when Eugene Meyer wrote a profile on Paul Tibbets in the *Washington Post*; then I learned that

> Suspicious of the museum's direction and motives, he had refused to have his recollections videotaped by the exhibit's curators. But quietly, on Nov. 1, Tibbets met for three hours at the Smithsonian with Under Secretary Constance Berry Newman, joined by Secretary I. Michael Heyman and others. Heyman was "pleasant, courteous." Tibbets urged Heyman to fire Air and Space Museum Director Martin Harwit, and two curators associated with the exhibit. To this request, Heyman did not respond.
>
> Tibbets's immediate mission was to assure that the section of the exhibit dealing specifically with the 509th and the mission were factually correct in every detail. Since the meeting, Tibbets and associates have offered corrections that have been accepted, Smithsonian officials say. Which is not to say he has endorsed the entire exhibition. He has not. "I'd vote 100 percent to have the whole thing canceled," he says.[12]

Not until two months after this meeting did I get a call from Newman, on January 4, that Tibbets and some of his former crew had looked over the script and had comments that Tibbets's biographer George Hicks was coming to discuss with us the following day. I was told that Tibbets had dropped by, and had the impression that this had been a brief meeting, quite recently, before Christmas. Here, curator Joanne Gernstein was trying to work with members of Tibbets's crew, wondering why they were suddenly unwilling to commit themselves to letting us use the video-film Patricia Woodside had made, which all of them had previously liked. Now it is obvious that word had come down from the general that he was dealing with the Smithsonian on a higher level. The secretary had taken over.

Other meetings apparently also took place without my knowledge. Testifying before Senator Stevens months later, Bob Manhan, of the Veterans of Foreign Wars, stated that he had been present at three meetings in the Smithsonian Castle at the turn of the year. He mentioned that Newman had attended all three, Heyman had been at one. I knew only about the two meetings that I had attended.[13]

The Smithsonian secretary of course has the right to take over any activity he chooses, but he needs to inform people that he has taken that action. Otherwise confusion sets in; the organization cannot function.

LEAKS

From this, one might conclude that my relations with Heyman and Newman were poor. If so, I was unaware of it. I believe the main reasons for the secrecy were twofold. Heyman was hesitant to share information, whereas I felt that I needed to be open if I wished to delegate responsibility to senior members of the museum's staff.

The second factor was a series of leaks of documents from the museum to the Air Force Association. Clearly someone, or perhaps several people at the museum, felt disgruntled and passed documents on to the AFA. The July 1993 planning document was one of those, as was my April 16, 1994, memorandum commenting,

> Though I carefully read the exhibition script a month ago, I evidently paid greater attention to accuracy than to balance.... A second reading shows that we do have a lack of balance and that much of the criticism that has been leveled against us is understandable....[14]

I had discussed these leaks with Newman some months earlier, and felt that a witch hunt to determine who was responsible would be the worst thing for museum morale. In any case, as a public institution, under the Freedom of Information Act the museum generally had to respond positively to requests for any document. Most of our staff knew that and were careful in what they wrote.

Heyman frequently complained, "Your museum is like a sieve." Leaks were unacceptable to him.

HEYMAN'S PUBLIC COMMENTS

In late October 1994, Heyman told the *Washington Times*,

> The first draft of this exhibition received significant criticism from both outside and inside the Smithsonian. Many of the critics were right, and the exhibit team has taken what I believe are the necessary steps to address these concerns. The process is not finished. . . . We will continue to accept and consider observations from all sources, but we feel that we are very close to where we want to end up.[15]

This suggested that he was quite happy with the way the exhibition was going at that time, though he felt we should continue to listen to advice. Some months later, in a letter dated January 17, 1995, Heyman wrote to George A. Mannes, chairman of the California Department of Veterans Affairs,

> I believe the script for the exhibition now strikes the appropriate balance that provides visitors with an opportunity to learn more about this critical event while at the same time recognizing the sacrifice of those who served in the armed forces and the resoluteness of those who led our nation. . . . The development of this script has served as a catalyst for a national discussion about the legacy of the *Enola Gay* and the atomic bombings. Next May we will open an exhibition that I believe will make a positive and thoughtful contribution to this dialogue. . . .[16]

The very next day, after American Legion national commander William Detweiler had stopped by his office, Heyman decided to cancel the exhibition. I will return to relate how that came about. But first we need to see how the Japanese were reacting to the revisions the museum had made in the script.

27

Japanese Doubts

ALARMED BY THE CONGRESS

The August 10 letter from twenty-four members of Congress caused a stir in the Japanese press, where the attacks on the exhibition from the veterans' organizations had already made a strong impression. On August 23, 1994, Kenji Ohara, the new associate director at the Hiroshima Peace Memorial Museum, wrote to Tom Crouch,

> Would you inform us of recent claims by veterans and Congressmen?
>
> How will you re-examine the contents of the script? Will you make any changes in the script we have received? If you make any changes, we would like you to give us an outline and schedule. After you make some changes, we would like you to send us a Japanese script with all changes immediately.
>
> We are going to present to the press the script ... as a final script from [the] Smithsonian Institution, if the claims by veterans and Congressmen do not cause you to make any [further] changes ...[1]

A week later, Ohara telefaxed Crouch again, wanting to know more about the pressures on the museum from veterans and Congress. He complained that the local press was asking him questions and he was not in a position to offer any answers.[2] Crouch telefaxed back the same day, describing the display of about fifty photographs through which visitors would pass on their way to the exhibition. He added that all the changes to the older portions of the script would be apparent in the revised script that "went out for copying this morning. We will meet with our translator tomorrow morning to discuss the translation of those pages of the old script which have been changed."[3]

Given these concerns, I decided to also write a letter to Ambassador Kuriyama, who in any case was slated to receive the latest version of the script:

> I am attaching a revised script for the National Air and Space Museum's planned exhibition, *The Last Act: The Atomic Bomb and the End of World War II.*
>
> As you may have noted, the script has recently been taken to task in the press. Critics have urged us to include more material on the earlier history of the War in the Pacific. Originally, the Museum had seen no need to go into that history in great depth, because we assumed that most visitors to the Museum would remember what they had learned in school about the war. Recently, however, we undertook some man-in-the-street interviews, and found that many people with whom we spoke knew little about [it]. . . .
>
> We have now decided to add further historical background to the exhibition. . . .
>
> If you or your staff have any questions regarding the changes incorporated in the new script, I would be happy to respond.[4]

I talked with Connie Newman about the worries of the Japanese. We had just had a good meeting with the American Legion at Minneapolis, and now it was important to make certain the Japanese would not back away from working with us. We agreed it made sense for me to return to Japan to reassure the mayors of Hiroshima and Nagasaki in person, though Newman thought I should not leave for another six weeks or so. By that time, we would have met with the Legion's representatives and perhaps settled any remaining differences.

On September 6, I sent copies of the revised script to both mayors and attached almost identical letters to both. The letter to Nagasaki mayor Motoshima, in part, stated,

> Because the added historical material in the revised plans was not included in the script and its translation into Japanese, which the Museum had earlier sent to Nagasaki for your consideration, I am writing you now to point out these changes. . . .
>
> You may have questions that you would like to raise in connection with the more broadly historical script as it now has developed. At the same time, there is also some urgency for the Museum to receive the materials we wish to borrow, so that they can be installed in the exhibition in time. I would therefore like to propose that I visit Nagasaki . . . during the week of October 17 . . . so that we can discuss any of your concerns. . . .[5]

The Japanese were particularly worried about the unanimous nonbinding U.S. Senate resolution that Senator Kassebaum had sponsored in late September.

On October 13, 1994, Yoshio Saito, secretary-general of Nihon Hidankyo—Japan Confederation of A- and H-Bomb Sufferers Organizations—wrote to the Smithsonian Institution and enclosed an October 4 letter he had sent to

> the President of the United States and the U.S. Senate. We protested that the U.S. Senate had adopted a resolution on September 23 urging a change in the planning of an exhibition of the *Enola Gay* and witnesses of the atomic bombing of Hiroshima and Nagasaki, planned to be held at the Air and Space Museum. . . .[6]

The declaration he attached was signed by three cochairs, one of them Senji Yamaguchi. I had not heard of Yamaguchi before, but he would soon write to me again. The statement read,

> The atomic attack on Hiroshima and Nagasaki brought to the human community the first experience of nuclear war. It was an immeasurably atrocious massacre. The atomic bombs instantly turned the two cities into a "hell," killing hundreds of thousands of people, including infants, the elderly, and the sick, who were burnt in the conflagration and trapped under the collapsed buildings. The bombs even inflicted lasting agony on the survivors, from which they cannot escape until their lives end. We learned through our own experience that nuclear weapons are weapons of the devil which must never be used for any reason whatsoever. . . .
>
> That the Senate adopted a resolution trying to justify the atomic bombing as a "benevolent act" is unendurable for all of us who have lived and are still living in the "hell" caused by the atomic bombing.
>
> We strongly protest the resolution adopted by the U.S. Senate which virtually calls black white. We hope that . . . the Smithsonian Air and Space Museum will exhibit A-bomb material . . . from Hiroshima and Nagasaki, in defiance of all outside pressure.
>
> We demand that the U.S. government take effective actions for the abolition of nuclear weapons at the earliest possible date.[7]

Not long thereafter I received a note from Mayor Motoshima, forwarding a letter from the same Senji Yamaguchi who had cosigned this declaration.[8] The mayor described him as a lifelong friend. Mr. Yamaguchi wrote,

> I was 14 when the Atomic Bomb was dropped in Nagasaki. At that time I was working outside, shirtless, 1.1 km from the point of detonation. I passed out instantly and it took 7 months for my burns to heal. In July of 1946, I developed leukemia. Although I recovered, I haven't been in perfect health since. I still have a large keloid from the blast.
>
> I have dedicated my life to speaking throughout the world on the dangers of nuclear proliferation. . . . I feel the Smithsonian's upcoming exhibi-

> tion . . . will do much to raise awareness of the effects of nuclear war. I only hope that the exhibition fully demonstrates its horror.
>
> The cities of Hiroshima and Nagasaki are providing relics for the exhibition with the strongest hope that peace can be reached by showing what damage nuclear weapons can do. This exhibition must show more of the human suffering that these weapons caused. It is not enough to only show memorabilia and physical evidence such as melted glass or burned wood. The human toll as well as the physical toll must be represented if visitors of the Smithsonian are to understand the consequences of nuclear war.
>
> As long as there are nuclear weapons, nobody is safe. People in Washington or New York could suffer just as those in Hiroshima and Nagasaki did. That is why we want everyone to witness the full reality of what happened here.[9]

REQUESTS FOR A FACE-TO-FACE MEETING

While I had written the two mayors in early September that I hoped to come to Japan in the second half of October, the Senate resolution, the continuing onslaught from the veterans' organizations and the media, and the increasingly conservative attitude in the United States, soon made such a trip doubtful, at least until after the November elections. Republican congressman Peter Blute, chief architect of the August 10 letter cosigned by twenty-three other members of the House, was already making this a campaign issue. My going to Japan now would have given him and like-minded members of Congress additional cause to attack the exhibition.

When the results of the November election were in, my going to Japan became even more unthinkable. Connie Newman was increasingly worrying that the new, conservative Republican Congress might ask for hearings on the exhibition. Having been involved with the Congress for many years, this was the one possibility she seemed to genuinely dread. I was quite surprised at the strength of her concerns.

The Japanese now decided they should send a delegation to Washington if I could not come to Japan to answer their questions. They wished to express their dismay at the removal from the exhibition of items we had originally asked to borrow. Apparently, though they had never officially consented to lend us any objects, the two city museums had already contacted the families who had originally owned the objects to see whether they would consent to a loan. Now some of those families were deeply hurt that these objects, which often were the sole reminder of a lost relative, were being rejected.

To us it appeared that the Japanese sending a delegation to Washington would provoke even more ill will in Congress than my going to Japan. The publicity that would surround a delegation would be devastating. On November 23, 1994, my deputy Gwen Crider and I met with Heyman and

Newman. We all agreed that I could not go to Japan now, and that we could not afford to have the Japanese come either. But we could not put this in writing. The furor such a letter would raise would top everything. Heyman adamantly wanted to avoid a "paper trail." Whatever we did needed to be done verbally to leave no trace.

We agreed that I should call Ambassador Kuriyama's office and ask for an appointment for Heyman and myself. Perhaps the Ambassador could send a confidential message asking the two cities to withdraw their demands for a meeting. I also called Brian Burke-Gaffney, international advisor to the Nagasaki municipal government, at his Nagasaki home, told him about the need for confidentiality and asked him whether he could talk with Dr. Itoh, the Nagasaki museum director, to let him know I could not come now. On November 28, he telefaxed me his private response:

> I called Dr. Itoh and spoke with him about the question of your visit to Japan and the current situation in Nagasaki and Hiroshima.
>
> Dr. Itoh says that in the opinion of both the Nagasaki and Hiroshima museums, your visit to Japan and explanation concerning the reasons for the changes to the exhibition script is essential to obtain the understanding and cooperation of the committees presently reviewing the plans in the two cities. He pointed out that the Japanese side has had to rely almost exclusively on television and newspaper reports for information on the controversy in the United States and the pressure to revise the exhibition. The museums were assured that no major changes would be made, but the fact is that the latest script differs widely from the original. Although he understands the difficulty of your present situation, Dr. Itoh thinks that hesitation would be greatly diminished here if you could come and speak directly to the Nagasaki City Committee on Atomic Bomb Information and Exhibits and explain the reasons for the changes.
>
> An added problem is that Mayor Motoshima promised to lend the Nagasaki exhibits without the necessary approval of the above committee. With a mayoral election coming next spring, Mr. Motoshima is concerned about the possibility of political ramifications of a controversy in the atomic bombed cities. . . .[10]

In the meantime, I had obtained an appointment for Heyman and myself with the ambassador. At 8:00 A.M. on November 30, Heyman, Newman, Acting Provost Bob Hoffmann, and I met in Heyman's office to discuss a strategy for the meeting later that morning. Heyman did not want any direct negotiations with the Japanese, nor did he want any letters going out, but he did want to obtain the loan of the Japanese artifacts.

Later, Heyman and I were driven to the Japanese embassy and met with Ambassador Kuriyama and his designated contact, Mr. Seiichi Kondo. I introduced Heyman to the ambassador and began by apprising him of the situation, namely that we could not publicly confer with Hiroshima and Nagasaki representatives without risking the entire shutdown of the exhibition by Congress. The August letter from Congressman Blute and twenty-three other members of the House, the September resolution in the Senate, and other warnings from Capitol Hill, and the constant criticism from veterans' organizations in the press foreboded the worst.

Kuriyama understood the difficulty and asked Kondo to communicate these sentiments through diplomatic channels. He promised to let us know their response.

On December 5, I decided to also call Paul Blackburn at the American embassy in Tokyo, to tell him about our problems and to seek his help. I told him I feared the exhibition would be shut down by Congress if there was a lot of publicity giving the impression that we were under the influence of the Japanese. Blackburn was sympathetic, but could not suggest any solutions. He said the embassy took care never to interfere in such matters.

Blackburn, just then, was preoccupied with the row the postal service's mushroom cloud stamp had raised in Japan. Its caption, "Atomic bombs hasten war's end, August 1945," had deeply offended the Japanese and was threatening to precipitate a confrontation between the two countries. When I talked with Blackburn a few days later, the controversy had been settled through the intervention of President Clinton, who had asked the postal service to withdraw the stamp. We wryly agreed that the museum should not complain; we had our difficulties, but at least for the moment, we still had an exhibition that was going forward.

DEADLINE PROBLEMS

In the middle of December, in response to requests from the veterans' organizations and their congressional supporters, Newman asked me whether the museum might be able to have all items for the exhibition in hand by the middle of January. She wanted to be able to show them to the American Legion leadership just after that, so they would be in a position to withdraw their condemnation at an upcoming convention in Washington on February 5.

There were two problems with this. The exhibition schedule only called for the four video-films to be ready on respective dates between February and early April. I said I would check with Patti Woodside, to see whether she saw any way of completing all the work in one month instead of three.

The other problem was getting the Japanese to release their artifacts and film footage to us. I did not see how I could even ask them, given that Heyman had ordered a moratorium on written communications with Japan. To settle this difficulty, Newman and I had another meeting with Heyman on December 16. He now consented to my writing the mayors and museum directors to ask for all their materials by mid-January, but I was also to confirm that no personal visits could be arranged for now.

I wrote essentially identical letters to the two cities' mayors the next day. The letter to Nagasaki mayor Motoshima read,

> As you know, the National Air and Space Museum has for many months now been working on an exhibition, titled *The Last Act: The Atomic Bomb and the End of World War II*, which is to open in May 1995. I am writing you now to ask that all of the items we have requested from Nagasaki for this important exhibition be made available to the Museum by mid-January. I realize that this is a short time, but the urgency is great.
>
> I know that the city of Nagasaki would have preferred to arrange a face-to-face consultation before reaching final agreement, but unfortunately such a meeting is not possible. Therefore, I am requesting that agreement be reached on the basis of the written material that our Museum has provided. We believe that it gives the most detailed depiction we can give of the nature and contents of the planned exhibition at this time. It also lists the artifacts we wish to borrow. . . .
>
> It is important, however, for you to understand that even this latest version of the script may be changed. While the progress made thus far is both significant and evident, the process is not yet final. The Secretary of the Institution and I both are committed to the principle that we will continue to be open to observation and comment from all interested parties right up to the exhibition's opening and even thereafter.
>
> Dr. Itoh and his staff are fully aware of all the items that we hope to receive from Nagasaki. We would need by mid-January to actually receive, in Washington, the following:

I then went into detail on delivery dates, formatting and payments, and concluded,

> . . . I will be happy to work closely with Dr. Itoh or any member of your staff to assure a successful outcome. . . . I regret any inconvenience that might be caused by the urgency of this request. However, in order for the Museum to pursue our exhibition through its normal review process and for the exhibition to open on time next May, we need to have in our hands the set of items we have requested, at latest by mid-January. . . .[11]

A shorter letter also went to the museum directors, with copies of the letters to their mayors. Heyman asked me to send the letters over, so he could check them before they went out.

A SECOND VISIT TO THE EMBASSY

In mid-December we heard that the Japanese embassy had received a reply to their confidential inquiries, and Newman and I obtained an appointment to see Seiichi Kondo at 3:00 P.M. on December 19, 1994.

As we were being driven over, I mentioned to Newman that Mike Neufeld had now shown me a diary entry of Admiral Leahy for June 18, 1945. It showed that the label we had negotiated with the American

Legion, on casualties to be expected in an invasion of Japan, would have to be rewritten. It was now clear that we had misquoted Leahy. We had attributed to him a casualty estimate of a quarter of million, on the basis of an interpretation by historian Barton Bernstein of minutes of a meeting held that day. Bernstein had since pointed out to us that Leahy's diary for that same day showed this to be an error. The Admiral had figured that American casualties in the invasion would be around sixty-three thousand. I was concerned that we needed to make this correction and had to let the American Legion know about it. These were figures we had debated with them at length. We needed to let them know we had made an error that had to be corrected.

When we arrived at the embassy, Mr. Kondo informed us that the Japanese groups from Hiroshima and Nagasaki still wanted to come to Washington. We talked about whether Kondo could not once more try to dissuade them. Then Newman surprised me by suggesting that perhaps I could call the Hiroshima and Nagasaki museum directors, directly and confidentially, to tell them of the situation and to see whether I could dissuade a visit and arrange for the artifact loans and the videos without one. The important point was not to leave a paper trail that might be leaked. I was astonished by the suggestion, because I could have been doing that all along if Heyman had not been so apprehensive about contacts with the Japanese and word about them getting to the press, the veterans, and the Congress.

CALLING JAPAN

I did not call at once, because I had written just two days earlier and wanted the Japanese side to first consider that request. With the year-end holidays I also needed to give them more time. But on the morning of January 6, I called Dr. Itoh at his home. It was nighttime by then in Japan. I had arranged for an interpreter with the phone company.

We talked for about half an hour, but the conversation was somewhat confusing, and I called Itoh once more from my home the same evening. Our son Eric was visiting us that day, and acted as translator this time. We reached Itoh around 10:00 P.M. our time, around noon his time. Eric introduced himself, and explained once more all that I had tried to say to Itoh that morning. Itoh apologized for not having understood me, but said he would telefax me his response at once. For him it was January 7 by then. He had heard from the Japanese embassy in Washington what our problems were but had been asked not to tell anyone else. So he had tried to persuade the advisory committee without giving them a clear indication of our political situation. He continued: "The best way to solve the situation is, I think, to have a face-to-face meeting between you and our delegation."[12] Since I had said that was impossible, he suggested they could have such a meeting with Mr. Heyman instead. If they could get a full explanation it would be a "great help" for him to persuade the advisory committee to lend us the things we hoped to borrow.

Of course, we could not afford their delegation seeing Heyman any more than me. I kept finding myself surprised by the enormous cultural gap between our two countries, which prevented what was obvious to us from being fully understood over there—and, I imagined, vice versa.

I decided to telefax Hanako Matano and ask her to get in touch with Itoh and talk to him, since he still failed to understand the full implications of what he was requesting.

When I called her later, she said she had spoken with Itoh, who said I would need to talk with Dr. Shunzo Okajima, a physicist who chaired Nagasaki's committee in charge of approving the loan of artifacts.

I called Brian Burke-Gaffney to ask whether he would translate the conversation with Okajima, and then arranged the conference call. I explained to Okajima that the Smithsonian had terribly hostile press coverage, accusing us of being pro-Japanese and anti-American. We could not afford to have any more criticism without jeopardizing the entire exhibition with the Congress. We would appreciate their not insisting on either a visit by us to Japan or a visit by them to Washington. We simply could not afford the press coverage. Could we not settle matters simply by mail? In advance, I had prepared some notes so I could say all this to Brian succinctly over the telephone. They read,

> The Smithsonian Institution and the National Air and Space Museum, which is part of the Smithsonian, have been accused in the press of mounting an anti-American exhibition unduly sympathetic to the Japanese.
>
> We feel we must minimize any further hostile press coverage, particularly if it could be construed or misconstrued as reflecting Japanese attempts to influence the content of our exhibition. Such a depiction by the press would make it likely that the exhibition would be canceled by the Congress and that the Smithsonian Institution would suffer a terrible blow to its standing with the American public.
>
> This is why we have been so reluctant to have face-to-face meetings with your group. The only way that a meeting would not be viewed in this light, would be if there was an announcement in advance, stating that the Japanese side has read the script, that the contents of the exhibition are fully acceptable and that no changes would be sought by the Japanese side. A meeting without any press coverage could then follow.
>
> But we fully understand that this may be very difficult or even impossible for the Japanese side to state.[13]

Dr. Okajima replied that he could not give me a response to my remarks. It was his hope and intention to cooperate with us to the greatest extent. But his committee had internal, conflicting ideas. He would like to communicate with the committee and get back to me through Dr. Itoh.

The next day, January 10, I called Itoh and reported what I had said to Okajima the previous evening. Itoh repeated that Nagasaki felt the need to state their point of view in a meeting. Even if those views were rejected, they would be willing to lend us artifacts, but this would need to be preceded by their public statement. We went round and round on this, without too much progress. I asked whether they could not voice those objections in a letter. Itoh thought that they could write a letter, but would still need to visit. I said that this would not work. Itoh then asked could the Congress close the exhibition down, and I said since they controlled seventy-five percent of our budget they could influence the Institution to comply with their wishes.

Itoh then proposed that the delegation hold a press conference asking for these changes and that we could say we cannot make them. I countered that if they would agree to the loan without condition, but wished to recommend certain changes of fact or balance for us to consider, they could publish those in the press and we would probably not mind.

On this point, I told him, I could not speak for the Institution, but that this might be an alternative.

Itoh still explained that the information given him by the Japanese embassy in Washington had been portrayed as confidential. So he had not told the advisory committee our reasons for the changes in the exhibition, or for not wishing to meet with them, even though this was what they would have wanted.

I was frustrated. We had sent Itoh the confidential information through the embassy so he could work with it behind the scenes, not keep it to himself. Again, the gap in cultures appeared to have worked against us.

We approached the end of the conversation.

I summarized our position in three sentences: If they insisted on sending a delegation, in all likelihood we would not have an exhibition. If they did not come, we would at least have an exhibition in which we could show the effects at Hiroshima and Nagasaki with photographs. And, rather than risk the total cancellation of the exhibition because of a delegation's coming, our decision would have to be to withdraw our request for artifacts.

Itoh understood that. Even if the exhibition stood as it was, it would still be better to lend us the artifacts. But their procedure required their sending a delegation.

We had reached an impasse.

I had two more phone conversations the next day, January 11, one with museum director Harada in Hiroshima and one with Itoh in Nagasaki. We again talked past each other, and never got any closer to resolving the issue of their sending a delegation. The only concession I could get was that they would send us the videotape material we had asked for, so we could inspect it and see what, if any part of it, we wanted to use in the exhibition. This would not commit them in any way to make these items available for the exhibition. From my point of view, however, their sending us the tapes was useful because we could decide what items we wanted and show them to the American Legion, as Connie Newman had promised.

JAPAN EPILOGUE

For all practical purposes the last two calls on January 11, 1995, constituted the end of negotiations with the two cities. By then, Heyman was telling me he would have to cancel the exhibition for other reasons.

28

Cancellation

THE VETERANS AND CONGRESS

On December 8, 1994, when I was away from Washington at a meeting in Europe, Connie Newman and Mark Rodgers, head of Smithsonian government affairs, met with Herman Harrington, and Hugh Dagley, of the American Legion. Rodgers recalls that Newman was worried that the new Congress might wish to hold hearings on the exhibition, and hoped to enlist the Legion's help in avoiding them. The Legionnaires apparently were willing to help; but in return, and before they recommended withdrawal of the Legion's condemnation, they expected to see all video-film footage and other materials that would be part of the *Last Act* exhibition. Since the Legion was to convene in Washington on February 5, 1995, they would need to review those materials by mid-January.[1]

On December 15, a week after her meeting with the Legion's representatives, Newman scheduled another meeting, which I did attend. Monroe Hatch and John Correll represented the Air Force Association, Tom Kilcline and Charles Cooper were there for the Retired Officers Association, and Bob Manhan attended for the Veterans of Foreign Wars. The meeting was relatively friendly. Everyone had received the first draft script of the "War in the Pacific" section by now, but the group asked to still have an updated, fully assembled script circulated to them at some point. They also felt that American casualties were emphasized too little and asked us to include something on disabled American veterans, to emphasize that our side had taken enormous casualties before using the atomic bomb. They also wanted to eventually view all videotaped material intended for the gallery.

Almost in passing, Hatch advised us that we should take care, because Congressman Blute could be expected to put additional pressure on us. I interpreted that statement as a not-so-subtle threat. But Newman, with whom I discussed it later, told me she thought he was just stating a fact—that Hatch had heard about such a letter in the making.

Knowing now how closely the AFA was involved in drafting congressional letters and legislation, I am not surprised that Hatch would have known what was coming. A letter addressed to Heyman came from Republican congressmen Sam Johnson, Peter Blute, Henry Bonilla, Ralph Hall, Randy "Duke" Cunningham, Robert Dornan, and Duncan Hunter. It was dated December 13, 1994, though I had heard nothing about it at the time of the December 15 meeting with Hatch:

> We would like to take this opportunity to convey to you our deep disappointment with how the Smithsonian Institution has handled the *Enola Gay* Exhibit.
>
> This very disturbing ordeal has blemished the reputation of this world renowned institution. With the *Enola Gay* exhibit's script currently on its fifth revision, it is our view that the Smithsonian should promptly coordinate with the associated outside parties to resolve this embarrassing incident as soon as possible. . . .
>
> Due to the fact that past revisions have taken approximately two months, we are requesting that a final script, including any accompanying video material, be completed by the first week in February. At that time we will determine whether or not to support the *Enola Gay* Exhibit.
>
> In closing, we would like to reiterate our deep displeasure with the way in which the Institution has handled this exhibit and the criticism surrounding it. There is no excuse for an exhibit which addresses one of the most morally unambiguous events of the 20th century to need five revisions.[2]

I was astounded at their labeling the atomic bombings of Hiroshima and Nagasaki "one of the most morally unambiguous events of the 20th century."

This letter also reemphasized that the museum had needed five drafts to shape the script. We never did persuade our critics that this was not unusual for a complex exhibition. Edwin Bearss, who as chief historian for the National Park Service had been responsible for the controversial fiftieth anniversary commemorations at Pearl Harbor, told an interviewer that on the movie they prepared, they had gone through "about 25 [drafts] in two years"—perhaps not a precise comparison, but at least a benchmark.[2]

Herman Harrington and Hugh Dagley had left Newman's office that December 8 Wednesday afternoon. By the following Monday, the letter from the seven congressmen had been signed and was on its way. It asked for precisely what the Legion had told Connie they needed in time for their February 5 convention in Washington. The Legion can move fast.

Detweiler and his associates had quickly capitalized on the Institution's fear of hearings. In exchange for promising to help with the Congress, they had extracted a Smithsonian commitment to let them judge all further materials meant for the exhibition. And now they also had all-but-overt congressional backing to make that judgment stick.

Dagley felt confident now that the Legion would be able to get more concessions from the Smithsonian. An interview published in the January/February issue of the American Legion Auxiliary magazine, *National News,* issued in late December 1994, reported,

> If in fact the Smithsonian Institution will not make any additional changes to the script, Dagley feels the Legion will express opposition.
>
> What are the Legion's options? They can continue working with the National Air and Space Museum to iron out the remaining kinks in the exhibition. But if Smithsonian officials are saying they will make no more changes then the Legion will have to explore other avenues. Dagley says one of those options could be for the Legion to go before Congress and point out the fact that no one seems to be happy with the exhibit, and therefore it is eminently important to save the Smithsonian Institution from itself.[3]

This echoed Dagley's more detailed reasoning raised in a letter he had sent to the American Legion National Commander's Advisory Committee on December 2,

> The American Legion is rapidly approaching the point at which you will be asked to consider the National Commander's final recommendation on the National Air and Space Museum's exhibit concerning the end of World War II. It is important for you to understand what has transpired so far, where we stand now, and where we may be within the next few weeks, certainly by the seating of the 104th Congress in January. . . .
>
> The consensus among veterans is overwhelmingly against the exhibit. . . .
>
> Although I do not presuppose the Commander's recommendation, he, his representative, Herm Harrington, and I are in complete accord that our meetings with NASM officials and the changes in the exhibit script, though many, are not sufficient to ward off the American Legion's renewed opposition. . . . The only defensible position is beginning to come into focus.
>
> It appears The American Legion has won the publicity battle. We are commonly seen as the only organization with the expertise and credibility to speak for the opposition . . .
>
> Members of Congress have been in touch with us from time to time, to express their continued support of our position and to ask what, if anything they can do to help. We have asked them to stand pat until our work

> with the NASM is complete. It is unlikely that they can be held in check much longer, for we are almost alone among veterans groups in not now clamoring for the exhibition to be canceled ... and the curators to be fired....
>
> Finally, and of particular significance, is that there now is a growing number of scientists and historians who are condemning the Air and Space Museum for "caving in" to the American Legion and for changing the exhibition to what they consider to be one that is historically inaccurate....
>
> Clearly the Air and Space Museum is between a rock and a hard place....
>
> Now that NASM has an exhibit that pleases absolutely no one, and is suspect from both our perspective and our opponents' perspective, the museum may be seriously damaged by the administration's dogged determination to proceed with this ill-fated exhibit and irreparably damage its reputation for scholarship and reliability. It is our view that Members of Congress can be motivated to act against the exhibit on these grounds, even if some might not be moved by the revisionist history argument....[4]

In late October, when we had left Indianapolis after our last long negotiating session with the American Legion, all of us at the Smithsonian had felt we were close to agreement. But the pressures on the Legion from other veterans' organizations and individual veterans who had been aroused by the AFA's and the Legion's media campaigns, appeared now to be leading to a tougher stance. The Legion by this time also appeared to have the means of implementing a harder line through the veto power over the exhibit that they had seized by capitalizing on the Smithsonian leadership's almost morbid fear of hearings.

At the start of 1995, National Commander Bill Detweiler was ready to move. By the time his advisory committee met, on January 4, the second working day of the new year, he had produced a detailed strategy, which needs to be seen in its entirety:

> Memorandum to: National Commander's Advisory Committee
>
> From: William M. Detweiler
>
> Subject: *Recommended Final Position on NASM Display*
>
> Based on lengthy discussions with officials of the Smithsonian Institution, extensive independent research, and failure of the Institution officials to meet certain agreed-upon conditions of mutual cooperation, it is my recommendation that The American Legion actively oppose the National Air and Space Museum and its controversial exhibit, "The Last Act: The Atomic Bomb and the End of World War II."
>
> It is my recommendation that our position call for:
>
> 1. The exhibit to be canceled
> 2. NASM's role and intent in the controversy to be investigated by Congress.

3. The *Enola Gay* to be immediately re-assembled and loaned, or ownership transferred, to an entity willing and able to display it without controversy.

American Legion opposition should center on the primary motivations that led the organization to involvement in review of the exhibit script.

1. Inaccurate and unfavorable portrayal of the nation and its fighting forces, and inaccurate or incomplete historical data.
2. Continuing erosion of public confidence in the museum as a result of the controversy.

In the first instance, the script continues to dwell on the civilian casualties sustained in the bombings of Hiroshima and Nagasaki, without providing the historic, military and cultural context by which to judge Japan's responsibility for such casualties; it treats lightly (nine lines of text) the attack on Pearl Harbor and glosses over the horror and suffering of the Pacific campaigns and essentially ignores the atrocities committed by the Japanese against Asian populations; it continues to question the legitimacy of the decision to use the atomic weapon and second guesses the motivation of the Nation's leaders, and gives only brief mention to the probable casualties of a land invasion of the Japanese home islands.

In the second instance, the exhibition is suspect from all perspectives. The curators, the process, and the Institution's administration have been, and continue to be, soundly criticized for the way the exhibit was originally conceived, how it was presented, the lack of veracity on the part of officials, the role of the Japanese in the exhibit, and the role of dissident historians and veterans groups in creating or revising the exhibit text.

It is my recommendation that our opposition take the form of:

1. A face to face meeting with Smithsonian Officials to declare our final position.
2. A face to face meeting with selected members of Congress to explain our opposition and the nature of our critique and evaluation.
3. A Washington, DC press conference to announce our opposition
4. A letter to the President of the United States
5. A letter to Senate and Congressional leaders
6. A copy of the leadership letters, under separate cover to all members of Congress
7. A letter to the Board of Regents of the Smithsonian Institution
8. Continued maximum exposure of the Legion's position in the public media.
9. Maximum communication of our position and the status of the exhibit to our members in internal publications.

Finally, it is my recommendation that the opposition strategy be implemented immediately, and completed, to the extent possible, by the

1995 Washington Conference, and that a report of progress be made during the Legislative Conference.

Respectfully Submitted:

William M. Detweiler
National Commander[5]

The National Commander's Advisory Committee approved these recommendations the same day.[6] That was on January 4. The Legion would have to begin at once and hustle to complete all the required steps by their February 5 Washington Conference.

Dated the same day as Detweiler's letter was an unsuspecting Mike Heyman's response to the December 13, 1994, missive from the seven Republican congressmen. Promising Sam Johnson full compliance with the Congress and the veterans' associations, Heyman wrote:

> Let me assure you and your colleagues that we indeed are taking all the necessary and proper steps to coordinate with the associated outside parties in bringing this exhibition to a successful opening. . . . We are doing our utmost to have both the revised script which will reflect these changes and any related media material, such as videos, in a form that is acceptable for review by early February. I will contact you again in late January to apprise you of the status of the revised script and all related material.[7]

Little did Heyman know that he was already too late. As Ben Nicks might have put it, Detweiler was rolling down the runway now.

PREBLE STOLZ

On the first working day of the year, January 3, 1995, Mike Heyman had scheduled a meeting with me in his office. He told me he had decided to have an investigation of the *Enola Gay* exhibition history. His old friend and former colleague at the Berkeley Law School, Professor Emeritus Preble Stolz, would come to Washington for a week to conduct the study.

I said I would be glad to have Stolz take a look at anything he wanted to see, talk to anyone he wished to interrogate. I was proud of what we were doing at the museum and it would be a pleasure to show off what the staff had done. Privately I felt that if we could show Stolz how proficiently we had proceeded on the exhibition, he might be able to convince his friend. I never seemed able to persuade Heyman, but maybe a trusted colleague could.

I arranged at once to have all available material on the *Enola Gay* gathered from our files and organized into a coherent package with which Stolz could work. Some of this was mailed to him ahead of time. The rest was catalogued and arranged for him in an office made available for the duration of his visit. There he could interview anyone on the staff he wished and work on his report.

On January 9, 1995, a week before Stolz's anticipated arrival, Heyman circulated to all the museum directors a draft announcement for comments and response. It outlined Heyman's new intended policy, motivated apparently not only by the coverage given to the *Last Act* exhibition, but by vocal criticism of *Science in American Life,* at the National Museum of American History; *The West as America: Reinterpreting Images of the Frontier, 1820–1920,* at the National Museum of American Art; and several displays at the National Museum of Natural History:

> Prompted by several recent exhibitions which have elicited criticism from various sources, the Institution must set up procedures to ensure that all exhibitions and public programs are well-conceived, balanced and objective in their presentation.

I wrote back, suggesting an insertion (here italicized) so his announcement would read,

> Prompted by several recent exhibitions which have elicited criticism from various sources, *I believe the Institution must examine the procedures that were followed in those cases, to see whether a satisfactory system of reviews was in place or new guidelines need to be established* to ensure that all exhibitions and public programs are well-conceived, balanced and objective in their presentation.

I added,

> I have told our staff that you have invited Prof. Preble Stolz to look into the review process the National Air and Space Museum followed, and that he will report to you whether, in his eyes, our process was thorough. Our staff would be disappointed to think that you had already made up your mind before Prof. Stolz gives you his report. The issue is important for staff morale; I know that everyone at the National Air and Space Museum would appreciate your considering this recommendation.[8]

THIS IS ALL ABOUT MONEY

During the first week of January, Heyman, Newman, and I had a meeting to lay out a general strategy for the weeks ahead. Both still worried about hearings and the possibility that the Institution's budget would be slashed. Heyman suggested that perhaps the exhibition should be shut down.

I was aghast. The Smithsonian had been established to provide the public with knowledge about the world of science, of art, of history. And here, on facing his first real test, the Institution's new secretary was ready to give up. If he did, the future would be bleak indeed. We would have lost our last hope of support from like-minded people who also stood for education as an important national goal. I said I understood his fears, but our supporters, and particularly the academic community, would be outraged and accuse us of capitulating. In the long term these were the groups on whom we would need to rely for help.

Heyman and Connie looked exasperated and said to each other, "He doesn't get it!" Turning to me, they said, "This is all about money."

Both were totally consumed with the issue of congressional funding. Where were those supporters I talked about? What help had the academic community been?

In January, we learned that as a result of the November elections, Republican senator Mark Hatfield was to be the new chairman of the Senate Appropriations Committee. This would make him a powerful figure in determining whether the *Enola Gay* exhibition would affect the Smithsonian's budget. Heyman had not yet met with Hatfield, and I thought I might allay his fears about how to deal with the exhibition by sending him some reassuring news:

> Given your concerns about the *Last Act* exhibition, I thought you might like to know about a photo-exhibition in the Senate that Sen. Hatfield sponsored in 1980. As a young man he had been in Hiroshima just weeks after the war's end and had been greatly impressed by the destruction he witnessed.
>
> In case you had not heard about this, I attach a clipping from the *Washington Post* of June 10, 1980, as well as a clean transcript and some other information that Sen. Hatfield's staff made available to us some months ago.[9]

The *Washington Post* article praised this exhibition, showing the destruction and suffering at Hiroshima. What a change fifteen years had made!

I do not know whether Heyman made any use of the information. By that time, he may have been looking for a way out—rather than seeking new allies in Congress.

Heyman came over to the museum the following morning, January 10, to meet with the *Enola Gay* exhibition team. He spoke of the difficulties he was facing in Congress, and said he might need to take drastic action if forced.

TIBBETS'S AMBASSADOR

Newman, Rodgers, and I spent most of the morning of January 5, 1995, with George Hicks. I had not met Hicks before; he was director of the Airmen Memorial Museum in nearby Suitland, Maryland, and, I was told, was Paul Tibbets's biographer. Often he also acted as Tibbets's spokesman.

The day before the meeting, I had been given a copy of the exhibition script that Tibbets and his former crew had carefully marked up. I asked Mike Neufeld at once to go over it and give me his comments. With those as my guide, Newman, Rodgers, Hicks, and I went over the draft and discussed the recommended changes.

Many of Tibbets's suggestions dealt with details known only to those who had been with the 509th. These changes were easy to accept. A few others of a broader nature I said we could not incorporate. But the mood was generally accommodating. In the course of the meeting, Heyman popped in for a few minutes to greet Hicks. It was clear that he wanted to forge good relations through him with Tibbets. I had, of course, tried to get Tibbets on our side months earlier, and understood what Heyman was trying to do.

A LETTER TO THE LEGION

After lunch on January 9, having again briefly spoken to Newman about the Leahy diary entry in the course of one of the many phone conversations we were now having daily, I sat down and wrote Hugh Dagley at the American Legion. As I had said to Newman, I did not want to surprise him later, when we would have to be showing the Legion all the new exhibition materials. This was too important to be thrown in with everything else:

> As you will remember, last fall we spent a good deal of time discussing one of the labels in the script for *The Last Act: The Atomic Bomb and the End of World War II.* That label involved losses of lives that would have been expected in an invasion of Japan, and cited the casualty estimates given to President Truman by his most senior advisors on June 18, 1945. They were the last estimates given to him before the war's end in August.
>
> The highest of the figures cited for the invasion of Kyushu at the June 18 meeting appeared to be those of Admiral Leahy, who said that he expected loss rates comparable to those suffered at Okinawa, or around 30%. Prof. Barton Bernstein of Stanford University, in a paper he had published some years ago, interpreted that figure to mean 30% of the 766,700 "total assault troops" Marshall had mentioned earlier in the meeting. On that basis, Bernstein thought Leahy's remarks meant casualty levels around a quarter of a million for the Kyushu invasion.
>
> Our Museum accepted those figures, but in a more recent meeting with Bernstein, he took us to task for this, saying that he had, in the meantime, found Leahy's diary entry for that same day. We checked on that in the archives and found that Leahy's entry summarizes the entire June 18 meeting with these words,
>
> "From 3:30 to 5:00 P.M. the President conferred with the Joint Chiefs of Staff, the Secretary of War, the Secretary of the Navy, and Assistant Secretary of War McCloy, in regard to the necessity and the practicability of an invasion of Japan. General Marshall and Admiral King both strongly advocated an invasion of Kyushu at the earliest practicable date.
>
> "General Marshall is of the opinion that such an effort will not cost us in casualties more than 63,000 of the 190,000 combatant troops estimated as necessary for the operation."
>
> As Bernstein pointed out to us, 63,000 represents 30% of 190,000, and that evidently is the figure that Leahy had had in mind at the meeting that afternoon.

> Seeing that our earlier label text had been based on a misapprehension, we needed to revise it. I am sending you the text of the label as it now reads. It does not alter the figures Truman cited after the war, but gives a different interpretation of what he might have had in mind.
>
> If you have any concerns or comments, I'd greatly appreciate your letting me know.[10]

The last sentence invited comment and discourse and was meant to be taken at face value. I attached the new label draft, and sent copies of the entire package to Connie Newman, Mark Rodgers, and Michael Neufeld. The new label, which I cite in its entirety below, was not that different from the earlier version in its thrust to present the casualty figures Truman's advisors had provided and in trying to make understandable why Truman would have claimed much larger figures after the war.

Readers may wish to compare this version of the casualty label to the one given earlier, in Chapter 25, to see whether what happened next was warranted. I reproduce the two versions side by side; the newer one is on the right:

INVASION OF JAPAN—AT WHAT COST?

Estimates of the number of American casualties—dead, wounded, and missing—that the planned invasion of Japan would have cost varied greatly. At a June 18, 1945, meeting, General Marshall told President Truman that the first 30 days of the invasion of Kyushu could result in 31,000 casualties. Admiral Leahy pointed out that the huge invasion force could sustain losses proportional to those on Okinawa—about 35 percent—which would imply a quarter of a million casualties, or at least 50,000 dead.

Had the Kyushu invasion failed to force Japan to surrender, an invasion of Honshu, with the goal of capturing Tokyo, would have followed, and losses would have escalated. If even that failed, and Japan continued fighting in the home islands and in the captured territories in Asia, casualties conceivably could have risen to as many as a million (including a quarter of a million deaths). Added to the American losses would have been perhaps five times as many Japanese

INVASION OF JAPAN—AT WHAT COST?

Estimates of the number of American casualties—dead, wounded and missing—that the planned invasion of Japan would have cost varied greatly. In a June 18, 1945, meeting, General Marshall told President Truman that the first 30 days of the invasion of Kyushu could result in 31,000 casualties. But Admiral Leahy pointed out that the huge invasion force could sustain losses proportional to those on Okinawa, making the operation much more costly.

Had the Kyushu invasion failed to force Japan to surrender, the United States planned to invade the main island of Honshu, with the goal of capturing Tokyo. Losses could have escalated.

After the war, Truman often said that the invasion of Japan could have cost half a million or a million American casualties. The origin of these figures is uncertain, but Truman knew that Japan had some two million troops defending the home islands. He believed, along with the many casualties—military and civilian. The Allies and Asian countries occupied by Japan would also have lost many lives.

For Truman, even the lowest of the estimates was abhorrent. To prevent an invasion he feared would become "an Okinawa from one end of Japan to the other," and to try and save as many American lives as possible, Truman chose to use the atomic bomb.

Americans who would have had to invade Japan, that such a campaign might have become, in his words from June 18, 1945, "an Okinawa from one end of Japan to the other." Added to the American losses would have been several times as many Japanese casualties—military and civilian. The Allies and Asian countries occupied by Japan would also have lost many additional lives.

For Truman, even the lowest of the casualty estimates was unacceptable. To prevent an invasion and to save as many lives as possible, he chose to use the atomic bomb.

THE SMITHSONIAN PANICS

Mail travels slowly within the Smithsonian. The copy of my letter to Dagley reached Mark Rodgers on the morning of January 11, and he at once sensed a problem and notified Newman. At eleven o'clock, I got a call from Heyman. He said he might need to shut down the exhibition. I'd better get over there right away. This was the third time in a week he had made the same threat. He had mentioned shutting down the exhibition just the day before in meeting with our exhibition team. I felt he was looking for a way to justify a decision he had already reached.

When I arrived in Newman's office a few minutes later, Heyman, Newman, Rodgers, and Acting Provost Bob Hoffmann were already there. They were upset at my letter, and said I should at least have called Dagley ahead of time. Rodgers was going to try to call Dagley before he received the letter to try to warn him.

I acknowledged that a call to Dagley would have been a good thing, but I had, by then, written him rather frequently, and did not think this was all that different. What I did not know became apparent many months later, when I saw another internal memorandum from the American Legion's Bill Detweiler, dated January 18:

> On December 8, 1994, Constance Newman and Mark Rodgers informed Herman Harrington, Hugh Dagley and me that Harwit and his crew of curators had been removed from decision-making on the exhibit, and were not to communicate in any way on its content. Newman, Rodgers and Heyman alone were to make contact or comment.[11]

I have no reason to doubt Detweiler. His memorandum gives further details on the context in which these remarks were made, and corroborates his assertions.

Unfortunately, neither Heyman nor Newman had ever told me they had taken over the task of acting as the Institution's sole communications link with the Legion. Their panic that day gives added credence to Detweiler's claim.

The next morning, while I was in another meeting, Heyman called again. He had just phoned Detweiler, who appeared very upset. Heyman thought he'd now definitely have to shut down the exhibition. Mark Rodgers evidently had reached Hugh Dagley an hour after Dagley had received the letter. At the time, Dagley reportedly said, he already had three calls from the press waiting to be answered by him, allegedly about prospective changes in the script. Heyman now accused the museum of having leaked this information. I told him I had a hard time believing that, and I certainly never heard about those press calls or alleged leaks again. The press might have been inquiring about Dagley's own calls for script changes in the interview he had given in the January/February issue of the Legion's *National News*.[12]

DETWEILER COMES TO TOWN

The next few days were ominously quiet. The museum's filmmaker, Patricia Woodside, had worked heroically over Christmas and New Year to get the video-films ready on time for the Legion to examine, and the final pieces of footage were coming in from Japan, in response to the calls I had recently made. Then, at 4:30 P.M. on January 18, Heyman called again and wanted me to come over right away. When I arrived, he and Newman were in his office looking glum. Detweiler and three of his colleagues Dagley, Harrington, and Phil Budahn, the Legion's public affairs officer, had just met with them.

Detweiler had called earlier in the day to say he was in town, and asked to come by. Heyman had understood from their conversation the previous week that they would try to negotiate about the changes in the casualty label. But by the time Detweiler came, he and the others had already been on Capitol Hill. They had seen Republican majority leader senator Dole, to inform him of their views on canceling the exhibition.

Reportedly, the meeting lasted no more than ten minutes before Heyman stood up and showed his guests the door.

Detweiler's memorandum written that day stated,

> The Secretary offered no alternatives or compromise. He remarked to me that because we had briefed members of Congress, we had made it more difficult for him to act on the exhibit and deal with the criticism from the historian and academic communities....
>
> I have contacted the following Members of the Senate and House of Representatives, or key members of their staffs. I wanted to brief them in advance on our position. All have been supportive: Senators Robert J.

Dole, John W. Warner, Kay Bailey Hutchinson, Nancy Kassebaum, and Alan K. Simpson, and Representatives Peter Blute, Sam Johnson, Newt Gingrich, Bob Stump, and G. V. Montgomery.[13]

With the exception of Mississippi Democrat Congressman G.V. "Sonny" Montgomery, former chair of the House Veterans Affairs Committee, all were Republicans.

Detweiler also issued a *"Position Statement of the American Legion"* to the media the same afternoon.

> I have today conveyed to the Secretary of the Smithsonian Institution, Michael Heyman, The American Legion's renewed opposition to the National Air and Space Museum's planned exhibit on the end of World War II. . . .
>
> This exhibit, in our opinion, so closely parallels the design, content and conclusions of the Nagasaki Peace Museum as to defy coincidence. Moreover, we have been informed by the Director of the National Air and Space Museum that, despite an agreement reached between NASM and The American Legion in September 1994, NASM has decided to return doubtful casualty estimates to the exhibit, thereby resurrecting the suggestion that the President of the United States acted from racist and political motives.[14]

To the media, the Legion kept insisting that my letter to Dagley had violated an "agreement." But this was false. Dagley had already been calling for changes in his December 2 letter to the National Commander's Advisory Committee. And Detweiler had obtained the approval of the Committee for his plan to close down the exhibition, *more than a week before my letter had reached them.* Moreover, Heyman had since October been saying, "The process is not finished. We will continue to accept and consider observations from all sources. . . ."[15] As late as January 5, 1995, dozens of suggestions offered by General Tibbetts and his colleagues had, in fact, been adopted in a meeting Heyman, Newman, Rodgers and I attended with George E. Hicks, Tibbets's representative.

I am convinced that if I had never written him, Detweiler would have announced some other reason to call for closing down the exhibition—for example, any of the reasons cited in his January 4 memorandum, or his first-cited reason now, "This exhibit, in our opinion, so closely parallels the design, content and conclusion of the Nagasaki Peace Museum as to defy coincidence."[16]

At the time, nobody knew about Detweiler's January 4 plan of attack, and so the media took at face value his allegations that I had broken an agreement with the Legion. Heyman soon would also publicly blame me.[17]

THE REGENTS' EXECUTIVE COMMITTEE

Late the following morning, January 19, Heyman again called me to the secretary's office. He told me he was having lunch with the regents' executive committee, and would recommend to them that they cancel the exhibition. I would be given an opportunity to make a statement following his announcement at a press conference.

I tried to dissuade him, but he stood firm and told me he was sure the executive committee would go along with him. Seeing there was nothing more to be said, I stood up and left, saying "Let me know what they say."

Around four o'clock, Heyman called back to tell me that the executive committee, headed by Chief Justice William Rehnquist, had felt this was a matter for the entire board to take up. He also gave me the news that a letter asking for my resignation was making the rounds and gathering signatures in Congress. To the media he released a statement that the "*Enola Gay* Controversy" would be taken up by the full Smithsonian Board of Regents at its next meeting on Monday, January 30, 1995. Until then, the Smithsonian would have no further comments:[18]

> This issue clearly matters to Congress and it matters to a great many Americans, which is why the full Board of Regents has agreed to take up this issue at its next meeting on Monday, January 30. It makes good sense to wait until that decision before publicly addressing this issue again.

That day, following his January 4, 1995, strategy to its details, Detweiler wrote President Clinton asking him to cancel the exhibition and have Congress "investigate the process by which it was developed":

> National Air and Space Museum officials, despite an accord reached with our representatives in September 1994, and in defiance of their Smithsonian Institution superiors, have restored to the exhibit highly debatable information which calls into question the morality and motives of President Truman's decision to end World War II quickly and decisively by using the atomic bomb. The hundreds of thousands of American boys whose lives were thus spared and who lived to celebrate the 50th anniversary of their historic achievement are, by this exhibit, now to be told their lives were purchased at the price of treachery and revenge. This is an affront to all Americans."[19]

In addition, Congressmen Peter Blute and Sam Johnson called for my resignation and for congressional hearings on the exhibition. Johnson's office issued the statement,

> Dr. Harwit has tarnished the name of the Air and Space Museum and presented an exhibit that is historically inaccurate and not in line with the thinking of most Americans. The institution needs new leadership....
>
> This week the American Legion was effectively stonewalled from any further input into the *Enola Gay* exhibit. After saying that they would continue to include all interested parties in the formation of a balanced exhibit, the Smithsonian is now ignoring our input....

The statement added that "Congressman Blute (Republican, Massachusetts) also announced that the Committee on Government Reform and

Oversight's Subcommittee on Government Management, Information and Technology will hold hearings regarding the Smithsonian Institution on this issue. The hearing will address, among other issues, the museum's compliance with its charter."[20]

An Air Force Association press release supported the American Legion:

> AFA Blasts the Air and Space Museum on *Enola Gay* Reversal: The Air Force Association denounced the National Air and Space Museum's latest breach of faith on the *Enola Gay* exhibit regarding its intent to revise previously agreed language that cited a range of casualty estimates for a land invasion of Japan.
>
> ... We had been assured that no unilateral actions would be taken by the curators and officials of the National Air and Space Museum, in whom we lost faith long ago. It now appears that, on the side and behind the scenes, the curators are still working their political agenda. This is unacceptable. Museum officials have failed in their stewardship and responsibilities. We do not believe that a fair and balanced presentation of the *Enola Gay* is possible with the present director-curator team in charge. We conclude, therefore, that it is time to cancel this exhibit.[21]

They also called for a congressional investigation and transfer of the *Enola Gay* to another museum if the Air and Space Museum could not display it properly.

The following morning, I read about all this in the newspapers. The *Washington Post* wrote of the museum's "incredibly propagandist and intellectually shabby early drafts," spoke with disdain about the negotiations and mutual denunciations by opponents and proponents, and concluded,

> Forget the coalitions of negotiators, forget the representatives of interest groups and, especially, the team that bungled this so badly in the first place—from Mr. Harwit down. Get a couple of respected historians of the period, a military expert or two and some people who know about mounting good exhibits, and charge them with getting a reasonable commemorative exhibit to the museum.[22]

In the *Washington Times,* Rowan Scarborough wrote about Congressman Peter I. Blute, Republican of Massachusetts, joining Representative Sam Johnson, Texas Republican, "in calling on Mr. Harwit to resign:"

> "If Harwit does not resign and the exhibit remains in its current form, I will begin to look at future funding of the institution," Mr. Johnson said. "In 1994, the Air and Space Museum received approximately $13 million from Congress. Taxpayers' dollars should not be used to fund a museum that does not have the best interests of the United States in mind."[23]

Having woken up to these reports over breakfast, I thought I could use some dispassionate advice. Confidentiality at the Smithsonian left most people unaware of the main facts. Whom could I ask without risking further leaks to the newspapers? I decided to call two people, Hanna Gray, former president of the University of Chicago, and Homer Neal, a professor of physics and vice president for research at the University of Michigan. Both of them knew of the issue, having been on the board of regents for several years, and both came from the same academic background as I. I thought they would have useful advice. I didn't know either of them well, but had talked to both at various times, with Neal more than with Gray. At 9:00 A.M., I called both and left messages for them.

A couple of hours later, Neal returned my call. As a member of the regents' executive committee he was totally up to date. I told him of my concerns about the entire crisis and the potential cancellation of the exhibition. He mused that he was just preparing a talk on the history of the Smithsonian and past crises the Institution had weathered. He felt it would weather this one as well. In any case, he said, Heyman was calling around to consolidate regents' views before the January 30 board meeting.

Later in the morning, Connie Newman called to tell me Heyman was so furious, he had asked her to call me in his place. Hanna Gray had called him about my wishing to speak to her. Newman warned me not to speak to any regents. That was the secretary's function. I was not to talk with them.

Still the same day, Congressman Gerald B. Solomon, Republican of New York, chair of the House Rules Committee wrote to me personally (with a copy to Heyman):

> I am writing to express my deep dissatisfaction with the course of events surrounding the *Enola Gay* exhibit.
>
> To date, I and other like-minded colleagues here in Congress who were disgusted with the original plans for the exhibit have patiently awaited adequate corrective action from the Smithsonian. Having learned that the American Legion has called for the cancellation of the exhibit due to the Smithsonian's unwillingness to make adequate changes, I have run out of patience....
>
> .... The Smithsonian is getting on my nerves. Let me make a promise: if the Smithsonian cannot accommodate the wishes of the American Legion concerning the *Enola Gay* exhibit, I will personally take measures this year to zero out the Smithsonian's congressional appropriation. You can count on that.[24]

It is a reflection on the clout of the American Legion that an influential congressman and chair of one of the most powerful committees in the House of Representatives will threaten to "zero out" the budget of a national resource like the Smithsonian simply on the word of the Legion. What about the people from his district who had elected Solomon? Would they all want the Smithsonian's budget zeroed out?

Four days later, the actual letter from Congressmen Sam Johnson, Peter Blute and seventy-nine other members of the House was mailed to Heyman and received the next day:

> It is after serious consideration that we the undersigned call for the immediate resignation or termination of Mr. Martin Harwit, Director of the National Air and Space Museum.
>
> It has come to our attention that Mr. Harwit, by his own admission, recently reintroduced controversial and speculative accounts to the script of the *Enola Gay* exhibit which earlier had been removed. . . .
>
> . . . In effect Mr. Harwit's actions were a slap in the face of all the parties who contributed their time and expertise in creating an exhibit that best reflects the contributions that all Americans made to the culmination of World War II.
>
> It is our hope that his removal will bring a fresh and unbiased approach in finalizing a script which will produce a mutually acceptable and balanced exhibit.
>
> We expect this issue will be addressed and a decision will be made on Monday January 30th at the meeting of the full Board of Regents. Also due to this recent controversy and the high level of public interest, we ask that you keep us informed during the selection process of Mr. Harwit's replacement.[25]

STOLZ'S STRATEGY FOR HEYMAN

Having been invited several weeks earlier, Prof. Stolz, Heyman's special investigator, had arrived at the Museum late in the morning on January 17. Three of us on the staff had lunch with him to find out what he would like to see, whom he would want to question, and how he planned to proceed. Then I took him around to show him a number of exhibitions on military subjects that we had done in recent years in preparing for the *Enola Gay* exhibition, including the *Enola Gay* video that had been running in full view of visitors at the entrance of the World War II gallery for four continuous years, by then, without ever eliciting a single response to the museum or the Smithsonian. I felt this was remarkable, because it included many of the aspects we had planned for our larger exhibition, which our opponents now so heavily criticized.

Over the next week, I saw Stolz several times, and he apparently was satisfied to find the information he sought. On January 24, his last day at the museum, he came in to say goodbye. During our previous conversations he had given me his interim impressions, which I would hear echoed in separate meetings with Heyman the same day. Evidently, as old friends, the two were spending a fair amount of time together discussing the exhibition.

Now Stolz had come by to give me a verbal outline of his intended report to Heyman, and to thank me and our staff for having helped him in his work. In his opinion, the major problem with the exhibition was that we wanted to do a "historic analysis" at a time when others wanted a "commemoration." Neither Adams nor I, he remarked, had recognized this well enough.

At my next meeting with Heyman I heard precisely the same conclusion from him. The two friends had found a way out. As Mike would phrase it the following week in announcing his cancellation of the exhibition, "In this important anniversary year, veterans and their families were expecting, and rightly so, that the nation would honor and commemorate their valor and sacrifice. They were not looking for analysis. . . ."

In a meeting the previous day, Heyman had already told me he wanted to do a new exhibit of his own, using none of the material we had already discussed with the veterans' organizations. Instead, it would be a minimalistic display centered on the *Enola Gay*. The prospect of taking the aircraft out of the museum now, potentially with the full glare of publicity the press had accorded its arrival from our restoration shops two months earlier, did not appeal to him. What he now was describing sounded remarkably like the strictures of Title 20 §80 of the United States Code, "The valor and sacrificial service of the men and women of the Armed Forces shall be portrayed as an inspiration to the present and future generations of America. . . ."

I was deeply troubled that Heyman should reject a carefully designed historical exhibition as unsuitable for commemorating our servicemen. Did the scholarship the Smithsonian could bring to bear have to be rejected at times of commemoration and be replaced by straightforward celebration? For Stolz, after one week's examination, to reject the scholarly, meticulously discussed display that curators and external advisors had offered, and for Heyman to accept this advice, struck me as cynical—designed solely to placate the Legion and Congressman Sam Johnson, who was about to join the board of regents as one of Heyman's bosses.

THE VICE PRESIDENT'S CALL

Late in the afternoon on January 24, the day of Stolz's departure, I was discussing these matters with my deputy, Gwen Crider, when the phone rang. She answered, took down a number and said, "He'll call you right back." Hanging up, she said, "It's the vice president's office."

I went next door to my own desk and called the number. The secretary at the other end of the line identified herself as "Ann" and said that Vice President Gore was thinking of coming to the regents' meeting next Monday and would like me to come in to talk with him on Friday afternoon.

Al Gore and I had come to know each other on a first name basis while he was serving in the Senate. The museum had produced a wide-screen film, *Blue Planet,* showing the earth and its natural resources, as seen from space. Because it so well matched Gore's ecological interests, he occasionally asked for specially scheduled evening showings of the film for groups he wanted to impress with the importance of environmental concerns.

I said I would be glad to come, but remembering Heyman's outburst four days earlier when I tried to call Hanna Gray, I asked whether the vice president would be willing to invite Heyman to come along also, since Heyman reported to the vice president, who also serves as vice chancellor of the board of regents. Ann said she'd ask about that and let me know.

I picked up the phone again and called Heyman's direct line in his office. It was nearly six o'clock, but he was still there. I told him of the call, and asked if he would like to come along if the vice president invited him, or whether he wanted me to go without him if not. Heyman did not sound pleased; he told me he had been trying to reach the vice president all this time and could not get him to respond to his calls. He said he would try again and let me know.

At 8:00 A.M. the following morning, January 25, 1995, I went over to the Castle for another scheduled meeting with Newman, Hoffmann, and Heyman in Heyman's office.

I mentioned the State of the Union Address the president had just delivered, and his admonition to reconcile politically polarized views. Could we not ask the vice president to see whether the president would intercede with the Legion and ask us to sit down again to talk? I recalled that presidents had done that at times, when it was in the national interest. Heyman responded that it was too late for that. He seemed to want to get the project off his back.

Several days later, President Clinton did intercede, though not successfully, in another national issue, the ongoing baseball strike. The *Enola Gay* dispute could easily have merited similar intervention.

At the end of our meeting, I asked Heyman whether he had been able to get through to the vice president's office, but he had not.

Late the same morning and back in my office, I received another call from Ann. Could I see the vice president that afternoon at 3:00 P.M.? I called Heyman, but he was to be out all morning. I decided to call Jim Hobbins, who normally dealt on all regents' matters for the secretary. He said he had been talking with his contact in Gore's office. Gore apparently had several people working for him and Jim had a different contact. He said he would try to straighten the matter out.

Early that afternoon, Ann called back to say that the afternoon meeting with Gore was off. I called Hobbins to tell him.

A few days later, on January 28, I learned that Heyman had obtained an appointment with Gore and seen him by himself. Heyman reported that the president had asked Gore to handle affairs with veterans, because, having avoided the Vietnam draft, his own rapport with them was low. Heyman said the White House was relieved at his wishing to close down the exhibition. They wanted to avoid another confrontation with the veterans.[26] His claim seemed substantiated by President Clinton's statement, three months later, that virtually endorsed Heyman's cancellation of the exhibit in words almost identical to those the secretary himself was using:

> I do not believe that on the celebration of the end of the war, and the service and sacrifice of our people, that that is the appropriate time to be asking about our launching a major re-examination of that issue.[27]

Taken at face value, it suggests that the president had cashed in a lot of chips the previous month, asking the postal service to withdraw its mushroom-cloud stamp. Veterans had been furious with him. The *World War II Times* discerned two coupled threats:

> Something is terribly wrong in America when the *Enola Gay* B-29 bomber cannot be displayed properly at the Smithsonian, and the Postal Service is pressured by the White House to withdraw a commemorative postage stamp with a historically correct statement on it about the dropping of atomic bombs.[28]

A spokesman for the Non Commissioned Officers Association added,

> The politically correct win again. We can't rewrite history. I wish people would stop trying.[29]

THE AFA AND THE REGENTS

The Smithsonian regents count three senators and three members of the House among their members. Two always come from the majority party in each chamber and one from the minority. With the overwhelming Republican win in November's elections, the Republicans were able, in late January 1995, to appoint four new regents. Speaker Newt Gingrich appointed Sam Johnson, from Texas, on whose stationery the eighty-one members of the House had written to ask for my dismissal, and also Congressman Bob Livingston, about to become chairman of the House Appropriations Committee. In the Senate, Thad Cochran and Alan Simpson were to be the new regents.

R.E. Smith, the president of the Air Force Association, wasted no time. Writing to Cochran on January 25, he congratulated him,

> Dear Thad,
>
> I am absolutely delighted with your appointment as a regent of the Smithsonian Institute. As you may know, the Air Force Association was the first national organization to shine light on the distorted plans of the National Air and Space Museum for display of the *Enola Gay*. . . .
>
> Unfortunately, we—like a number of other veterans' organizations—have now concluded that a fair and balanced presentation of the *Enola Gay* is not possible with the present director-curator team. These officials simply will not give up on their radical political agenda despite the massive crit-

icism that has fallen on them from the American public, the news media and your colleagues in Congress. . . .

. . . . We have a comprehensive record and file of our own work, as well as other various documents, including papers and memos written by National Air and Space Museum officials, that leave little doubt about the outrage that is going on over there. We will be pleased to make this material available to you and your staff, and to answer any questions that you may have.

Penned at the bottom is a personal note:

Thad, you can be assured that this subject is on the mind of every person I talk with in Mississippi when they find out I'm national pres. of AFA. We need your help in getting this on track.[30]

It was signed "Gene." The postscript was a gentle reminder to the senator from Mississippi.

THE HISTORIANS' ASSOCIATIONS RESPOND

The nation's historians were appalled at Heyman's apparent intention to cancel the exhibition. In a last-minute attempt to change the regents' minds before they met, Brigadier General Roy K. Flint (USA Ret.), president of the Society for Military History and a former dean of the faculty at West Point, wrote to the chairman of the board of regents, Chief Justice Rehnquist, on January 26, with a copy to Heyman. The full text of his letter read,

Dear Mr. Chief Justice,

I am writing to you in your capacity as Chancellor and head of the Board of Regents of the Smithsonian Institution on behalf of the members of the Society for Military History to share a concern about the exhibit of the *Enola Gay*, planned by the National Air and Space Museum.

Please do not think I pretend to represent the interpretations of two thousand military historians. But while our members possess varying views about the history of World War II, we share a passionate commitment to freedom of speech and to providing the best scholarship with integrity. Overwhelmingly, we believe the past should be represented honestly, sensitively, and with verisimilitude. We also believe that ending the war against Japan by employing atomic weapons was perhaps the most significant historical event of the time and therefore deserves just such careful treatment. Moreover, the Smithsonian's prominence as the leading museum of our nation and its possession of the *Enola Gay* demand a full presentation of the context and history of those events.

Even more importantly, the Smithsonian must stand publicly against the politicizing of scholarship in public discourse, and it must resist all efforts to impose conformity in the rendering of history. If the Institution cancels this exhibit, or directs the display of the artifacts without the history of its use or a discussion of the significance of the events, the Smithsonian will deal the presentation of honest history by publicly funded institutions a crippling blow. We believe the Smithsonian must lead in presenting the best scholarship on the most important subjects to the American people; to do otherwise is to forfeit its leadership in American scholarship and education.

Therefore, as a matter of principle, we respectfully ask the Regents not to cancel the exhibit of the *Enola Gay*. If there remain questions about the quality of the textual support, we recommend further consultation with qualified scholars rather than with political or special interest groups. In this regard, the Society for Military History will gladly suggest names of scholars in whom we have full confidence.[31]

The following day, the immediate past presidents of the Organization of American Historians, Eric Foner and Gary B. Nash, and the organization's president-elect, Michael Kammen, also wrote to the Chief Justice, with a copy to Heyman. I present it also in its entirety:

Mr. Chief Justice:

We are writing to you and your colleagues on the Board of Regents on behalf of the Organization of American Historians, the largest organization of historians of the United States. Our membership includes historians who teach at the college, university, and pre-collegiate levels; historians who work at museums of many kinds; and public historians who work for government at all levels, and in the private sector.

We earnestly hope that the Board will not decide to cancel the forthcoming exhibition planned for the National Air and Space Museum—the controversial exhibition concerning the end of World War II in the Pacific theatre.

We are concerned about the profoundly dangerous precedent of censoring a museum exhibition in response to political pressures from special interest groups. Such a precedent is likely to invite subsequent attempts to cancel other exhibitions at the Smithsonian Institution.

Moreover, the attendant publicity if this exhibit is canceled will send a chilling message to museum administrators and curators throughout the United States because the Smithsonian is the most visible and most public of all American museum complexes. Doing so would send the explicit message that controversial subjects cannot be examined openly as a part of our democratic civic life. More specifically, it would send the message that certain aspects of our history are "too hot to handle," so susceptible to contested points of view that they must be excluded from the public mind. Differences of opinion about the study of the past have long been an inescapable part of the social and cultural process. We cannot hide that from the public.

History museums should not be confined only to exhibitions about subjects for which a perfect consensus exists. Where consensus already exists, there is the least need for the presentation of information and the opportunity for members of our diverse society to be educated and formulate opinions.

In sum, cancellation of "The Last Act" would be a disservice to the American people, to thousands of history museums across the country, and to the Smithsonian Institution itself. It is an act that could later haunt the Institution because its professional integrity and intellectual freedom will have been compromised.

Sincerely yours,

Eric Foner, DeWitt Clinton Professor of History, Columbia University; President, Organization of American Historians, 1993-94
Gary B. Nash, Professor of American History, University of California, Los Angeles; President, Organization of American Historians, 1994-95
Michael Kammen, Professor of American History and Culture; President Elect, Organization of American Historians, 1995-96; Member of the Smithsonian Council [32]

OTHER RESPONSES

Syndicated columnist George Will saw the matter differently. His January 26 article claimed that the Smithsonian was imbued with a type of anti-American crankiness that characterized the nation's campuses. In Will's view this left-wing movement was unwilling to face its own irrelevance. Instead, it was now concerned with teaching the humanities as an indictment of America and more broadly all of Western civilization—both being portrayed as a blemish on our planet.

Will went on to recommend that President Clinton stride into the White House press room and, with great emphasis, declare that there would be retaliation, with firings and slashed funding at the Smithsonian unless this kind of "cloth-headed" condescension and "insulting rubbish" was stopped. The president should make it clear that the Smithsonian must stop using public funds to tell the nation how nasty it is and that its citizens are Philistines who need to be tutored on the nation's sins.[33]

An accompanying cartoon by Oliphant also accused us of revisionism.[34]

The new Speaker of the House Newt Gingrich, who had just appointed his close associates Sam Johnson and Bob Livingston to the board of regents, echoed these words. On January 28, only two days before the regents' meeting, he announced that the *Enola Gay* exhibit would be dramatically pared down in response to protests, and added,

> Political correctness may be okay in some faculty lounge, but ... the Smithsonian is a treasure that belongs to the American people and it should not become a plaything for left-wing ideologies.[35]

'We've rewritten the war with Japan; now I suggest we rewrite the history of the war with Germany from the Nazi view. Poor things have been misunderstood lately.'

SHIFTING PLANS

That same January 28, Saturday morning, Connie Newman, Bob Hoffmann, Jim Hobbins, and I met with Heyman in another marathon session on the exhibit. Heyman had asked us at the museum to let him handle all relations with the press and to make no public statements. I had been pressing him, however, to permit the museum to speak out on the accusations that were heaped on us. In particular, the curators should be given an opportunity to clear their names. At this meeting, Heyman withdrew his earlier offer to have me participate in the press conference following the Monday morning regents' meeting. He asked me not to attend because the press would certainly single me out, and he feared my responses to their questions would probably conflict with what he would say.

This was to be only the first of several promises Heyman would make to the museum, only to withdraw them again when the time came to realize them. As a concession to the museum, he had arranged with Smithsonian regent Homer Neal to hold a one-day forum on controversial exhibitions, centered on the *Enola Gay* experience. It would be held at the University of Michigan, in Ann Arbor, in the spring. The exhibition script would be circulated there to participants, so they could discuss the lessons that could be learned, the role of the Smithsonian Institution, and the ways in which we might mount difficult exhibitions without losing legitimacy in the eyes of the country. Here would be the place for the museum to speak out.

MORE VOICES IN THE PRESS

On January 30, the morning of the regents' meeting, a *New York Times* editorial argued,

> To reduce the complexities or painful ambiguities of the issue to slogans or historical shorthand is wrong. To let politicians and groups with a particular interest frame the discussion and determine the conclusion is worse. But that is precisely what is happening as the Smithsonian Institution tries to prepare an exhibit at the National Air and Space Museum of the *Enola Gay*. . . . The Smithsonian effort, while not without its own missteps, is in danger of being highjacked by a band of Congressmen and veterans outraged that the exhibit does not tell just their side of the story. . . . Scaling back the exhibit to eliminate information in dispute, which seems to be the compromise plan of Smithsonian Secretary Michael Heyman, only ducks the issue. . . .[36]

And an editorial, in *USA Today* commented,

> . . . And now the Smithsonian seems ready to give up on using the *Enola Gay* to teach any historical lesson at all. That would be a loss for everyone. . . . to learn any lessons, people need to get more than a narrow, one-eyed view of the world. Different perspectives on events are what enliven history and help us learn. The Smithsonian's role is to provide such perspectives. Members of Congress should butt out and let the curators do their jobs. When Congress, which controls 77% of the Smithsonian's budget, threatens jobs and calls for hearings, it only cuts off the debate needed to keep history alive. Then the sacrifice of veterans becomes a dusty memory, and the *Enola Gay* a million-dollar hunk of junk.[37]

Following the board of regents' meeting that morning, Mike Heyman faced the press, with the entire board literally backing him in front of the cameras. The vice president and chief justice, who were not able to make the meeting, were not part of this tableau, but the members of Congress and all others who had attended were. Heyman announced that the exhibition "The Last Act: The Atomic Bomb and the End of World War II" would be replaced with a smaller, simpler display. He explained,

> I have concluded that we made a basic error in attempting to couple a historical treatment of the use of atomic weapons with the 50th anniversary commemoration of the end of the war. Exhibitions have many purposes, equally worthwhile. But we need to know which of many goals is paramount, and not to confuse them.
>
> In this important anniversary year, veterans and their families were expecting, and rightly so, that the nation would honor and commemorate their valor and sacrifice. They were not looking for analysis, and, frankly, we did not give enough thought to the intense feelings such an analysis would evoke.
>
> Once the controversy was upon us, our staff made a sincere effort to create a more balanced exhibition. Within a month of my becoming Secre-

> tary of the Smithsonian last fall, plans for the exhibition were substantially revamped. . . . However . . . there was . . . a fundamental flaw in the concept of the exhibition. In retrospect, I now feel strongly that despite our sincere efforts to address everyone's concerns, we were bound to fail. No amount of re-balancing could change the confusing nature of the exhibition. . . .[38]

So much for the formal statement. The question and answer period that followed permitted Heyman to be more direct. Distancing himself from the museum, he referred to my January 9 letter to Hugh Dagley and told the assembled media,

> They made a terrible error in judgment sending a letter saying, "This is the way we're going to do it."[39]

Coupled to this, he promised "an extensive management review" at the Air and Space Museum.

Admittedly, Heyman did not yet know of American Legion commander Bill Detweiler's January 4 detailed plan of attack to close down the exhibition, nor that it had predated my letter to Dagley by many days.[40] But he did know of Dagley's call for further changes in the exhibition, as reported in the January/February issue of the Legion's *National News*.[41] Even Heyman's close friend and confidential advisor, Preble Stolz, whom he had asked to investigate the museum, had reported to him just days earlier,

> [Harwit's letter to Dagley] gave the American Legion a way to do something that some of its leaders probably had been wanting to do for a long time—call for the cancellation of the show. . . .[42]

In his conclusions on the museum's approach to the exhibition, Stolz added,

> I find no fundamental flaws in the processes used by NASM in the design and planning for this exhibition. . . .[42]

Why then had the American Legion decided as early as January 4 that the exhibition needed to be canceled?

I believe the Legion's negotiators on the script, Hugh Dagley and Herman Harrington, were talking a different language from their public affairs staff, and publicity is what forges a membership's will. Even as we were sitting across the table from each other at American Legion headquarters in late October, the *American Legion* magazine's November 1994 issue was going to press, with Paul Tibbets describing the exhibition as "a package of insults."[43]

How should Dagley and Harrington have answered that?

The letter I had written Dagley was really a nonissue. The Legion took great care to suppress the new label I had written, claiming, instead, that it now cited only sixty-three thousand expected casualties in place of an earlier quarter of a million. The label had *never even mentioned* that number, sixty-three thousand. If I had not written that day, the Legion would have found an excuse in the film footage we had promised to show them, or picked on the pictures of civilian casualties, as Detweiler's January 4 letter had already announced.

The Legion ultimately was after one major prize. They wanted their membership to feel heroic and have the nation's enthusiastic approval for the atomic bombings, just as, a year earlier, the nation had thanked and applauded its veterans of World War II on the fiftieth anniversary of the Normandy invasion. Speaking on the day the Legion had commanded the exhibition's termination, Hugh Dagley summed up,

> The number of casualties is not really that important. The issue remains: Was the use of the weapon a morally defensible act that shortened the war, saved lives and was done for a good purpose?[44]

The Legion did not trust a straightforward history to honor them sufficiently. Is honest history really so cruel that America had to fear it?

I am confident that it is not.

29

The Immediate Aftermath

MUSEUM MORALE

Smithsonian regents' meetings are always held on a Monday. I had spent the previous weekend writing letters of thanks to all our many supporters on the now canceled exhibition, both in the United States and in Japan. As soon as Heyman had delivered his announcement, those went into the mail. The difficult problems, I knew, would be internal to the museum. My main concern was with morale after this tremendous setback on an exhibition toward which so many members of the staff had devoted their efforts for years.

I could take three steps at once.

I met with the entire exhibition staff within an hour of the secretary's announcement; thanked them for all they had done; answered the many questions they had about recent weeks, during which the secretary had imposed a curtain of silence; and sought to have them understand they were not to blame.

Reconciliation among members of the staff whose opinions had mirrored those of the world outside was going to be difficult. I had to take the first step and set an example. I wrote Frank Rabbitt, the docent I had dismissed:

> Dear Mr. Rabbitt,
>
> As you may have heard, the Secretary of the Smithsonian Institution today decided to terminate the exhibition *The Last Act: The Atomic Bomb and the End of World War II*, to which you had so strongly objected early last year.
>
> Since this exhibition now is no longer going forward, and since you had previously given many years of loyal service to the National Air and Space Museum as a volunteer and docent at the Museum's Garber facility, I felt I should write you to invite you back if you should still feel inclined to return to your former duties. Your fellow docents at Garber have spoken very highly of you and have mentioned that you might wish to be reinstated.
>
> If you would like to return, I would be pleased to hear from you.[1]

Rabbitt accepted within a few days and returned to his volunteer duties.

In anticipation of his announcement, I had also asked Heyman whether he would be willing to come over to the museum early on the morning after the regents' meeting for a town meeting with our entire staff. His appearance at one of these periodic meetings shortly after his inauguration in September had made a good impression on the staff. I hoped he could now help raise their morale. He agreed to come, though he had an appointment shortly thereafter and could stay only ten minutes. I felt that even a short pep talk might raise spirits and did not think the length of his stay would matter much.

The meeting was held in the museum's planetarium, which could seat the entire staff. Heyman stepped up to the podium, pulled out a few sheets of paper, and read the statement which he had delivered at his press conference the day before. Everyone had seen that repeatedly on the previous night's news. He then added that he did not believe in exhibitions that are books on the walls, suggesting that the trouble with the *Enola Gay* exhibition was the length of its script. (In fact, its script was no longer than that of many of the museum's most popular galleries; and this criticism didn't make sense.) Having finished his remarks, Heyman accepted no questions, stepped down from the podium, and walked out through a pall of stunned silence.

I waited till he was out the door. Somehow, I had to undo his damage. I had prepared some notes, but spoke off the cuff, roughly saying that

> I was disappointed the exhibition had been canceled, but could understand why the Secretary had acted as he did.
>
> Our original objective had been to present an exhibition that would lead our visitors to better understand a pivotal event in this century. Instead, controversy and passions had erupted, and—most disappointingly of all—led some American veterans to believe that we did not fully appreciate their sacrifices. Certainly there was no point in going forward with an exhibition that was going to divide the country.

> The Museum had undoubtedly suffered a major blow. But this was the first setback in nearly nineteen years of operation. If we did not have an occasional failure it probably meant we weren't trying hard enough. We had attempted a difficult exhibition, that I felt was worth undertaking, and we had admittedly not managed to succeed. Greatness was not measured by how well one did in times of success, but how well one could recover from a downfall. I felt that we were a great museum, with a powerful staff, and we would learn from the experience and come back all the stronger.
>
> I expressed deep appreciation to the curators, the exhibition staff and the restoration staff, for the difficult, professional job they had done. Certainly, they were not to blame for what had happened. I said that I appreciated the Secretary's announced proposal for a forum at Ann Arbor in the spring to shed greater light on the process we had followed in putting together the exhibition. It would give us an opportunity to set the record straight after all that had been said about the Museum in the media.
>
> We had many other exciting activities and projects coming up, to which we should now turn and go forward.
>
> Above all, we needed to join ranks and not give way to rancor, which in many ways simply mirrored a dispute that had gone on across the country. As average Americans, it was not surprising that we too were divided over a national dispute on Hiroshima and Nagasaki. To be able to go forward with other pressing projects, we had to let these differences go.
>
> I reminded people that almost everyone who worked at the Museum had probably been asked about the exhibition more times than they would wish, by friends and strangers, and that these stresses would have also spilled over on their families. All of us and our families had been affected by this exhibition; we all needed to recover.[2]

We still had a few questions and answers, talked about several other matters and then broke up. I hoped it had helped, but knew full recovery would take a long time.

EXTERNAL REACTIONS

The first reactions came from Japan. Mayor Hiraoka of Hiroshima released a statement the same day:

> It is extremely regrettable that the National Air and Space Museum of the Smithsonian Institution has reduced the scale of its exhibit and revised it to commemorate a tone of the victory of the U.S. in World War II, contrary to its original plans.
>
> Today, when the world is still facing the threat of nuclear weapons, it is not an appropriate time to display the *Enola Gay* that dropped the bomb solely under the concept of an exhibit displaying U.S. victory in war.
>
> We, the cities of Hiroshima and Nagasaki, had no intention of criticizing the United States by lending our A-bombed materials for this exhibit. We simply hoped to heighten public opinions toward the building of a world

> free of nuclear weapons by letting U.S. citizens know the cruel aftermath brought by the bomb. There are no words to describe our disappointment in not realizing our hopes. . . .[3]

Given the difficulties I had experienced in mid-January in getting the agreement from Hiroshima and Nagasaki to support our exhibition, I could not help but feel the mayor might be somewhat relieved that the exhibit had been closed down by others without his city being first to pull out.

A note from the Hiroshima Peace Memorial Museum director, Mr. Harada, expressed regrets:

> I was hoping that the exhibit would convey the wishes of the citizens of Hiroshima to as many people as possible in your country and I am therefore very disappointed. I understand, however, that you were put in a difficult situation, receiving requests that were completely different from ours.[4]

I did appreciate a perceptive note from Ambassador Shinichiro Asao, who headed the Japan Foundation and had provided us with our first contacts, two years earlier, on our first trip to talk with the museums in Japan. He now wrote to me that he wished us well and that he understood our difficulties. In Japan, people remembered Hiroshima, whereas Americans remembered Pearl Harbor. In every country, he felt, the sense of having been a victim dies hard, particularly in recollecting the atrocities of wars. He thought it would take time for our two countries to narrow the wide gap in their perceptions of World War II but also hoped that this experience would nevertheless deepen our mutual understanding.[5]

Many friends and colleagues from all over the United States and abroad also wrote thoughtful and supportive letters. I circulated some of them to the staff to raise their spirits.

General Tibbets was quoted as endorsing Secretary Heyman's plan:

> Just as now expressed by Secretary Heyman, separate the horror of nuclear war from the fighting it took to get the bases close enough to Japan to fight their heartland with conventional air power. Portray the *Enola Gay* separately as the ultimate development of the B-29 and the first airplane to strike an enemy with an atomic weapon, and after that, period.[6]

W. Burr Bennett, Jr., commented,

> I'm pleased with the outcome. I think this is what we wanted in the first place. To have the plane displayed properly like any other airplane, like Lindbergh's *Spirit of St. Louis* or the Wright *Flyer*. Just say what it did—which was to end the war in nine days.[7]

But Ronald H. Spector, professor of history and international relations at George Washington University's Elliott School of International Affairs, whose book *Eagle Against the Sun* is a study of the Pacific war, regretted the decision to scale back the exhibition:

> I think everybody lost. Basically, the public has lost the opportunity to understand the great range and complexity of this whole issue and, certainly, to understand the great differences of opinion about this. . . . The mind-set of 1945 is not the mind-set of 1965 or 1995. The historian's job is to understand the mind-set of the people at the time, not to pass judgment on them from the vantage point of 50 years. It's hard to do, but that's what we get paid for.[8]

A particularly thoughtful letter to the editor in the *Whittier Daily News* showed that there were even World War II veterans who genuinely regretted the cancellation of our exhibition. Dell Herndon, of Whittier, California, wrote,

> As a World War II veteran who participated in the bombing of Japan, I have been saddened by the attitude of some veterans regarding the planned Smithsonian exhibit to commemorate the 50th anniversary of the atomic bombing of Hiroshima and Nagasaki.
>
> I was one of 11 crew members on a B-29 arriving on the island of Tinian the day after Christmas 1944. Our crew was one of many which repeatedly bombed many Japanese cities. Of the original 15 plane crews in my squadron, only five survived. I was one of the lucky ones who returned home after the war. . . .
>
> From early June 1945 until the Hiroshima bombing, my B-29 crew was assigned to weather observation. . . . Just prior to the Hiroshima bombing, our plane spent six hours over Japan, flying back and forth over several cities. No anti-aircraft guns shot at us and no fighter planes challenged us. This experience led me to believe that Japan was defeated and that no invasion would be necessary. Declassified information released in the past 20 years has strengthened my belief that Japan would have surrendered without an invasion or the use of atomic bombs. I know many veterans agree with me.
>
> In our democratic society, it is the patriotic duty of every citizen to question the actions of our government. Even if it is true that the atomic bombings saved thousands of Americans, it is our patriotic duty to acknowledge the results of those bombs—instant death for many innocent children and the long-term effects on the descendants of the survivors.
>
> My hope is that such reflection will encourage our leaders to seek better solutions and make this world a better place for my three granddaughters and all children.[9]

Prime minister Tomiichi Murayama was quoted as saying "This is regrettable (in terms of) the Japanese people's feelings," while President Clinton and Vice President Gore, conveying their message through press secretary Michael McCurry, said, "while believing firmly that academic freedom has its place" they also were "very sensitive to the concerns expressed by veterans' groups."[10]

The nation's newspapers were divided:
An editorial in the *Washington Post* commented,

It is not, as some have it, that benighted advocates of a special-interest or right-wing point of view brought political power to bear to crush and distort the historical truth. Quite the contrary. Narrow-minded representatives of a special-interest and revisionist point of view attempted to use their inside track to appropriate and hollow out a historical event that large numbers of Americans alive at that time and engaged in the war had witnessed and understood in a very different—and authentic—way.[11]

But columnist Coleman McCarthy, also writing in the *Post,* countered,

Whining and bellyaching—long the specialty of the American Legion and similar pensioner warrior groups—have prevailed. The Smithsonian Institution is canceling its planned *Enola Gay* exhibit.[12]

A Steve Benson cartoon in the *Arizona Republic* had already made the same point.[13]

The *Philadelphia Inquirer* expressed disbelief:

The nation has flunked a history test. The Smithsonian Institution's long nervous breakdown over how to display the Enola Gay . . . was a test of whether this nation could seriously re-examine a watershed event, not just commemorate it.

Smithsonian Secretary I. Michael Heyman said that the exhibit's "basic error" was trying to "couple a historical treatment of the use of atomic weapons with the 50th anniversary of the end of the war." Imagine

> that, trying to get people to think about the meaning of atomic weapons by exploring the first and only use in actual warfare. That would require a dedication to sincere and open historical inquiry. Positively un-American.[14]

This was echoed by cartoonists Stahler,[15] Auth,[16] Sargent,[17] and Wasserman.[18] Political figures also expressed opinions:

Speaker of the House Newt Gingrich told the National Governors' Association,

> You are seeing a reassertion and a renewal of American civilization. The *Enola Gay* fight was a fight, in effect, over the reassertion by most Americans that they're sick and tired of being told by some cultural elite that they ought to be ashamed of their country.[19]

And his colleague congressman and Smithsonian regent Sam Johnson offered,

> We've got to get patriotism back into the Smithsonian. We want the Smithsonian to reflect real America and not something a historian has dreamed up.[20]

A HIROSHIMA EXHIBITION

I had been unable in mid-January to get a definite commitment from Hiroshima and Nagasaki for the loan of artifacts from their museums. But as soon as the National Air and Space Museum's exhibition was canceled, an outcry erupted in Hiroshima for some sort of display of their artifacts in the United States.

On February 13, 1995, just two weeks after the cancellation, I had a call from Hiroshi Harada, who asked what Hiroshima might be able to do to exhibit their artifacts in other museums or otherwise get publicity for their cause in the interests of peace. I said that I was not really well equipped to help him on these matters, but suggested a few people with whom he might talk.

By March 31, the *Washington Times* was reporting that

> American University in Washington and the city of Hiroshima have agreed in principle to jointly hold an exhibit on the U.S. atomic bombing of Japan at the university's campus. . . . The exhibit will replace the one planned by the Smithsonian Institution but canceled under pressure from American veterans' groups. . . . The exhibit will open July 8 as part of a summer seminar sponsored by the university on the history of nuclear power. It will show photos of damage and radiation injuries suffered by people in the atomic bombing of Hiroshima on Aug. 6, 1945.[21]

And on April 6, the *Washington Post* reported,

> Emotionally laden artifacts and vivid images from the atomic bombing of Japan that were to have been part of the canceled Smithsonian exhibit on the end of World War II may have found another home in the nation's capital: on the campus of American University.[22]

The following day, the *Washington Times* added,

> American war veterans, weary from their battle over a national exhibition on the atomic bombing of Hiroshima, are watching vigilantly as American University and a Japanese museum plan a display on the same divisive topic this summer in Washington. Veterans question the involvement of the Hiroshima Peace Memorial Museum because of the role Japan played in World War II. . . .

Phil Budahn, spokesman for the American Legion, was cited in the same article as saying,

> The Hiroshima museum is at the philosophical heart of the effort to portray the Japanese as victims of World War II. They identify the Japanese with the Jewish victims in Europe.

Air Force Association spokesman Jack Giese was quoted to have said,

> Although we're very interested in fairness, this is a university program.... Our major concern with the Smithsonian exhibit was that it used public funds.

And American University history professor Peter Kuznick reportedly said,

> This is not going to be what the *Enola Gay* exhibit would have been.... Anything we do is going to be balanced.[23]

By July 9, Hiroshima's exhibit at the American University in Washington had opened. The speed with which these arrangements had been made was impressive. One might be tempted to chide the Japanese for not having committed themselves to the National Air and Space Museum after nearly two years of discussions, and then turning around with such speed to do an exhibit at American University. But the circumstances were different. At the museum, we had given the Japanese little control over the context in which their artifacts would be displayed. At the American University they had far more free rein, and they did not have to worry about displaying their artifacts in the overwhelming presence of the *Enola Gay* fuselage.

30

The Last Act

ANN ARBOR FORUM

When Heyman, on January 30, 1995, canceled the exhibition the museum had planned, he also publicly announced that a forum on the exhibition would be held later, in spring, at Ann Arbor. There, historians and museum scholars would be able to openly discuss the role of museums in the presentation of sensitive topics and debate the means by which such issues could best be aired. This broad, scholarly forum would attempt to discern the lessons taught by the *Last Act* controversy. Professor Homer Neal, vice president for research at the University of Michigan and a Smithsonian regent, had offered to host the forum at Ann Arbor, far away from the political pressures of Washington, so all aspects of the exhibition might be openly discussed.

On February 13, I sent Heyman a memorandum outlining my recommendations for the museum's participation. I emphasized the need for circulating the exhibition script, addressing the care we had taken in talking with different constituencies, and showing the videotape we had made of *Enola Gay* and *Bockscar* veterans recalling their wartime missions. I wrote that one of the curators and I would need an hour out of the day's proceedings to make such a presentation:

For the forum you are organizing in Ann Arbor, which is to deal with controversial exhibitions and use the *Last Act* exhibition as a point of departure, I would like to request the following participation by the National Air and Space Museum.

We would like to circulate to participants of the Forum the final version of the exhibition script as it stood on January 30. It differs from the October 26 / December 6 version in relatively minor ways, primarily in incorporating a number of recommendations that were made by Gen. Tibbets, military historians and members of the exhibition's advisory committee.

To give participants an account of factors that are not apparent in the script, I would like to request two twenty-minute slots for presentations, at the start of the day, with additional time for discussion from the floor at a suitable point on the agenda. The first presentation would be by one of the exhibition's curators, Tom Crouch, and would augment the written script with slides that would provide a better feel for what the exhibition would have been like. The script's written words do not adequately serve that purpose; the quality of the images we can include unfortunately is poor. In the second of these two talks I would like to give a summary of the procedures we followed in compiling successive versions of the script, the groups we consulted, and the criteria we applied in reaching decisions on changes we would make.

Finally, we would like to complement these two talks with a presentation of the video recounting the recollections of the *Enola Gay* and *Bockscar* crew members. The script we now have primarily addresses the historical approach we took. The videos brought out the commemorative approach which was to balance the historical accounts. To accurately portray what we did, and to show the balance between history and commemoration that we were planning, this thirteen minute film needs to be shown.

In summary, we would need an hour to make the overall presentation, with any additional time for discussion allotted separately, whenever appropriate during the day.[1]

On March 1, Acting Provost Bob Hoffmann asked me to see him about the forum. Before going over, I called in Neufeld and asked whether he would be willing to join me in making the presentation at the forum. Crouch, whom I had initially asked, had said he was reluctant, and I did not wish to force him if he felt awkward. Neufeld was willing to go, but said he would probably wish to differ with me on some points. That didn't matter, I said; the differences between us were not that important, and if they came out in the course of the forum, so be it.

An hour later, I met with Bob Hoffmann and an aide. They showed me the preliminary plans. Half a morning was to be devoted to the *Last Act* exhibition. An as yet unnamed neutral commentator was to make a presentation on the museum's activities, followed by a panel discussion. Each panelist would be given five minutes for an opening statement before audience participation began. Tom Crouch was one of the panelists and the only representative for the National Air and Space Museum in the entire forum.

I strongly objected to this plan, saying that I had been told by Heyman, in Hoffmann's presence, that this would be a forum where the curators and I could present what we had done. I reminded him of my memorandum to Heyman, which he had seen, and said that the proposed program was in clear violation of the promise to the museum that we would be permitted to make a full presentation at the Ann Arbor forum. We talked for an hour. Hoffmann felt he did not have the authority to make the changes, and said he would have to take it up with the secretary. We left it at an impasse.

The next day, after a meeting with Connie Newman on another matter, she asked me to stay on and called Bob Hoffmann to join us. They now were willing to allot the museum more time at the forum than Hoffmann had specified the previous day and to place that session first in the morning, as I had suggested, but they would not tell me how much time we would get, certainly not the full hour I had asked for. They wanted to have Crouch represent the museum. *I would absolutely not be permitted to contribute to the forum.* The video also could not be shown. When I asked why, I kept hearing, "because the secretary doesn't want it."

I reminded them that I had been told in their presence that the museum would have an opportunity to present its case at Ann Arbor, and that they were now shackling us. The secretary had been publicly criticizing us with innuendos for lack of care, for books on the wall, for not paying enough attention to "stake holders—veterans of the Second World War";[2] here was our chance to show how well we did, and we were not being permitted to. I bluntly called it a "betrayal." The forum as structured rescinded the most important commitment the secretary had made to the museum.

Newman and Hoffmann said they would talk with the secretary.

The next day I had a note from Heyman, saying he wanted to talk with me about a number of matters including

> . . . how the sessions at the Michigan conference should be handled. After discussions with the Michigan group, we definitely concluded that we should change your plan in some significant ways. Primarily, someone else, probably Tom Crouch, should make a 20–30 minute presentation, the central purpose of which would be to describe the initial conception of the exhibition and the one that emerged in the final draft. Thus we will have two detailed examples for the larger questions that are being addressed.[3]

Three days thereafter, Heyman and I had a long session late in the afternoon. I will return to that later, but in the course of our conversation, he reiterated his decision to have only Crouch participate, saying Crouch would be restricted to talking about just those subjects Heyman had determined.

In deciding to entrust the task to Crouch, Heyman was picking someone who was being portrayed as sharing his views. Crouch's July 1993 memorandum to me, though not my reply to him, had been widely cited by then, and a comparison of Heyman's and Crouch's views had been published just a few days earlier in the *Cleveland Plain Dealer*:

> Smithsonian Secretary I. Michael Heyman said the exhibit had two irreconcilable goals—to honor the valor of the soldiers who fought the war and to reassess the *Enola Gay*'s heritage....
>
> Heyman's conclusion was foreshadowed early in a note from curator Tom Crouch to museum director Martin Harwit, a memo promptly leaked by internal critics. "Do you want to do an exhibit intended to make veterans feel good?" Crouch asked, "Or do you want an exhibition that will lead our visitors to think about the consequences of our atomic bombing of Japan? Frankly I do not think we can do both."

The *Plain Dealer* continued:

> Why not? If a high school history book can offer a noncontroversial account of Hiroshima, why couldn't the Smithsonian?[4]

Crouch's memorandum to me had also been quoted in the detailed report on the exhibition Heyman had commissioned and received in January from Preble Stolz.[5]

While I, who was responsible for the exhibition, was prevented from participating, Monroe Hatch, who had led the attack on the Smithsonian for so long, received a personal invitation from Heyman to attend.[6]

At the Ann Arbor forum, "Presenting History: Museums in a Democratic Society," on April 19, Crouch gave the presentation that Heyman had restricted to just the script contents. Predictably the museum was roundly criticized by colleagues from other institutions for not having built up a sufficiently large group of "stake holders" and for not giving enough attention to the voices of veterans, as contrasted to the historians' views.

This was just what I had feared. Since I had handled them, Crouch was not aware of many of our dealings with veterans, the military, the pacifists, or the politics in the United States and Japan. He therefore could not speak to our efforts at building support. And, denied permission to show our video footage of *Enola Gay* and *Bockscar* veterans, we were robbed of the opportunity to show that we had indeed given veterans' recollections a central place in the exhibition.

Heyman had left little to chance. To serve as introductory speaker and chairman of the panel debating *The Last Act* he had once again called on his trusted friend Preble Stolz.

The only unexpected moment may have come with the statement by panelist Vice Admiral Tom Kilcline, president of the Retired Officers Association, who spoke of his disappointment that *The Last Act* had been canceled:[7]

> There was frustration early on, because we did identify some things that were obviously not going to be accepted. But by the time we got to the fifth script, and by the time the first video was prepared . . . we were enthused!
>
> Recognize that the media was still talking about script number one. That was the unfortunate part. How could you turn them off? We intentionally did not have any press conferences. We tried to stay quietly behind the scenes while things were happening. Maybe that was our mistake; but I'm not really sure. I still think we did the right thing. . . .
>
> I was very sorry . . . I think we should have gone ahead and had the exhibit, because I think people would have loved it. And the veterans would have accepted it, because the basic idea was a good idea. And as it was modified it got even better. . . . I wish we could turn on the camera here and show you the thirteen minute video—a beautiful sense of emotion. And real people—four crewmen—nobody setting them up, nobody interviewing them; just talking about what they did and how they felt about it. That's the kind of a tone they had come to. And I feel badly; but I also feel badly that at first it needed so much correction.
>
> I live in a political world, more than I do anything else. And in politics we talk about the government's special interests these days. . . . Special interests are causing you problems in museums too. . . . It started with a couple of small associations and then it got to a bigger association. . . .

When Kilcline finished, his fellow panelist, Rem Rieder, from the *American Journalism Review*, questioned,

> Maybe it is a shame that you were not more aggressive going public with how impressed you were with the way it had evolved. . . . Here's a wonderful chance for what the newspapers call a "stunning reversal." Suddenly the Smithsonian has fixed the thing! Some veterans groups are happy with it! They've been responsive! . . . It *is* possible to turn a story around.
>
> Very often mistakes are repeated . . . Because nobody really brings it to the attention of the media. . . . Here is a case where a lot of good might have been accomplished by getting the word out very aggressively that "Hey, they might have gotten off on the wrong foot, but we are really happy with it now."

To this Kilcline replied:

> I hear you. . . . But we had a reason for not doing it. Frankly . . . We wanted to keep on monitoring the progress of the development. I wasn't sure that if we had backed off and said it was great, it would not have gone back to where it was.
>
> So that's it. Tactical mis . . . [his voice trails off; the last word is unclear.]

I appreciated Kilcline's generous praise, but found it bittersweet.

Ironically, the museum had produced the video-film he liked so much without any pressure from veterans organizations or the media. Its production had entirely escaped their notice. The museum had been able to fashion, reshape, and revise the footage, quietly going through many, many revisions until we were satisfied and released it.

This was how we normally worked—without outside interference, without the tumult and glare of the media. In producing this film, there had been no need for Kilcline's "monitoring progress," nor for the veterans' "corrections," to which Kilcline attributed such importance. Nor was there any fear the museum might "have gone back" on the film. Once we at the museum had felt satisfied and shown the veterans the video, everyone had agreed on its quality. The process we had followed in making the film was an example of how every other film or exhibition in the National Air and Space Museum had always been produced.

Much of the veterans' advice indeed *had* been useful and had resulted in changes. That was why we had in the first instance asked them for comment. But their insistence and their attacks in the media and Congress to force their views on us had hurt the museum badly.

Though the video that had so "enthused" Kilcline had been first shown at the American Legion's 1994 Labor Day convention, four full months before the exhibition's cancellation, neither he nor any other representative of a veterans' organization had ever publicly praised it in the intervening months or spoken up for the exhibit. Praising the exhibition now, so many weeks after its cancellation, was no help; but at least Tom Kilcline had the grace and character to belatedly admit that "people would have loved it." Others were still sniping.

A month later, at congressional hearings chaired by Senator Ted Stevens of Alaska, the Retired Officers Association's Colonel Charles D. Cooper expressing similar sentiments, testified that

> . . . There was a growing consensus, at least among the reviewers of the Air Force Association, the Retired Officers Association, and the VFW, that the exhibit would have been found acceptable by most veterans. This sense was conveyed to Under Secretary Newman and Dr. Harwit at a joint meeting at the Smithsonian on December 15, 1994.[7]

At these same hearings, the VFW's Robert Manhan similarly testified that

> The [final] script was a fairly decent package. . . . After all the time, effort, and money that had been expended up to this point on presenting a balanced exhibit, [Secretary] Heyman's decision was a surprise, at least to the VFW. . . . This action, in the VFW's opinion was not justified. . . .[7]

Cooper's and Manhan's testimony came on May 11, 1995. I had left the Smithsonian and the museum the previous week.

AVOIDING HEARINGS IN THE HOUSE

Following the script of his January 4, 1995, plan of attack, American Legion national commander Bill Detweiler had written to Congressman

William F. Clinger, Jr., chairman of the House Committee on Government Reform and Oversight: "We adamantly believe the interest of the American people, as well as the Institution's long term welfare, lies with Congress proceeding with hearings.... We believe only Congressional hearings can elicit the needed answers from Institution principals."[8] His letter, dated February 10, listed twenty-three specific questions. A similar request went to the Senate.

In the House the Smithsonian was able to avoid hearings through the intervention of congressman and regent Sam Johnson, who instead expanded Detweiler's list of questions to twenty-eight and sent it to the Institution asking Secretary Heyman to answer them in writing.[9]

Several of the questions went to the heart of the freedoms that academic institutions need to enjoy in a democratic society. I present these, together with the answers Heyman provided:[10]

Johnson's question 7 is one of the most important, because it emphasizes the difference between practical politicians and academics. Academic institutions judge the merits of scholars by the quality of their work, not by their personal beliefs. Heyman's response to this question, particularly his last two sentences, were correct and important to the future of the Institution:

Q.7. Why was Michael Neufeld, a Canadian National, hired by NASM? What are his philosophical and political underpinnings?

A.7. I am informed that Michael Neufeld was hired for his broad knowledge of World War II, as displayed in his prize-winning book on the development of the V-2 rocket, *The Rocket and the Reich*, published in 1994. The book won the "best book of the year" award from the American Institute for Aeronautics and Astronautics, and was critically acclaimed in the *New York Times Book Review*. . . .

Before embarking on the exhibition of the *Enola Gay*, Dr. Neufeld had already curated a World War II commemorative exhibition on the Republic P-47 Thunderbolt, affectionately known as "The Jug." In the same commemorative series he curated a display on the German Arado, the first operational jet bomber which was used also for reconnaissance.

I do not know Dr. Neufeld's political affiliations or philosophical propensities. These are not matters that the Smithsonian inquires about.

Questions 15 and 16 concern a matter that was repeatedly raised by Detweiler and Johnson. It regards the *Last Act* exhibition catalogue. The museum had long planned to print a catalogue that contained word for word the entire label text for the exhibition. So much had been written and said about this exhibit, by people who had never read the script or who quoted it out of context, that I wanted to have the text published. That way, long after the event, we would be able to defend its actual, rather than alleged, contents. The *Last Act* catalogue was ready to go to the publisher on January 30, when Heyman canceled the exhibition; the Smithsonian Institution Press had already announced it in their Spring catalogue.

The Legion and Congressman Sam Johnson desperately wanted the catalogue withdrawn from publication and suppressed. Sam Johnson's first inquiry was conveyed to me by Newman as early as February 13, probably the day that Detweiler's February 10 letter had reached him. She told me the Institution was reassuring Johnson that the catalogue would never be published. She wanted to know its status. I provided the information in a memorandum to Heyman the same afternoon.[11]

To the two questions posed by Johnson, Heyman gave alarming answers:

Q.15. What is the status of the companion volume on the now canceled exhibit? How does Secretary Heyman intend to stop this? What will happen to the 10,000 copies said by an unidentified spokesperson at the Press to exist? Do they? If so, have they been distributed to anyone? How will they be recalled?

A.15. There has been no publication of the catalogue that was to have accompanied the *Enola Gay* exhibition. No catalogue has been printed, published or distributed that must be recalled. The Acting Director of the Smithsonian Press reports that the Press did not even keep copies of the draft manuscript.

In response to my decision regarding the *Enola Gay* exhibition and the decision to cancel the publication of *The Last Act*, the Smithsonian Press took the following steps:

- Sent a memorandum to all sales representatives throughout the world informing them of the cancellation. The same memorandum went to key wholesale accounts.
- Placed a message in the system at the warehouse that automatically informs customers who order the book that it has been canceled.
- Sent a letter to book reviewers and other media that informed them that the publication has been canceled. That information was included in a routine letter from the publicist to the key media contacts.

Q.16. Does Secretary Heyman intend to honor his promise to cancel all related materials? More important, is he positioned properly to effectively control the actions of NASM personnel?

A.16. I fully intend to honor my promise to cancel all related materials. The only material that may be in the public purview are copies of the various draft scripts that were out for comment prior to the decision to cancel the exhibition. The Institution has received some requests for the first and last scripts. We are referring those requests to the Office of General Counsel. There is, however, no basis to deny people access to the documents that had already been made public. To date there have been very few requests of the General Counsel to supply copies of any of the draft scripts.[12]

Worried that the exhibition would undermine the public's appreciation for their sacrifices in World War II, countless veterans had strenuously objected. Given these fears, the intense pressures on the Smithsonian to cancel the exhibition could be understood. But it was quite another matter to suppress a catalogue—a book that people must find sufficiently attractive to actually spend their own money to buy. Every democratic society abhors the suppression of books, and here we had a congressman asking for just that and the secretary of the Smithsonian Institution acquiescing.

Finally, The Legion and Congressman Johnson also raised the issue of Title 20 §80 of the United States Code, authorizing a National Armed Forces Museum that had never been built in the thirty-four years since the legislation was passed. Here Heyman, after consulting general counsel, gave an answer that corresponded to the way I had also been advised to interpret the legislation. Unfortunately, however, he added his intention to have the mission statement of the museum reviewed by the board of regents.

If the regents decide to place the added restrictions of this legislation on the museum, they will eliminate all possibility of free inquiry into military aviation at the Smithsonian, because important aerospace activities that might not be "an inspiration to future generations" would systematically have to be kept out of exhibits. And who would judge what was inspirational and what was not? Such restrictions would curtail dispassionate discourse, so fundamental to our democratic society. Johnson's question and Heyman's response were as follows:

> Q.1. To what extent did the now canceled exhibit conform to the charge of the Smithsonian Institution, as stated in 20 USC, Ch. 3 Para #80? NASM officials respond that the requirement to present "the service and sacrifice of America's service men and women as an inspiration to the future generations" applies only to the National Armed Forces Museum—which was never built. However, the language in the cited section clearly states that "*The Smithsonian Institution* shall. . . ." Absent case law to clarify the intent of the legislation, no prevailing interpretation of that language exists, it appears that NASM is citing an interpretation designed to free their hands from responsibility as probably intended by Congress.
>
> A.1. The legislative language quoted pertained to a National Armed Forces Museum which was authorized but never funded. The statute containing it specifically provided that that statute was not intended to apply to the National Air and Space Museum. As stated in 20 USC §80:
>
> The provisions of this subchapter [Subchapter X—National Armed Forces Museum Advisory Board] in no way rescind subchapter VII of this chapter, which established the National Air and Space Museum of the Smithsonian Institution, or any other authority of the Smithsonian Institution. . . .
>
> As you know from my statement on January 30, I am undertaking a management review of the National Air and Space Museum, and one of our

> goals is to review a mission statement for the Museum to make sure that it is responsive to this statutory provision. I will discuss with the Regents the parameters of this management review on May 8, 1995, and I expect to have the review completed by September 1995.[13]

MANAGEMENT REVIEW

In 1994 the Smithsonian Council, the Smithsonian's highest-level blue-ribbon committee, provided the Institution a report evaluating the museum's entire operations. Council members are internationally recognized scholars, research scientists, museum specialists, and educators. They come from many different establishments and are familiar with a variety of management practices. Heading the council was Dr. Maxine Singer, president of the Carnegie Institution of Washington. A subcommittee of the council had spent the latter half of 1993 examining the museum's activities and operations in great detail. Late in 1993, the entire council met at the museum for further briefings and discussions. Their report, issued to then secretary Adams early in 1994, and through him to the regents and members of Congress, was a strong, positive evaluation of the museum's work.

Thus, on January 30, 1995, the day Secretary Heyman decided to cancel *The Last Act* exhibition, he had at hand a carefully crafted report on the museum prepared barely a year earlier by a group thoroughly familiar with all Smithsonian operations. Heyman had also a second, rather favorable opinion in the Preble Stolz report, which he had just received on January 26. As already mentioned, Stolz had concluded, "I find no fundamental flaws in the processes used by the NASM in the design and planning for this exhibition." With two such recent positive reports in hand, Heyman nevertheless ordered yet a third study. He told me Congressman Sam Johnson had insisted on it at the regents' meeting.

The Smithsonian commissioned the National Academy of Public Administration (NAPA) to conduct this study. NAPA is one of several reputable Washington management consulting firms. Under Secretary Constance Berry Newman serves as a member of their governing board.

When the NAPA report finally appeared, months after I had left the Institution, it correctly identified many of the problems facing the museum, and also offered a number of reasonable recommendations. However, it failed to note that the museum had already been working, for some years, with Resolution Dynamics, another Washington management consulting firm, often used by the Smithsonian. We had already identified these same problems; begun to reallocate resources, particularly now that Congress had authorized us to begin planning the construction of a Dulles museum extension; and initiated strategic planning, in 1994, to systematize these efforts.

Listing the museum's problems and recommending solutions while failing to note any of the actions the museum had already taken, the NAPA report gave the impression that the museum's management was unaware of its problems and incompetent. To the contrary, I had met with the NAPA project staff and the investigating panel for a total of roughly ten hours over a period of five months, had identified for them the problems that lay ahead, corrected some misapprehensions, told them about management changes we had already instituted, and outlined the steps we were pursuing.

How can one account for these omissions from so reputable a firm? The structure of the NAPA report itself provides a possible explanation. The report clarifies, right at the outset, that "the central administration of the Smithsonian posed a series of questions [that] form the basis for the main findings and principal recommendations [of this report]." Thus, the very specific questions the report was asked to answer appear to have constrained the report's contents and shaped the overall impression it conveyed.[14]

I was proud of the museum staff. They were serving millions of visitors a year, mounting exciting new exhibitions, devising improved ways of preserving aerospace artifacts, restoring aircraft to the highest standards, conducting forefront research, producing informative films that ranked among the top fifteen revenue-earning documentaries of all time, and making all these activities possible, in part, by supplementing the museum's annual federal budget by fifty percent through energetic fund-raising efforts. To manage all this with a federal staff of just 220 was no easy feat. But not a hint of these accomplishments appeared in NAPA's ten-thousand-word report.

I was disappointed. It felt like Ann Arbor all over again.

BASIC DIFFERENCES

I had strongly disagreed with Heyman's decision to cancel the *Last Act* exhibition, particularly at a time when even major veterans' organizations were showing signs of being satisfied. I disagreed, though I could understand his decision. He was under great pressure in Congress, he had been in office only four months, and he had been brought in by the regents to raise money, which he otherwise feared he might not be able to do.

But there were deeper basic differences. One had shown up in the way Heyman had organized the Ann Arbor forum. It was billed as a scholarly meeting, but he prevented the museum from providing a full report. I saw the forum as an unfortunate exercise, aimed at satisfying the Institution's critics in Congress, but misleading the entire family of sister museums, which might have been able to learn from the National Air and Space Museum's true experiences rather than from a dismembered report.

The suppression of the exhibition catalogue was another example of an action totally at variance with American, and more broadly, democratic approaches to scholarship.

One other incident underlined our different approaches:

Late in the afternoon on March 6, Heyman and I spoke frankly and informally for almost an hour in his office. I had written him a note taking him to task for his many public statements accusing the museum of not taking sufficient care. In particular, I had cited his statements to the National Press Club on February 23, 1995, where, in the question and answer period he had said,

> I think there are a lot of ways that we at the Smithsonian ... can search through and evaluate complex historical events. I'm not sure that exhibitions with great plaques of words are the best way.... I don't think we should eschew controversial exhibitions.... I think we can do quite controversial exhibitions if we are very careful about how we do them. And I think one of the problems was—with the *Enola Gay* exhibition—that not sufficient care was taken to do it in that careful way that would have avoided quite a lot of the collisions of emotion that otherwise occurred....
>
> In my view ... there should always be room at the Smithsonian to explore important contemporary issues lest our great potentialities as an educational institution be wasted. However, when we create an exhibition, a program or even an individual label, we need to do so with great care. We need to distinguish between opinion and fact. We need to contribute to light rather than heat....
>
> The Air and Space Museum was not very, very careful in permitting all of this to go public [at] what I believe was a premature time....
>
> I think that what ... in my view happened with regard to the original rendition of the *Enola Gay* script, is that there was very little attention paid to what stakeholders would feel—in this case ... veterans of the Second World War.[15]

I told Heyman I could not understand these allegations. It was precisely because we had worked with "stakeholders," that our script had been prematurely released. The Air Force Association, whom we had asked to comment so we could take their views into account, had instead widely circulated the script—violating our explicit request.

Heyman had never informed himself about the amount of care we had taken and about all the liaisons we had built up. I had never had an opportunity to list for him all the people we did talk with and all the stakeholders—in the military services, in the State Department, among scholars and veterans, and in Japan—who, in fact, had been asked for opinions and had made significant contributions. There were more than six hundred grassroots B-29 veterans who had helped us restore the *Enola Gay* over the years by sending in small checks that eventually totaled over $22,000; there were all the members of the 509th who sent in memorabilia for us to use, crew members of the *Enola Gay* and *Bockscar* who helped us make the video for the exhibition, all the members of our advisory committees, and members of Congress and their staffs the museum and Institution had kept informed. The list could go on and on.

Heyman said he understood, but that he really would prefer if I left the handling of politics up to him.

I sensed that he might mean just Washington politics, so I said that I would be glad to, but that neither Bob Adams nor Connie Newman had ever seemed particularly interested in the Japanese part of the politics. So I had left the Washington politics up to them and handled the Japanese part myself.

Heyman did not appear to find that objectionable.

We were about to break up when, almost as an afterthought, Heyman said, "There is just one thing I've never understood. Why were you so set on borrowing artifacts from the Japanese? I would have thought you could have done the exhibition without them."

Where Heyman's own replacement exhibit was going to avoid all contact with Japan, I had always seen *The Last Act* as joining the histories of our two countries in a world event that may have the gravest consequences for the human race. But I thought this would be too hard to explain when Heyman's mind-set appeared so different. I replied with the most practical, one-sentence rationale I could give him—the only one that might make political sense to him: "I feared right from the beginning that leaving out the Japanese would put us in exactly the place where the postal service ended up with the mushroom-cloud-stamp fiasco. I wanted the Japanese to buy into this exhibition early."

Bob Adams might have put it differently, quoting the words he had written Bill Rooney nine years earlier:

> I think it would be wrong . . . to deal with the *Enola Gay* merely as a significant step in the development of several vital aerospace technologies. "A decent respect for the opinions of mankind" . . . requires us also to touch on the demonstrated horror and yawning future risk of the Age that the *Enola Gay* helped inaugurate.[16]

Yet here his successor was setting out to do just that—exhibit the *Enola Gay* as a significant development in aerospace technology while disregarding the horror and risk symbolized by her mission. Three months later, when Heyman's own exhibit opened, to a packed news conference, in late June 1995, he was asked why he had failed to treat the destruction and devastation at Hiroshima. He responded,

> I really decided to leave it more to the imagination.[17]

Such differences between Heyman and me led to mounting difficulties as veterans' organizations and Congress continued to press their advantage.

A Smithsonian museum director serves at the pleasure of the secretary. That is as it should be; but unless they work well together, the strains interfere with the Institution's functioning. The time was approaching when I would have to resign.

TIMING

The timing of my resignation had not been my choice. While I had quietly begun looking around for other positions, I was surprised at the speed of events that actually unfolded.

On the morning of Thursday, April 20, 1995, the day after the Ann Arbor forum, I was asked to come to Newman's office. When I arrived, she and Hoffmann were already there. Newman began, saying she was sorry it had come to this: the secretary wanted my resignation by next Monday or at the latest Tuesday—giving me four or five days to resign.

I mentioned I was scheduled to leave town the next hour, and would not be back until late Saturday. That made it a little tight. Newman explained Heyman was in a hurry. He wanted to have a clear announcement well before the next regents' meeting on May 8, and before the mid-May Senate hearings right after that.

My contract with the Institution entitled me to stay on as a senior researcher if ever I chose to step down from the museum's directorship. But while I was now offered this option, I saw no purpose in staying and chose instead to sever my relations and resign from the Smithsonian. I had enjoyed eight hectic but happy years working with the museum's outstanding staff. The Institution's new secretary had an outlook entirely alien to me. It was time to leave.

RESIGNATION

By the afternoon of Monday, May 1, the Castle and I had agreed that I would officially resign the next day. At the museum, I had kept only my deputy, Gwen Crider, fully informed, because the weight of running the museum would be on her shoulders. Three others on the staff had to know now in order to effect a sensible transition on my departure. At 3:30 P.M. on that Monday afternoon, Gwen distributed an announcement for a town meeting of the entire staff for the next morning at 9:00 A.M. She gave no reason. I wanted to tell the staff myself, before they read about it in the newspapers.

The next morning I arrived at the office, turned in my badge, parking sticker, library card, and keys, and then went down to the planetarium.

It was already packed. People must have guessed. I walked up to the podium and said,

Well, The Last Act!

Some months ago, on a visit to Garber, someone asked me whether I intended to resign over the issue of the *Enola Gay*. I said then that it would depend on whether I felt I could continue to provide effective leadership to the Museum.

I have concluded now that I cannot. I have therefore written the following letter to Secretary Heyman:

May 2, 1995

Dear Mike,

I am writing to let you know that I am today stepping down from the Directorship of the National Air and Space Museum.

I do so with deep regret. During my eight years at the Museum, I have gained a profound respect for the men and women on the staff. Their high professional standards, dedication to excellence, and devotion to public service have made this the most visited museum in the world. In the past five years, the attendance at Air and Space rose, while all other Smithsonian museums on the National Mall saw a decline.

During my time at the Museum we acquired a number of truly significant donations: the Air Force SR-71 reconnaissance aircraft, which on delivery broke the transcontinental speed record, by flying coast to coast in 68 minutes, and the recently declassified Corona camera for high-precision photography from Earth orbit. Both played an important role during much of the Cold War. To these we added a Soviet SS-20 and a U.S. Army Pershing 2 rocket, harbingers of the dramatic reduction in nuclear armaments set in motion by the Intermediate-Range Nuclear Forces (INF) treaty; a Vietnam era Navy A-6 attack aircraft; and a rare World War II Shturmovik Il-2, the Soviet ground-attack aircraft, which helped turn the tide on the Eastern front.

The Museum has also attracted the highest caliber scientific and historical researchers, archivists, art historians, exhibition specialists and educators, to bring greater depth and breadth to the production of exhibitions and films and the provision of archival services. A newly created development department has substantially augmented Federal funding with private contributions in support of new galleries, films, and collections care. Working together, as teams, this staff has produced a series of informative galleries, including the futuristic "Where Next Columbus?"; the World War I gallery, "Legend, Memory and the Great War in the Air"; "Star Trek," the most popular gallery ever produced at the Smithsonian; and "Beyond the Limits," the story of computers and flight. To these, they have added two recent, wide-screen IMAX films, "Blue Planet," about Earth seen from space; and "Destiny in Space," a vision of the future of space exploration.

The Museum's craftsmen have restored for display a wealth of aerospace treasures, including a World War I French Voisin bomber; the first operational jet aircraft, the German World War II Arado reconnaissance/bomber; the Hubble Space Telescope Structural Dynamic Test Vehicle; and

> the World War II U.S. Army Air Force B-29 bomber *Enola Gay*—the largest restoration project ever undertaken by the Museum—which is to go on display in a few weeks.
>
> Initial plans for a historical display of this aircraft have persistently provoked controversy and divisiveness. Three months after the cancellation of that planned exhibition, the controversy still continues. I believe that nothing less than my stepping down from the directorship will satisfy the Museum's critics and allow the Museum to move forward with important new projects, such as the Extension to be built at Washington's Dulles International Airport to provide better care for the collections.
>
> There is no choice but to resign: The Museum's welfare and future are too important.
>
> Sincerely yours,
>
> Martin Harwit
> Director [18]

Turning from the letter of resignation, I thanked the staff for all the work they had done these years, and for all it had meant to the museum, the Institution, and the public. Then I stepped down from the rostrum and walked out.

I was no longer their director.

SENATE HEARINGS

While the Smithsonian had successfully avoided hearings in the House, hearings in the Senate still loomed.

Senator Ted Stevens, chairman of the Senate Committee on Rules and Administration, had announced the hearings for May 11 and May 18, 1995. I wanted to have an opportunity to answer the charges that I knew would be made—and that did get made—about my handling of the *Enola Gay* exhibition. But when I called Christine Ciccone, the staff person in charge, she told me I could not testify in person, because neither the veterans' organizations nor the Smithsonian had included me on their list of witnesses. I was entitled, however, to submit testimony in writing, which I did.[19]

Perhaps the most discouraging aspect of the hearings was Senator Stevens's wish to have Title 20 §80 of the United States Code apply to the National Air and Space Museum.

> I am still disturbed . . . that [the Smithsonian believes] the law applies to a museum that was never funded when it specifically says the Smithsonian Institution shall take action . . . . I am going to ask the Congress to modify the statute to make clear our intent because I believe it might help settle this controversy if we did have a fulfillment of the original instruction of the Congress and the President to the Smithsonian to take the action required to

> commemorate the service of the men and women of the armed forces, and to portray them as an inspiration to present and future generations....

Just like the hundreds of veterans who had written the museum telling us that the bomb saved their lives and that they wanted their grandchildren to be proud of what they had done for their country, Senator Stevens felt uncomfortable:

> I have a Japanese daughter-in-law and I have a grandson. I know that in her country they are teaching that in the history of World War II, we were, in fact, the aggressors.... I do not want my grandson to walk out of that museum and ask me why I was the one who was the aggressor, and why did I try to kill Japanese babies.[20]

If Senator Stevens has his way, he may negate the Smithsonian Institution legal counsel's long-standing interpretation of the law, namely that the functions of any individual Smithsonian museum specified by its enabling legislation have "always been understood to be demonstrative, rather than restrictive, since they cannot diminish the basic Smithsonian trust responsibility for the increase and diffusion of knowledge."[21] In other words, congressional legislation authorizing a new bureau cannot restrain that bureau or others from dedicating themselves to "the increase and diffusion of knowledge," demanded by the legislation that originally established the Smithsonian Institution.

But that is precisely what tampering with the National Air and Space Museum's enabling legislation along the lines proposed by Senator Stevens and people like Congressman Sam Johnson would do. If they succeed, they will unrecognizably alter the Institution. The Smithsonian would no longer be an establishment dedicated to knowledge, but the government's organ for disseminating propaganda.

James Smithson would shudder!

Epilogue

In all the years that we at the National Air and Space Museum were preparing *The Last Act,* I felt that the museum would prevail in mounting a powerful exhibition. We had worked with friends in the military and in academic institutions for many years. We had consulted with the secretary and the Institution's board of regents. We all knew that this would be a difficult project. But I was confident that by listening carefully to colleagues and critics and reasoning with them, we could surmount our difficulties and steadily progress toward a successful opening.

This did not happen.

I asked myself why?

Then, in June 1995, only weeks after I had left the Smithsonian, the Air Force Association published its voluminous files of correspondence. This sudden, unexpected opportunity to see what had gone on behind the scenes, gave me the impetus to reconstruct the events of this controversy and record them for others.

In writing this book, I have tried to faithfully present each side's views by letting proponents speak in their own words, through their letters, memoranda and articles. What emerged was a portrayal of strongly dedicated individuals who feared that the exhibition could cast into doubt a hallowed, patriotic story that they believed was essential to our national self-image. It was the story of America's sacrifices and ultimate victory in World War II—the nation's fight for freedom and democracy for all the peoples of the world.

The Air Force Association and the American Legion were determined that newly available, declassified information must not be permitted to change the way this story had always been told. It didn't matter that some of this new material came from the diaries of President Truman, Secretary of War Stimson, or Chief of Staff Fleet Admiral Leahy. There could be no changes. The traditional story would have to be told the way it always had been told for the past fifty years.

An essential feature of the veterans' story was that the bombings of Hiroshima and Nagasaki, while regrettable, were a small price to pay for the millions of lives that the atomic bombs had saved, Japanese as well as American. When I maintained that *The Last Act* could not present this conjecture as fact, because we had no way of knowing how the war might have ended without atomic bombs, American Legion National Commander Bill Detweiler wrote President Clinton that the exhibition needed to be shut down. He could not understand why the veterans' oft-repeated claims should now be challenged—why a story they had always known to be true should be cast into doubt. But the museum had a responsibility to check its facts before presenting them to the public, and recognize different points of view where differences of opinion prevailed.

To have the exhibition changed to suit their aims, the AFA pursued a strategy of denigrating the planned exhibition, distorting its thrust and intent, and fanning the fears of aging veterans by telling them that *The Last Act* would dishonor their wartime service to the nation. Coupled to this was a deliberate campaign to gain the support of the media and Congress. To capture the press, the AFA "fed" reporters and editorial writers sound bites they could use, among them, "It was a war of vengeance. We [are] portrayed as the bad guys." . . ."We are worried that American youth will get a distorted view of what America did in the War."

The media took over this message. *The Washington Post* criticized the exhibition as narrow-minded and representative of a special-interest and revisionist point of view, and columnist George Will, referring to his reading of the *Post,* characterized the Smithsonian Institution as "besotted with . . . cranky anti-Americanism."

In the Senate, the AFA was able to gain the passage of a resolution admonishing the Smithsonian to produce a properly patriotic exhibition. In the House, they persuaded Republican Congressman Sam Johnson of Texas to spearhead an attack that called the exhibition historically inaccurate and not in line with the thinking of most Americans. Johnson had the backing of Speaker of the House Newt Gingrich who saw the fight over the exhibition as an assertion by most Americans that they were "sick and tired of being told by some cultural elite that they ought to be ashamed of their country."

This strategy worked. The onslaught by the veterans, the media and the Congress eventually took its toll. A Smithsonian secretary, barely four months in office, released a series of statements in which he concluded that the museum had made a basic error in attempting to couple a historical treatment of the use of atomic weapons with the 50th anniversary commemoration of the end of the war; that veterans and their families were not looking for analysis in this commemorative year; that we needed to distinguish between opinion and fact; and that we needed to contribute to light rather than heat.

How could the Smithsonian abandon such an important exhibition only four months before its opening? How could it declare, after seven years of careful planning, that *The Last Act* was so fundamentally flawed?

A confluence of two factors had come into play.

In selecting a new secretary in 1994, the Smithsonian regents had made clear that financial stability was now the Institution's highest priority. The secretary's mandate was to raise contributions from private sources and persuade Congress to provide adequate federal funding.

With money the highest priority of the Institution, academic integrity began to take second place. That Secretary Heyman was not fully behind the exhibition was apparent from statements he issued on his inaugural day in office. As he continued to make such statements to the press and television, the veterans associations took notice and I became increasingly alarmed.

The elections of November 1994, brought a conservative Republican leadership to power in both the House and Senate, and the Legion and AFA gained renewed influence. The combination of Congressional support and a wavering Smithsonian secretary were decisive. Where the Legion and the AFA had previously been willing to sit down with the museum and work out a reasonable exhibition, they now saw an opportunity to dictate better terms. They came in with their demands, and found the secretary ready to abandon *The Last Act* and substitute an exhibition of his own.

Once it was canceled, those who had opposed *The Last Act* claimed that they had rescued from possible harm an important aspect of our national self-image.

I do not believe this.

Heroic stories, no doubt, are important for national morale, but clear thought and sound information will more surely sustain us. James Smithson understood this, when he sent to America his bequest "for the increase and diffusion of knowledge." Those who diffuse such new knowledge will always find opponents who accuse them of having an ideological, narrow-minded, special interest—of dispensing opinion rather than fact. But a hundred and fifty years after the Institution's founding, Smithson would have deplored the secretary's decision to suspend an exhibition because it was "analytic" in a 50th anniversary commemorative year. Analysis is, after all, basic to the Institution's mandate to increase knowledge. It cannot be turned on an off for special occasions.

In one way or another, each of *The Last Act*'s opponents was attacking the basic charter of the Smithsonian Institution. Once a Congressman Johnson tells the National Air and Space Museum that it has no business teaching history, or orders one of its exhibitions to be shut down, or bans the publication of this exhibition's catalogue, it becomes difficult to see where his concern for patriotism and national self-image will stop. It becomes a dangerous game.

Do I have regrets?

I am dismayed that now, nobody will ever be able to see an exhibition that I am convinced every visitor to the National Air and Space Museum, especially American veterans of World War II, would have found thought-provoking and inspiring. Despite months of controversy, the public remains unaware of what this exhibition would have shown

I was prepared to accept criticism once the exhibit opened. But with *The Last Act* closed down without ever opening, and with its catalogue suppressed, every trace of the Museum's work has been wiped out. Only the caricature disseminated by the media remains; its distortions now serve those who seek to justify the closing of the exhibition. I believe our opponents wanted to erase what we had really done so that nobody could challenge their decision to deny the public the opportunity to judge the exhibition and its catalogue for themselves.

The nation is the poorer for what they did.

I still stand behind the final plans of the exhibition as they stood on January 30, 1995, the day the Smithsonian Board of Regents decided to cancel *The Last Act*. Though the new Smithsonian secretary and a newly reconstituted board of regents did not, I am convinced that James Smithson would have stood by me to see the exhibition open as planned. The museum was, after all, doing what he had founded his Institution to do.

Some day, some year, perhaps *The Last Act* exhibition will be reassembled. I shall be ready to defend it when it is.

Chronology of Significant Events

August 6, 1945	Colonel Paul Tibbets, commander of the 509th Composite Group, pilots the *Enola Gay* on her mission to Hiroshima to drop the atomic bomb.
July 3, 1949	The Air Force transfers ownership of the *Enola Gay* to the Smithsonian.
December 2, 1953	The *Enola Gay* is flown to Andrews Air Force Base, Maryland.
August 10, 1960	The staff of the National Air Museum begins disassembly of the *Enola Gay* at Andrews Air Force Base.
July 21, 1961	Disassembled components of the *Enola Gay* are moved to the National Air Museum storage facility in Suitland, Maryland.
Summer 1984	509th veterans Donald C. Rehl and Frank Stewart launch a campaign to restore the *Enola Gay* and found the *Enola Gay* Restoration Association, EGRA.
Early 1985	The museum begins to restore the forward fuselage of the *Enola Gay.*

1986	WWII veteran William A. Rooney begins a letter writing campaign for the "proud display of the *Enola Gay*."
August 17, 1987	Martin Harwit becomes Director of the National Air and Space Museum.
October 26, 1987	The museum's Research Advisory Committee meets and discusses the potential exhibition of the *Enola Gay*.
April 8, 1988	WWII veteran Benjamin A. Nicks begins a campaign to have the *Enola Gay* restored and displayed.
June 10, 1988	General Paul Tibbets and Frank Stewart visit the museum to discuss ways of accelerating the restoration of the *Enola Gay*.
Summer 1988	WWII veteran Lucius Smith III offers to help the museum restore the *Enola Gay* by launching a fund raising campaign.
Fall 1989	The museum launches a sixteen-month-long series of talks, panel discussions, and films on "Strategic Bombing in WWII."
Fall 1990	The museum opens a display outside the World War II gallery of a short video-film on the restoration of the *Enola Gay* and her mission.
February 4, 1991	Lead curator Michael Neufeld prepares "A Proposal to Exhibit the *Enola Gay* on the Mall."
September 5, 1991	A debate on the museum extension puts planning for the exhibition on hold.
July 23, 1992	WWII veteran W. Burr Bennett, Jr. starts a letter writing campaign for the "proud display of the *Enola Gay*."
August 5, 1992	The museum establishes contact with the Japan Foundation on the exhibition of the *Enola Gay*.
December 21, 1992	Planning for the exhibition of the *Enola Gay* officially resumes.
March 29, 1993	The museum staff meets with General C. M. Kicklighter, Executive Director of 50th Anniversary of World War II Commemoration Committee and his staff.
April, 1, 1993	Harwit and Aeronautics Department chair, Tom Crouch, travel to Japan.
May 4, 1993	The museum decides to seek no external funding for the exhibition.
June 10, 1993	The museum submits its exhibition proposal "50 Years On" to Smithsonian Secretary Adams.

July 12–22, 1993	Adams, Harwit, and Crouch debate the thrust of the 16-page exhibition planning document.
August 1993	Bennett, Rooney and Rehl amass more than 5,000 petitions from veterans for "the proud display"of the *Enola Gay.*
November 19, 1993	General Monroe Hatch (USAF Ret.) and John Correll of the Air Force Association (AFA) visit the museum to discuss the proposed exhibition.
January 14, 1994	The museum completes a first draft script for the *Enola Gay* exhibition.
February 7, 1994	The exhibition's external advisory committee meets to discuss the script.
March 1994	The AFA widely distributes copies of the script despite the museum's explicit request on transmittal to not "circulate the material at this time, since it is not yet in suitable form."
April 1, 1994	The AFA publishes "War Stories at Air and Space" in *Air Force* Magazine.
April 13, 1994	The exhibition team members meet with historians of the military services.
April 16, 1994	Harwit writes to curators about questions raised on the script's balance.
April 26, 1994	Harwit appoints an independent "Tiger Team" to review the script for signs of imbalance.
May 25, 1994	The "Tiger Team" issues a report recommending numerous changes.
May 26, 1994	General Tibbets and members of the 509th are filmed at the museum's Garber facility and are invited to work with the museum on the exhibition.
May 31, 1994	Adams and Harwit agree on the exhibition's title, "The Last Act: The Atomic Bomb and the End of World War II."
June 9, 1994	General Tibbets calls the proposed exhibition "a package of insults."
July 13, 1994	The museum staff briefs representatives of veterans' organizations on the exhibition.
August 10, 1994	Under Secretary Constance Newman and Harwit meet with members of the U.S. House of Representatives on the proposed exhibition.
	24 members of Congress send Secretary Adams a letter expressing their "concern and dismay."

September 3, 1994	Newman and Harwit address the national convention of the American Legion in Minneapolis, Minnesota.
September 8, 1994	Secretary Adams, Secretary-elect Heyman, Newman, and Harwit meet with the editorial board of the *Washington Post.*
September 19, 1994	I. Michael Heyman is installed as Secretary of the Smithsonian Institution.
September 21, 1994	Newman, Harwit, and members of the exhibition team meet with representatives of the American Legion to review the script. Two additional meetings are held during the next four weeks.
September 22, 1994	The Smithsonian Institution and the American Legion call a press conference to announce a "relationship" between the organizations to review the script and all related materials. The U.S. Senate passes a non-binding resolution expressing concern about the exhibition of the *Enola Gay*.
October 3, 1994	An "interim" draft of the exhibition script is prepared and distributed only to representatives of select veterans organizations.
October 7, 1994	Heyman reads the "interim" draft and suggests minor changes to two or three labels.
October 13, 1994	The Executive Board of the Organization of American Historians passes a resolution asking for the protection of the Smithsonian from political interference.
November 17, 1994	Delegation of writers and historians, organized by the Fellowship of Reconciliation, meets with museum staff.
November 22, 1994	The forward fuselage of the *Enola Gay* is moved from Suitland, Maryland, to the National Air and Space Museum.
December 6, 1994	The first full draft of "The War in the Pacific," an introductory photo-display for the exhibition, is issued.
December 13, 1994	Seven members of Congress express "deep disappointment with how the Smithsonian Institution has handled the *Enola Gay* exhibit."
January 4, 1995	American Legion National Commander William Detweiler sends his Advisory Committee a detailed plan for cancelling the exhibition and calling for congressional hearings.

January 9, 1995	Harwit sends a letter to Hubert Dagley of the American Legion suggesting changes to the exhibition's "invasion casualties" label.
January 18, 1995	The American Legion calls for the exhibition's cancellation.
January 24, 1995	81 members of the House call for Harwit's resignation and for congressional hearings on the exhibition.
January 26, 1995	The President of the Society for Military History writes to Chief Justice Rehnquist, Chancellor of the Board of Regents, asking the regents not to cancel the exhibition.
January 27, 1995	The President of the Organization of American Historians and two immediate past presidents also write to the Chief Justice asking the regents not to cancel the exhibition.
January 30, 1995	Secretary Heyman announces that "The Last Act: The Atomic Bomb and the End of World War II" has been cancelled and will be replaced by a small display devoid of historical content.
April 19, 1995	The Ann Arbor Forum on "Presenting History: Museums in a Democratic Society" takes place.
May 2, 1995	Harwit resigns from the Smithsonian Institution.
May 11, 18, 1995	The U.S. Senate holds hearings on the exhibition of the *Enola Gay.*

Notes

Preface

1. I. Michael Heyman, press release of statement made at his press conference following the regents' meeting, Monday, January 30, 1995. *NASM/MH*. Also reported in portions in national newspapers the following day.
2. R.S. §5579 derived from the Act of Congress of August 10, 1846, ch. 178, §1, 9 Stat. 102.
3. *Title 20, U.S. Code*. Chapter 3, Subchapter VII §77a.
4. I. Michael Heyman to Congressman Sam Johnson, April 4, 1995. Hearings before the Committee on Rules and Administration, United States Senate, 104th Congress, first session, in "The Smithsonian Institution, Management Guidelines for the Future," May 11 and 18, 1995, Appendix IV, 170–171.
5. I. Michael Heyman, letter to George A. Mannes, January 17, 1995, *NASM/MH*.
6. Roy K. Flint, letter to Chief Justice William H. Rehnquist, January 26, 1995, *NASM/MH*.
7. Eric Foner, Gary B. Nash, and Michael Kammen, letter to Chief Justice William H. Rehnquist, January 27, 1995, *NASM/MH*.

Chapter 1

1. I am indebted to Frank B. Stewart and Ken Eidnes for providing me this information in telephone conversations on December 21, 1995.
2. Paul W. Tibbets, *Flight of the Enola Gay* (Reynoldsburg, Ohio: Buckeye Aviation, 1989), 155 ff, 187.
3. Ibid. p. 170.
4. Document dated 8-20-90, *Carmen Turner File tab. 10,* 1991, *NASM/MH*.
5. Tibbets, *Flight of the Enola Gay,* 198.
6. Ibid., 200.
7. Ibid., 208, 218.
8. Ibid., 206; *Carmen Turner File, tab. 10*, 1991, *NASM/MH*.

9. Patricia Woodside, *Enola Gay: In Their Own Words* (19 June 1995), videotape.
10. Ibid.
11. Tibbets, *Flight of the Enola Gay,* 201.
12. Woodside, *Enola Gay.*
13. Frank Morring, Jr., Scripps Howard News Service, "Smithsonian restoring 'Enola Gay,' " Franklin, Indiana, *Daily Journal*, Aug. 6, 1985. *(See Carmen Turner File tab 3, NASM/MH.)*
14. Woodside, *Enola Gay.*

Chapter 2

1. Defense Nuclear Agency Report DNA 6037F, "Operation Redwing: 1956," 1 August 1982, 38.
2. Paul W. Tibbets, *Flight of the Enola Gay* (Reynoldsburg, Ohio: Buckeye Aviation, 1989), 245, 252.
3. Robert Standish Norris and Thomas B. Cochran, "United States Nuclear Tests, July 1945 to 31 December 1992," NWD 94-1, Natural Resources Defense Council, Washington, D.C., February 1, 1994.
4. Mike Wallace, "The Battle of the *Enola Gay,*" *Museum News* 74, no. 4 (July/August 1995): 45.
5. Aug. 12, 1946, ch. 955, §1, 60 Stat. 997.
6. Wallace, "Battle of the *Enola Gay*": 45.
7. Fred D. Cavinder, "2 Hoosier veterans aim to see *Enola Gay* restored, displayed," *Indianapolis Star*, Aug. 4, 1985. Also, Associated Press release N105, August 5, 1985.
8. Ibid.
9. Cavinder, "2 Hoosier veterans."
10. Kevin Cook, "Here lie pieces of history—named *Enola Gay,*" *Washington Times*, August 6, 1985.
11. Fred D. Cavinder, ibid.
12. Ibid.
13. Stewart, conversation, 1995.
14. Cook, "Here lie pieces of history."
15. Robert McC. Adams, letter to William A. Rooney, June 13, 1986. *Carmen Turner file tab 5, NASM/MH.*
16. William A. Rooney, letter to Robert McC. Adams, January 22, 1987. *NASM/MH.*

Chapter 3

1. Committee on House Administration, Subcommittee on Libraries and Memorials, United States House of Representatives, 91st

Congress, 2nd Session, "General Hearings: Smithsonian Institution—General Background—Policies, Purposes, and Goals from 1846 to Present" (July 21, 1970), 173, 185. See also, Mike Wallace, *Mickey Mouse History and Other Essays on American Memory* (Temple University Press, 1996), 287–8.

2. R.S. §5579 derived from the Act of Congress of August 10, 1846, ch. 178, §1, 9 Stat. 102.

Chapter 4

1. Wendy Stephens to Martin Harwit, private conversation, around 1992.
2. Martin Harwit, handwritten note dated July 7, 1987, in the author's possession.
3. David C. Beeder, of the Omaha World-Herald Bureau, "42nd Anniversary of Hiroshima Blast—Nebraska-Made *Enola Gay* Draws Few Visitors," *Evening World-Herald* (Omaha, Nebraska), August 6, 1987.
4. Handbill dated August 6, 1987, *NASM/MH.*
5. Martin Harwit, letter to Bill Winter, December 2, 1987, *NASM/MH.*
6. Peter Powers, letter to Martin Harwit, with a handwritten note, dated November 18, 1987, *NASM/MH.*
7. Arthur Veik, letter to Rep. Virginia Smith, August 25, 1987, *NASM/MH.*
8. Margaret Gaynor draft of a letter to Honorable Virginia Smith, hand-dated October 6, 1987, *NASM/MH.*
9. National Air and Space Museum, Smithsonian Institution, Minutes of the Research Advisory Committee Meeting, October 28, 1987, *Carmen Turner File, tab 2, NASM/MH.*
10. William D. Leahy, *I Was There* (New York: Whittlesey House, 1950), 441.
11. Dwight D. Eisenhower, *Crusade in Europe* (Garden City, New York: Doubleday, 1948), 443.

Chapter 5

1. Von Hardesty, through Don Lopez, to Martin Harwit, Memorandum on "Strategic bombing exhibit and the proposed display of the B-29 *Enola Gay,*" November 18, 1987, *NASM/MH.*
2. Ibid.
3. Lin Ezell to Martin Harwit and six other members of the museum staff, Memorandum on "5-year Restoration Schedule," December 7, 1987, *NASM/MH.*

4. Lin Ezell, through Don Lopez, to Martin Harwit, Memorandum on "Exhibiting the *Enola Gay*," December 11, 1987, *NASM/MH*.
5. Ronald Wagaman to Nadya Makovenyi, Memorandum on "Temporary Structures," December 30, 1987, *NASM/MH*.
6. Adrian J. Ayotte, Wasco Products, Inc., Sanford, Maine, letter to Victor Govier, January 22, 1988, *NASM/MH*.
7. Martin Harwit, letter to General Charles MacDonald [*sic*], March 14, 1988, *NASM/MH*.
8. Charles C. McDonald, Lt. Gen., USAF, letter to Martin Harwit, March 31, 1988, *NASM/MH*.
9. Von Hardesty to Martin Harwit, Memorandum on "Exhibition of the *Enola Gay*," December 28, 1987, *NASM/MH*.

Chapter 6

1. Fred D. Cavinder, "2 Hoosier veterans aim to see *Enola Gay* restored, displayed," *Indianapolis Star*, August 4, 1985, *NASM/MH*.
2. Ibid.
3. Paul Filipowski, letter to Congressman Buddy McKay, January 22, 1988.
4. Robert McC. Adams, letter to Honorable Buddy McKay, March 14, 1988, *NASM/MH*.
5. Paul Filipowski, letter to Mr. Adams, March 26, 1988, *Carmen Turner File, tab 5, NASM/MH*.
6. Martin Harwit, letter to Paul Filipowski, April 27, 1988, *Carmen Turner File, tab 5, NASM/MH*.
7. Paul Filipowski, letter to Martin Harwit, May 3, 1988, *NASM/MH*.
8. Frank B. Stewart, letter to Martin Harwit, May 15, 1988, *NASM/MH*.
9. Elbert Watson, letter to Robert McC. Adams, March 14, 1988, *Carmen Turner File, tab 5, NASM/MH*.
10. Frank B. Stewart, letter to Martin Harwit, May 15, 1988, *NASM/MH*.
11. Martin Harwit, letter to Frank B. Stewart, May 20, 1988, *NASM/MH*.
12. Frank B. Stewart, letter to Martin Harwit, June 2, 1988, *NASM/MH*.
13. Frank B. Stewart and Rolland S. Nail, letter to Martin Harwit, June 20, 1988, *NASM/MH*.
14. Martin Harwit, letter to Frank Stewart, July 7, 1988, *NASM/MH*.
15. Frank B. Stewart, letter to Martin Harwit, August 18, 1988, *NASM/MH*.

16. Linda Neuman Ezell, letter to Frank Stewart, September 1, 1988, *NASM/MH.*
17. Frank Stewart, *"Enola Gay,"* letters section, *American Legion,* October 1988, 4.

Chapter 7

1. Von Hardesty to Martin Harwit, Memorandum on "Exhibition of the *Enola Gay,*" December 28, 1987, *NASM/MH.*
2. Von Hardesty, through Don Lopez, to Martin Harwit, Memorandum on "Strategic bombing exhibit and the proposed display of the B-29 *Enola Gay,*" November 18, 1987, *NASM/MH.*
3. Steven Soter, Memorandum to Von Hardesty, Martin Harwit, and Dom Pisano on "Strategic Bombing Exhibit," Draft, dated April 18, 1988, *NASM/MH.*
4. Michal McMahon, "The romance of technological progress: a critical review of the National Air and Space Museum, *Technology and Culture* 22 (1981): 281- 296.
5. Steven Soter, Draft Memorandum on "Strategic Bombing Exhibit," April 18, 1988, *NASM/MH.*
6. Morihisa Takagi, letter to Martin Harwit, November 28, 1988, *NASM/MH.*
7. Robert McCormick Adams, "Smithsonian Horizons," *Smithsonian* 19 (July 1988): 4.
8. Martin Harwit to Robert McC. Adams, Memorandum on "*Enola Gay* Column," June 10, 1988, *NASM/MH.*
9. Martin Harwit, "*Enola Gay,*" Viewport column, *Air & Space Smithsonian* 3 (August/September 1988): 4.
10. National Air and Space Museum Research Advisory Committee Report, October 24–26, 1988, *Carmen Turner File, tab 2, NASM/MH.*
11. Martin Harwit logbook, October 18, 1988, *MH.*
12. Martin Harwit, letter to Ruth Adams, November 1, 1988, *Carmen Turner File, tab 4, NASM/MH.*
13. James M. Furman, letter to Martin O. Harwit, March 22, 1989, *Carmen Turner File, tab 4, NASM/MH.*
14. Daniel S. Greenberg, "Smithsonian space museum exhibits add truth to labeling," *Houston Post*, December 4, 1990.
15. "The Talk of the Town—Notes and Comment," *New Yorker*, August 13, 1990, 23.
16. Martin Harwit, "Smart Versus Nuclear Bombs," Viewport column, *Air & Space Smithsonian* (June/July 1991): 4.
17. Hank Burchard, "Plane Truths During WWII," *The Washington Post,* November 22, 1991.

18. Michael Kilian, "Grounded in reality," *Chicago Tribune,* November 26, 1991.

CHAPTER 8.

1. Ben Nicks, letter to Robert C. Mikesh, April 8, 1988, *NASM/MH.*
2. Robert C. Mikesh, letter to Ben Nicks, June 24, 1988, *NASM/MH.*
3. Ben Nicks, letter to Jay P. Spenser, September 26, 1988, *NASM/MH.*
4. Linda Neuman Ezell, letter to Ben Nicks, October 6, 1988, *NASM/MH.*
5. Martin Harwit to Bob Adams, Memorandum on "*Enola Gay* letter from Mr. Nicks," October 5, 1988, returned with a handwritten note by Bob, and dated the next day, *NASM/MH.*
6. Martin Harwit, letter to Ben Nicks, October 14, 1988, *NASM/MH.*
7. Benjamin A. Nicks, "The *Enola Gay*: Report to the Ninth Bomb Group Association and Resolution Calling for Its Speedy Restoration and Display," October 13–16, 1988, *NASM/MH.*
8. Ben Nicks, letter to Martin Harwit, November 3, 1988, *NASM/MH.*
9. Ben Nicks, "*Enola Gay* Restoration Committee Report—Nov. 5, 1988," *NASM/MH.*
10. Ben Nicks, letter to Martin Harwit, December 3, 1988, *NASM/MH.*
11. Linda Neuman Ezell, letter to Ben Nicks, December 9, 1988, *NASM/MH.*

CHAPTER 9

1. Frank H. Murkowski, letter to Lucius Smith III, June 28, 1988, *MH;* see also a letter from Lucius Smith III to Elbert Watson, August 15, 1988, *NASM/MH.*
2. "*Enola Gay* crewman's wife wins radiation-death claim," Associated Press, Sunday, August 6, 1988, *NASM/MH.*
3. Steven Soter, letter to Lucius Smith III, August 19, 1988, *NASM/MH.*
4. Robert S. Faron, letter to Elbert L. Watson, August 26, 1988, *NASM/MH.*
5. Lucius Smith III to Steven Soter, August 30, 1988, *NASM/MH.*
6. Martin Harwit to George Robinson, Memorandum on "*Enola Gay,*" September 2, 1988, *NASM/MH.*
7. Lucius Smith III, letter to Martin Harwit, September 20, 1988, *NASM/MH.*

8. Lucius Smith III, "Memorandum of intent to support and help defray cost of restoring the '*Enola Gay*,' " October 1, 1988, *NASM/MH*.
9. Lucius Smith III, letter to Martin Harwit, November 21, 1988, *NASM/MH*.
10. Martin Harwit, letter to Lucius Smith III, November 22, 1988, *NASM/MH*.
11. Lucius Smith III, letter to Martin Harwit, December 13, 1988, *NASM/MH*.
12. Lucius Smith III, letter to Martin Harwit, (date of writing obliterated), date received, January 10, 1989, *NASM/MH*.
13. Martin Harwit, letter to Lucius Smith III, February 6, 1989, *NASM/MH*.
14. "Memorandum of Understanding between Smithsonian Institution, National Air and Space Museum and Atomic Library and Technology Foundation," signed by Martin Harwit, February 22, 1988; and cover memorandum from George S. Robinson to Martin O. Harwit on "*Enola Gay* Fund Raising Agreement with ALTF," March 1, 1989, and a further Memorandum, "Your questions re: ALTF/NASM Agreement for *Enola Gay* restoration," February 22, 1989, *NASM/MH*.
15. Martin Harwit, letter to Lucius Smith III, April 21, 1989, *NASM/MH*.
16. Lucius Smith III, letter to Linda Neuman Ezell, June 19, 1989, *NASM/MH*.
17. Linda Neuman Ezell, letter to Lucius Smith III, July 3, 1989, *NASM/MH*.
18. Andrew H. Anderson, letter to Luke (Smith), July 15, 1989, *NASM/MH*.
19. Lucius Smith III, letter to Martin Harwit, July 21, 1989, *NASM/MH*.
20. Lucius Smith III, letter to Martin Harwit, September 16, 1989, *NASM/MH*.
21. Lucius Smith, letter to Linda Neuman Ezell, August 25, 1989, *NASM/MH*.
22. Terrence R. Ward, letter to Martin Harwit, September 27, 1989, *NASM/MH*.
23. Lucius Smith III, letter to Martin Harwit, November 1, 1989, *NASM/MH*.
24. *Atomic Veteran's Newsletter*, ed. Dr. Oscar Rosen, Vol. 12, No. 3 (fall 1990).
25. Bill Dalton, staff writer, "Atomic Museum 'up in smoke'?" Kansas City *Star*, March 17, 1991.
26. Ben Nicks, letter to Dr. Oscar Rosen, March 17, 1991, *NASM/MH*.

27. Linda Neuman Ezell, letter to Ben Nicks, March 21, 1991, *NASM/MH*.
28. Ben Nicks, "*Enola Gay* Status Report to 9th Bomb Group Association," May 11, 1991, *NASM/MH*.
29. Lin Ezell to George Robinson, Memorandum on "*Enola Gay* Memorandum of Understanding," April 1, 1991, *NASM/MH*.
30. F. Mark Miller, letter to Martin O. Harwit, June 12, 1991, *NASM/MH*.
31. Martin Harwit, letter to F. Mark Miller, August 5, 1991, *NASM/MH*.
32. Lucius Smith III, letter to Martin Harwit, May 3, 1995, in author's possession.
33. Dr. Oscar Rosen, telephone conversation with Martin Harwit, March 9, 1996.
34. Mayor Rondell F. Stewart, telephone conversation with Martin Harwit, February 13, 1996.

Chapter 10

1. Richard Horigan, through Lin Ezell and Don Lopez, to Martin Harwit, Memorandum on "*Enola Gay* Restoration Update," June 2, 1988, *NASM/MH*.
2. Lin Ezell, through Don Lopez, to Martin Harwit, Memorandum on "*Enola Gay* Restoration at NASM," May 2, 1988; Lin Ezell and Steven Soter, Memorandum on "*Enola Gay* Restoration Display," to Martin Harwit and eleven other members of the NASM staff, May 16, 1988, *NASM/MH*.
3. Lin Ezell to Von Hardesty and Bob Mikesh, Memorandum on "*Enola Gay* Restoration Project," June 15, 1988, *NASM/MH*.
4. Robert C. Mikesh, through Von Hardesty, to Lin Ezell, Memorandum on "*Enola Gay* Restoration Project," June 29, 1988, *NASM/MH*.
5. Richard Horigan to Lin Ezell, Memorandum on "*Enola Gay* Restoration," July 8, 1988, *NASM/MH*.
6. Richard Horigan, through Lin Ezell, to Martin Harwit, Memorandum on "*Enola Gay* Restoration Update," August 4, 1989, *NASM/MH*.
7. Linda Neuman Ezell, letter to David C. Knowlen, July 24, 1991, *NASM/MH*.
8. Martin Harwit, letter to David C. Knowlen, July 26, 1991, *NASM/MH*.
9. Martin Harwit, letter to Walter Roderick, August 23, 1991, *NASM/MH*.

10. Richard Horigan, through Lin Ezell, to Martin Harwit, Memorandum on "*Enola Gay* Restoration Update," October 10, 1991, *NASM/MH.*
11. Lin Ezell, informal note to Martin Harwit and Wendy Stephens on "*Enola Gay* Facts and Figures," December 11, 1991, *NASM/MH.*
12. William Reese, through Lin Ezell, to Martin Harwit, Memorandum on "*Enola Gay* Restoration Update," September 1, 1993, *NASM/MH.*
13. E.T. Wooldridge, letter to John Cornet, May 20, 1986, *Carmen Turner File, tab 7, NASM/MH.*
14. Don Ofte, letter to Walter J. Boyne, June 26, 1986, *Carmen Turner File, tab 7, NASM/MH.*
15. Richard E. Malenfant, letter to Robert Mikesh, March 7, 1988, *Carmen Turner File, tab 7, NASM/MH.*
16. Richard E. Malenfant, letter to Robert C. Mikesh, October 3, 1988, *Carmen Turner File, tab 7, NASM/MH.*
17. Meeting with George Robinson and Dominic Pisano, July 12, 1989, 9:30 A.M.
18. Martin Harwit, letter to Dr. Siegfried Hecker, August 16, 1989, *NASM/MH, Carmen Turner File, tab 7, NASM/MH.*
19. S.S. Hecker, letter to Martin Harwit, September 19, 1989, *Carmen Turner File, tab 7, NASM/MH.*
20. Dom Pisano, "Memorandum for The Record on 'Atomic Bomb,' " November 9, 1989, *Carmen Turner File, tab 7, NASM/MH.*
21. Tom Crouch and Dom Pisano, letter to Martin Harwit, Memorandum on "Atomic Bomb," February 7, 1991, *NASM/MH.*
22. Meeting held March 26, 1991, 2:00 P.M.
23. Martin Harwit, "Notes for Carmen Turner on the Enola Gay's Proposed Exhibition," August 11, 1991, *Carmen Turner File, Cover Note, NASM/MH.*

Chapter 11

1. Michael Neufeld, "A Proposal to Exhibit the Enola Gay on the Mall," February 4, 1991, *NASM/MH.*
2. Martin Harwit, letter to Richard West, May 1, 1991, *NASM/MH.*
3. Martin Harwit, letter to Robert McC. Adams, May 23, 1991, *NASM/MH.*
4. Kim Masters, "Arts Beat—The Enola Gay, on the Mall," Washington *Post,* June 3, 1991.
5. Richard West, letter to Martin Harwit, June 13, 1991, *NASM/MH.*
6. Martin Harwit, letter to Carmen Turner, July 23, 1991, *NASM/MH.*

7. Black loose-leaf binder with ten sections and cover page, "Notes for Carmen Turner on the *Enola Gay*'s Proposed Exhibition," dated August 11, 1991, in author's possession, with a copy archived at NASM.
8. Martin Harwit, letter to Robert McC. Adams, October 22, 1991, *NASM/MH.*
9. Minutes of the National Air and Space Museum advisory board meeting held December 11, 1991, dated and distributed December 31, 1991, *NASM/MH.*
10. Theodore C. Barreaux, memorandum to members of the National Air and Space Museum Advisory Board, January 3, 1992.

Chapter 12

1. Meeting with Rusty Mathews and Charles D. Estes, Martin Harwit log book, November 21, 1989.
2. Margaret Gaynor, letter to Robert McC. Adams, November 18, 1992, *NASM/MH.*
3. Martin Harwit, letter to Bob Adams and Connie Newman, November 19, 1992, *NASM/MH.*
4. Michael Neufeld, to Martin Harwit, through Tom Crouch, "A Proposal for Fiftieth Anniversary Hiroshima Exhibit in the West End (Gallery 104)," December 1, 1992, *NASM/MH.*
5. Martin Harwit to Robert McC. Adams, Memorandum on "World War II, Strategic Bombing and the Enola Gay," December 3, 1992, *NASM/MH.*
6. Jim Hobbins to Martin Harwit, Memorandum on "World War II, Strategic Bombing and the Enola Gay," December 11, 1992, *NASM/MH.*
7. Martin Harwit to Jim Hobbins, Memorandum on "World War II, Strategic Bombing and the Enola Gay," December 21, 1992, *NASM/MH.*
8. SEF Planning Grant Proposal, "Hiroshima and Nagasaki: A Fiftieth Anniversary Exhibit at the National Air and Space Museum," January 4, 1993, *NASM/MH.*
9. Martin Harwit to Tom Freudenheim, Memorandum on "Special Exhibition Fund," February 19, 1993, *NASM/MH.*
10. Martin Harwit, log book entry, March 29, 1993; Luanne Smith, Memorandum for the Record on "Meeting at Smithsonian Air and Space Museum, 29 Mar 93," April 1, 1993, *MH.*
11. Martin Harwit, letter to C.M. Kicklighter, March 31, 1993, *NASM/MH.*
12. Martin Harwit, letter to Robert McC. Adams, October 22, 1991, *NASM/MH.* The text of this letter appears in Chapter 11.

CHAPTER 13

1. William A. Rooney, separate but identical letters to William H. Rehnquist, David C. Acheson, and Samuel C. Johnson, May 11, 1990, *NASM/MH*.
2. Robert McC. Adams, letter to William A. Rooney, June 25, 1990, *NASM/MH*.
3. William A. Rooney, letter to Hon. Sidney Yates, September 10, 1990, *NASM/MH*.
4. W. Burr Bennett, Jr., letter to Robert McC. Adams, July 23, 1992, *NASM/MH*.
5. Robert McC. Adams, letter to W. Burr Bennett, Jr. August 27, 1992, *NASM/MH*.
6. W. Burr Bennett, Jr., letter to Robert McC. Adams, September 6, 1992, *NASM/MH*.
7. Tom L. Freudenheim, letter to W. Burr Bennett, Jr., October 9, 1992, *NASM/MH*.
8. W. Burr Bennett, Jr., letter to Robert McC. Adams, October 12, 1992, *NASM/MH*.
9. W. Burr Bennett, Jr., letter to Robert McC. Adams, October 20, 1992, *NASM/MH*.
10. W. Burr Bennett, Jr., letter to John Edward Porter, October 22, 1992, *NASM/MH*.
11. Martin Harwit, letter to W. Burr Bennett, Jr., November 23, 1992, *NASM/MH*.
12. Donald C. Rehl, letter to Martin Harwit, December 2, 1992, *NASM/MH*.
13. Linda Neuman Ezell, letter to Donald C. Rehl, December 9, 1992, *NASM/MH*.
14. Martin Harwit, letter to Donald C. Rehl, December 18, 1992, *NASM/MH*.
15. Donald C. Rehl, letter to George C. Larson, December 3, 1992, *NASM/MH*.
16. George C. Larson, letter to Donald C. Rehl, December 14, 1992, *NASM/MH*.
17. W. Burr Bennett, Jr., letter to George C. Larson, December 11, 1992, *NASM/MH*.
18. George C. Larson, letter to W. Burr Bennett, Jr., December 15, 1992, *NASM/MH*.
19. Martin Harwit, letter to W. Burr Bennett, Jr., December 15, 1992, *NASM/MH*.
20. W. Burr Bennett, Jr., letter to Robert McC. Adams, March 12, 1993, *NASM/MH*.
21. Robert McC. Adams, letter to W. Burr Bennett, Jr., March 23, 1993, *NASM/MH*.

22. Donald C. Rehl, letter to Robert McC. Adams, April 26, 1993, *NASM/MH.*
23. Martin Harwit, letter to Donald C. Rehl, May 17, 1993, *NASM/MH.*
24. Alfred A. Yee, letter to W. Burr Bennett, Jr., May 10, 1993, *NASM/MH.*
25. W. Burr Bennett, Jr., letter to Robert McC. Adams, July 9, 1993, *NASM/MH.*
26. Ben Nicks, letter to Don Rehl, August 10, 1993, *NASM/MH.*
27. William A. Rooney, letter to Ben Nicks, August 13, 1993, *NASM/MH.*
28. Petition on the *Enola Gay* circulated by the Committee for the Restoration and Display of the *Enola Gay*. For copies see attachments to notes 27 and 29, *NASM/MH.*
29. Donald C. Rehl, letter to Ben Nicks August 18, 1993, *NASM/MH.*
30. Ben Nicks, letter to Don Rehl, October 23, 1993, *NASM/MH.*
31. Martin Harwit, letter to Ben Nicks, September 3, 1993, *NASM/MH.*
32. Martin Harwit, identical letters to Donald C. Rehl, William A Rooney, and others, September 3, 1993, *NASM/MH.*
33. Donald C. Rehl, letter to Martin Harwit, September 21, 1993, *NASM/MH.*
34. Martin Harwit, letter to Donald C. Rehl, October 7, 1993, *NASM/MH.*
35. William A. Rooney, letter to Martin Harwit, October 13, 1993, *NASM/MH.*
36. Donald C. Rehl, letter to Martin Harwit, November 27, 1993, *NASM/MH.*
37. Martin Harwit, letter to Donald C. Rehl, December 21, 1993, *NASM/MH.*
38. Martin Harwit, letter to Donald C. Rehl, February 10, 1994, *NASM/MH.*
39. W. Burr Bennett, Jr., letter to Barber B. Conable, January 13, 1993, *NASM/MH.*
40. Peter G. Powers to I. Michael Heyman, Memorandum on "Effect of National Armed Forces Museum Advisory Board Legislation on the National Air and Space Museum," January 24, 1995, in the author's possession.
41. Martin Harwit, letter to W. Burr Bennett, Jr., December 15, 1992, *NASM/MH.*

CHAPTER 14

1. Morihisa Takagi, letter to Martin Harwit, November 28, 1988, *NASM/MH.*
2. "White House Bomb Ends the Mushroom-Cloud War," Associate Press article, *International Herald Tribune*, December 9, 1994.
3. Martin Harwit, letter to Jun Wada, July 24, 1991, *NASM/MH.*
4. Martin Harwit to Carmen Turner, Memorandum on *Enola Gay*, July 23, 1991, *NASM/MH.*
5. Martin Harwit, letter to Michael J. Mansfield, August 19, 1991, *NASM/MH.*
6. Note from Gregg [Herken] to Martin [Harwit], with an attachment Gregg had received from Mr. Itonaga in Tokyo, September 5, 1991, *NASM/MH.*
7. Martilla & Kiley, Inc., Boston, Massachusetts, "A Survey of Public Attitudes in the U.S. and Japan," conducted in Japan, in the U.S., and in Detroit between November 2 and 5, 1991.
8. Steven R. Weisman, "Japanese Think They Owe Apology and Are Owed One on War, Poll Shows," *New York Times,* December 8, 1991.
9. ABC News Polling Unit and the Culture Research Institute of NHK Broadcasting, "U.S.-Japan Relations 50 Years After Pearl Harbor," for release after December 6, 1991.
10. Sadao Ishizu, letter to Gregg Herken, November 15, 1991, *NASM/MH.*
11. Martin Harwit, letter to Shinichiro Asao, August 11, 1992, *NASM/MH.*
12. Jun'etsu Komatsu, letter to Martin Harwit, November 11, 1992, *NASM/MH.*
13. Martin Harwit, letter to Itsuzo Shigematsu, January 18, 1993, *NASM/MH.*
14. Itsuzo Shigematsu, letter to Martin Harwit, March 4, 1993, *NASM/MH.*
15. Martin Harwit, letter to Itsuzo Shigematsu, February 26, 1993, *NASM/MH.*
16. Akira Iriye, letter to Martin Harwit, February 25, 1993, *NASM/MH.*
17. J.W. Thiessen, letter to Martin Harwit, March 5, 1993, *NASM/MH.*
18. Martin Harwit, letters to Takashi Hiraoka and Hitoshi Motoshima, March 10, 1993, *NASM/MH.*
19. *Asahi Shimbun,* 17 March 1993.
20. Michael J. Neufeld, *The Rocket and the Reich—Peenemünde and the Coming of the Ballistic Missile Era* (New York: The Free Press, 1995).

21. Martin Harwit, letter to Takakazu Kuriyama, March 17, 1993, *NASM/MH.*
22. Martin Harwit, letter to Bob Adams, March 29, 1993, *NASM/MH.*
23. Norma Field, *In the Realm of a Dying Emperor—Japan at Century's End* (New York: Vintage Books, 1993), 177–178.
24. Takashi Hiraoka, letter to Martin Harwit, December 1, 1993, *NASM/MH.*
25. Hitoshi Motoshima, letter to Martin Harwit, December 7, 1993, *NASM/MH.*
26. The *Japan Times,* December 1, 1993.

CHAPTER 15

1. Martin Harwit to Tom Freudenheim, Memorandum on "1995 Exhibition on Hiroshima and Nagasaki at the National Air and Space Museum," May 4, 1993, *NASM/MH.*
2. Wendy Stephens to Lin Ezell, Tom Crouch, and Alice Adams, Memorandum on "Spending on the *Enola Gay* Exhibit," April 30, 1993, *NASM/MH.*
3. Tom L. Freudenheim to Martin Harwit, Memorandum on "Special Exhibition Fund," February 28, 1994. *NASM/MH.*
4. Martin Harwit, note to Jim Hobbins, dated June 10, 1993, *NASM/MH.*
5. "Fifty Years On" (A draft proposal for the exhibition), June 10, 1993, *NASM/MH.*
6. Jim Hobbins to the record (cc: Bob Adams, Connie Newman, Tom Freudenheim, and Martin Harwit), Memorandum on Barber Conable's Reactions to "Fifty Years On," June 21, 1993, *NASM/MH.*
7. Ben Nicks, letter to Dr. Michael J. Neufeld, June 26, 1993, *NASM/MH.*
8. Gregg Herken, E-mail to Mike Neufeld, Tom Crouch, Martin Harwit, Laurenda Patterson, Subject: Re *Enola Gay*, June 28, 1993, *NASM/MH.*
9. National Air and Space Museum, Smithsonian Institution, Exhibition Planning Document, "Ground Zero: The Atomic Bomb and the End of World War II," Projected Dates: May 1995 to January 1996, Draft #2, June 30, 1993.
10. Martin Harwit to Tom Crouch, Memorandum on "The latest Draft of '50 Years On,' " July 2, 1993, *NASM/MH.*
11. Original and cover-sheet for a telefax from our secretary, Toni Thomas, to Beth, in Hobbins's office, July 6, 1993, *NASM/MH.*
12. Martin Harwit, letter to Bob Adams, July 8, 1993, *NASM/MH.*
13. Martin Harwit, letter to Bob Adams, July 13, 1993, *NASM/MH.*

14. Bob Adams, letter to Martin Harwit, July 17, 1993, *NASM/MH.*
15. Tom Crouch to Martin Harwit, Memorandum on "A Response to the Secretary," July 21, 1993, *NASM/MH.*
16. Martin Harwit, letter to Tom Crouch and Michael Neufeld, July 22, 1993, *NASM/MH.*
17. Martin Harwit to Bob Adams and Tom Freudenheim, Memorandum on "Crossroads," July 22, 1993, *NASM/MH.*
18. Michael Neufeld, letter to Martin Harwit, Memorandum on "The Crossroads," July 23, 1993, *NASM/MH.*
19. Tom Crouch to Martin Harwit, Memorandum on "Crossroads," July 23, 1993, *NASM/MH.*

CHAPTER 16

1. "Collings Foundation Newsletter," The Collings Foundation, River Hill Farm, Stow, Massachusetts 01755, January/ February 1993.
2. Ibid., March/April 1993.
3. Arthur H. Sanfelici, "Is NASM Thumbing Its Nose at Congress While No One's Watching?" *Aviation,* July 1993.
4. United States General Accounting Office, Report to the Honorable Kay Bailey Hutchison, U.S. Senate, *Smithsonian Institution—Better Care Needed for National Air and Space Museum Aircraft*, October 1995, 3, 15.
5. Claude M. Kicklighter, letter to Martin Harwit, May 10, 1993, *MH.*
6. Martin Harwit, log book, May 19, 1993; Phyllis E. Phipps-Barnes, Memorandum for Record, "Meeting on Air and Space Hiroshima and Nagasaki Exhibit." May 26, 1993, *MH.*
7. Martin Harwit, letter to Luanne Smith, July 13, 1993, *MH.*
8. *Air Force* magazine, August 1993.
9. W. Burr Bennett, Jr., letter to John Correll, August 6, 1993, *AFA, The Enola Gay Debate, Key Documents*, May 1995.
10. John Correll to Monroe Hatch, Memorandum on "Info Letter to McCoy," August 10, 1993, *AFA, The Enola Gay Debate, Key Documents*, May 1995.
11. Monroe W. Hatch, letter to Martin Harwit, September 10, 1993, *NASM/MH*; this letter can also be found in *AFA, The Enola Gay Debate, Key Documents*, May 1995; however the copy reproduced there inexplicably bears the date September 12, 1993. Otherwise it appears to be identical to the original.
12. Richard P. Hallion, letter to Lt. Gen. McInerney, 8 September, 1993.
13. Martin Harwit, letter to Monroe Hatch, October 7, 1993, *NASM/MH.*
14. Sandy Rittenhouse-Black to Distribution, Memorandum on "Department of Defense Funds for Crossroads," September 24, 1994, *NASM/MH.*

15. Dick Hallion, handwritten note to General Kicklighter, September 24, 1996, *MH*.
16. Luanne Smith to Edward T. Linenthal, undated, but written August 10, 1995, and given to the author on February 2, 1996, *MH*.
17. Russell E. Dougherty, letter to Martin Harwit, October 12, 1993, *NASM/MH*.
18. Martin Harwit, letter to General Russell E. Dougherty, October 20, 1993, *NASM/MH*.
19. Joanne Gernstein to Martin Harwit, through Tom Crouch, Memorandum on "Air Force Association," October 27, 1993, *NASM/MH*.
20. Joanne Gernstein, update on her October 27 memorandum, November 1, 1993, *NASM/MH*.
21. John T. Correll, Memorandum for the record on "Meeting at Air & Space Museum, November 23, 1993," *AFA, The Enola Gay Debate, Key Documents*, May 1995.
22. Martin Harwit, letter to Monroe Hatch, November 26, 1993, *NASM/MH*.
23. Martin Harwit, letter to Monroe Hatch, December 16, 1993, *NASM/MH* and *AFA, The Enola Gay Debate, Key Documents*, May 1995.
24. Monroe Hatch, letter to Martin Harwit, January 5, 1994, *NASM/MH* and *AFA, The Enola Gay Debate, Key Documents*, May 1995.
25. John Correll, letter to Monroe Hatch, December 16, 1993, *AFA, The Enola Gay Debate, Key Documents*, May 1995.
26. Martin Harwit, letter to Monroe Hatch, January 31, 1994, *NASM/MH*.

Chapter 17

1. Philip Nobile, *Judgment at the Smithsonian—The Bombing of Hiroshima and Nagasaki*: The Uncensored Script of the Smithsonian's 50th Anniversary Exhibit of the Enola Gay (New York: Marlowe & Company, 1995).
2. Edward T. Linenthal, *Sacred Grounds: Americans and Their Battlefields*, 2nd Edition (Urbana: University of Illinois Press, 1993), 234.
3. Barton J. Bernstein, "A postwar myth: 500,000 U.S. lives saved," *Bulletin of the Atomic Scientists*, June/July 1986, 38.
4. Martin J. Sherwin, *A World Destroyed—The Atomic Bomb and the Grand Alliance* (New York: Vintage Books, 1987).
5. Richard Rhodes, *The Making of the Atomic Bomb (New York:* Simon & Schuster, 1986).

6. Akira Iriye, telefaxed letter to Michael J. Neufeld, February 7, 1994.
7. Edwin C. Bearss, letter to Dr. Tom Crouch, February 24, 1994, *NASM/MH*.
8. Richard Hallion and Herman Wolk, "Comments on Script, 'The Crossroads: The End of World War II, the Atomic Bomb and the Origins of the Cold War,' " February 7, 1994, *NASM/MH*.

Chapter 18

1. Albert Einstein to F.D. Roosevelt, August 2, 1939. A copy of this letter is reprinted in *Leo Szilard: His Version of the Facts*, eds. Spencer R. Weart and Gertrude Weiss Szilard (Cambridge: MIT Press, 1978), 94.
2. Henry Stimson, letter to President Truman, April 24, 1945. Harry S. Truman Library, Independence, Missouri.
3. Henry Stimson, "Memo Discussed with the President," April 25, 1945. Harry S. Truman Library, Independence, Missouri.
4. Paul W. Tibbets, "Our Job Was to Win." Excerpts from a speech given to the Air Force Sergeants Association, June 1994, reprinted in *The American Legion,* November 1995, 29.
5. Henry Stimson, diary entry for May 15, 1945. Stimson diary, Yale University.
6. "The Last Act: The Atomic Bomb and the End of World War II," exhibition label script, National Air and Space Museum, October 26, 1994, 200–59.
7. Ibid., page 200–58ff (Original letter in Harry S. Truman Library, Independence, Missouri).
8. Ibid., page 200–60.
9. Ibid., page 500–8 (and Truman Library), letter from Truman to Richard Russell, August, 9, 1945.
10. Ibid., page 500–7.
11. Ibid., page 200–45 (Original in the National Archives).
12. William D. Leahy, *I Was There* (New York: Whittlesey House, 1950), 441.
13. Dwight D. Eisenhower, *Crusade in Europe* (Garden City, New York: Doubleday, 1948), 443.
14. John Dower, *War Without Mercy—Race and Power in the Pacific War* (New York: Pantheon Books, 1986).
15. *Life* Magazine, May 22, 1944, 35.

Chapter 19

1. Martin Harwit to Gwen Crider, "Notes for Gwen," February 11, 1994, *NASM/MH*.

2. John T. Correll, "The Smithsonian and the Enola Gay," Air Force Association Special Report, March 15, 1994. *AFA, The Enola Gay Debate, March Report*, May 1995.
3. David McCullough, *Truman*, Simon & Schuster, 1992.
4. See also John T. Correll, "The Decision That Launched the *Enola Gay*," *Air Force* (April 1994): 30, for a slightly modified text of half of the Special Report, note 2.
5. National Air and Space Museum, Smithsonian Institution, Exhibit Planning document, Tentative Exhibit Title: "The Crossroads: The End of World War II, The Atomic Bomb, and the Onset of the Cold War," July 1993, *NASM/MH; AFA, The Enola Gay Debate, Key Documents*, May 1995.
6. See also John T. Correll, "War Stories at Air and Space," *Air Force* (April 1994): 24, for a slightly modified text of half of the Special Report, note 2.
7. "The Crossroads: The End of World War II, the Atomic Bomb, and the Onset of the Cold War," National Air and Space Museum script, January 1994, page EG: 100–5.
8. Ben Nicks, "Keep Moralizing Out of Museums," Letters, *Air & Space* (December 1990/January 1991).
9. Hank Burchard, "Plane Truths During WWI," *Washington Post*, November 22, 1991; Michael Kilian, "Grounded in Reality," *Chicago Tribune*, November 26, 1991.
10. Elisabeth Kastor, "At Air & Space, Ideas on the Wing," *Washington Post*, October 11, 1988.
11. Arthur H. Sanfelici, "Is NASM Thumbing Its Nose at Congress While No One's Watching?" *Aviation*, July 1993.
12. Stephen P. Aubin, in an interview with Tony Capaccio, editor of *Defense Week*, May 12, 1995.
13. Jack Giese, in an interview with Tony Capaccio, editor of *Defense Week*, April 17, 1995.
14. United States Senate, Committee on Rules and Administration, signed by Wendell Ford, Ted Stevens, Robert Dole, Thad Cochran, Jesse Helms, and Mitch McConnell, letter to Martin Harwit, March 24, 1994, *NASM/MH*.
15. Senator Nancy Landon Kassebaum, letter to Robert McCormick Adams, March 30, 1994, *NASM/MH, AFA The Enola Gay Debate, Congress*, May 1995.
16. Martin Harwit, letter to the individual senators listed in note 14, identical letters dated April 4, 1994.
17. John McCaslin, "Rewriting History," *Washington Times*, p. 9A, March 28, 1994, Inside the Beltway.
18. John McCaslin, "Naked Brutality," *Washington Times*, March 31, 1994, Inside the Beltway section.

19. Martin Harwit, reply to article by John Correll, *Air Force* (May 1994).
20. Martin Harwit, letter to Monroe Hatch, *NASM/MH*, April 7, 1994, *AFA The Enola Gay Debate, Key Documents*, May 1995.
21. "Enola Gay Coverage 1994," *Air Force Association* (May 1995): 1, 3, 5; *The Enola Gay Debate, Air Force Association* (May 1995).
22. Stephen Aubin, letter to Ed Bolen, April 4, 1994, *AFA The Enola Gay Debate, Congress.*
23. Stephen Aubin, letter to Ron Stroman and Marty Morgan, April 8, 1994, *AFA The Enola Gay Debate, Congress.*
24. Kenneth A. Goss to Dan Stanley, July 21, 1994, *AFA The Enola Gay Debate, Congress.*
25. John T. Correll, "The Smithsonian Plan for the Enola Gay: A Report on the Revisions," June 28, 1994. *AFA, The Enola Gay Debate, Script Analyses,* May 1995.
26. John T. Correll, "'The Last Act' at Air & Space," *Air Force* (September 1994): 58.
27. An article on this exhibition appeared in the *Washington Post*, June 10, 1980.
28. Tom Lewis, "Defending America from Scholars," *Air Force Times* (September 12, 1994): 33.
29. Act of Congress, August 10, 1846, Ch. 178, §1, 9 Stat. 102.
30. Congressman Tom Lewis, news release, "Lewis Disappointed with Smithsonian Director's Explanation of Japanese Bias in Upcoming Exhibit of Atomic Bomb," August 10, 1994, *AFA The Enola Gay Debate, Congress,* May 1995.
31. Lisa Metheny, note to John Correll, August 10, 1994, *AFA The Enola Gay Debate, Key Documents,* May 1995.
32. Congressman Sam Johnson, news release, "Purpose of *Enola Gay* Exhibit Disputed," August 10, 1994, *AFA The Enola Gay Debate, Congress,* May 1995.
33. Tom Lewis, letter to Martin Harwit, August 10, 1994, *NASM/MH.*
34. Martin Harwit, letter to Honorable Tom Lewis, August 12, 1994, *NASM/MH.*
35. Peter Blute and twenty-three other members of the House, letter to Secretary Robert McCormick Adams, August 10, 1994, *NASM/MH, AFA The Enola Gay Debate, Congress*, May 1995.
36. John T. Correll, " 'The Last Act' at Air & Space," *Air Force* (September 1994): 58.
37. Stephen Aubin to Rob Gray, Office of Congressman Blute, Memorandum on "Comments on the Enola Gay Exhibit," August 22, 1994, *AFA The Enola Gay Debate, Congress*, May 1995.
38. Monroe W. Hatch, letter to Martin Harwit, August 24, 1994, *AFA The Enola Gay Debate, Key Documents,* May 1995.

39. Stephen Aubin, letter to Ed Bolen, September 15, 1994, *Enola Gay Debate, Key Documents,* May 1995.
40. Senate Resolution 257, 103rd Congress, Second Session, September 19, 1994, *AFA The Enola Gay Debate, Congress.*
41. Ed Bolen, letter to Steve Aubin, September 19, 1994, *AFA The Enola Gay Debate, Congress.*
42. David W. Davis, letter to John Correll, September 19, 1994, *AFA The Enola Gay Debate, Congress.*
43. Monroe Hatch, letter to Senator Kassebaum, September 20, 1994, *AFA The Enola Gay Debate, Congress.*
44. House Resolution 531 and HRpt 103–740 Conference Report on HR 4602, 103rd Congress, Second Session, September 19, 1994, *AFA The Enola Gay Debate, Congress.*
45. Stephen Aubin, letter to Ron Stroman and Marty Morgan, April 8, 1994, *AFA The Enola Gay Debate, Congress.*

Chapter 20

1. Luanne J. Smith, Memorandum for the Record on "January 12 Smithsonian Air and Space Museum Meeting," February 7, 1994.
2. L.J. Smith, Memorandum for Director of Support and Outreach on "Update and Activities," April 18, 1994.
3. Richard Hallion, letter to Merrill A. McPeak, April 8, 1994, *AFA The Enola Gay Debate, Key Documents*, May, 1995.
4. Richard Hallion, letter to Lt. Gen. Claude M. Kicklighter, April 19, 1994, *AFA, The Enola Gay Debate, Key Documents*, May 1995.
5. Harold W. Nelson, letter to Lt. Gen. Claude M. Kicklighter, Ret., April 19, 1994, *AFA, The Enola Gay Debate, Key Documents,* May 1995.
6. Kathleen Lloyd, to LCDR Smith, World War II Commemorative Committee. Memorandum on "Review of National Air and Space Museum Script, 'The End of World War II, The Atomic Bomb and the Origins of the Cold War,' " April 25, 1994, *NASM/MH.*
7. L.J. Smith, Memorandum "thru Director, Support and Outreach for Executive Director," on "Smithsonian Air and Space Museum's Enola Gay Exhibit," April 25, 1994.
8. W. Burr Bennett, Jr., letter to Lt. General Claude M. Kicklighter, USA Retired, April 15, 1994.
9. William A. Rooney, letter to Lt. Gen. C. M. Kicklighter, USA Retired, April 20, 1994.
10. Draft letter prepared by Commander L.J. Smith for Claude M. Kicklighter to send to General McInerney, undated; Courtesy of Commander Smith, provided to the author February 2, 1996, *MH.*

11. Claude M. Kicklighter, Memorandum for the assistant vice chief of staff, United States Air Force on "Smithsonian Air and Space Museum Enola Gay Exhibition," May 11, 1994, *MH.*
12. Thomas G. McInerney, letter to Claude M. Kicklighter, June 30, 1994, *MH.*
13. Lt. Col. Albers, handwritten note to Col. Epifano, September 19, 1994, *MH.*
14. Colonel Epifano, letter to Lt. Col. Albers, Subject: Lt. Gen. Kicklighter's proposed letter, September 28, 1994.
15. L.J. Smith, letter to Martin Harwit, April 22, 1996, *MH.*
16. Martin Harwit, letter to General Merrill A. McPeak, July 15, 1994, *NASM/MH.*
17. Martin Harwit, letter to Secretary of the Air Force Dr. Sheila Widnall, July 18, 1994, *NASM/MH.*
18. Martin Harwit, letter to Ambassador Winston Lord, July 19, 1994, *NASM/MH;* Winston Lord to Martin Harwit, undated, but marked received July 28, 1994, *NASM/MH.*
19. Luanne J. Smith, Memorandum for the Record on "Enola Gay Exhibit Meetings," July 26, 1994.

CHAPTER 21

1. Sandy Murdock to Martin Harwit, unsigned, undated document, telefaxed February 14, 1994, in the author's possession.
2. Martin Harwit, letter to Frank Rabbitt, March 14, 1994, *NASM/MH.*
3. See (Biloxi, Mississippi) *Sun Herald*, Bob McHugh, "Space Museum has a bomb planned for Aug. 6," May 6, 1994; Opinion column, "Museum's exhibit should upset World War II vets and the rest of us," May 24, 1994.
4. Martin Harwit, letter to Frank Rabbitt, June 17, 1994, *NASM/MH.*
5. Tom Crouch to Martin Harwit, Memorandum on "Reaction to Crossroads," March 31, 1994, *NASM/MH.*
6. Hugh Sidey, "War and Remembrance," *Time* (May 20, 1994): 20.
7. Hugh Sidey, letter to Frank Rabbitt, May 24, 1994, in the author's possession.
8. Kathy Boi, letter to Martin Harwit, April 5, 1994, *NASM/MH.*
9. William G. Hulbert, letter to Robert McC. Adams, undated, but received April 12, 1994, *NASM/MH*; Martin Harwit, letter to William G. Hulbert, April 14, 1994, *NASM/MH.*
10. Martin Harwit, letter, to William G. Hulbert, April 14, 1994, *NASM/MH.*
11. Robert McC. Adams, letter to Honorable G.V. "Sonny" Montgomery, April 12, 1994. *NASM/MH.*

12. Martin Harwit, "Comments on Crossroads," April 16, 1994, *NASM/MH.*
13. Monroe W. Hatch, Jr., letter to Martin Harwit, September 27, 1994, *AFA, The Enola Gay Debate, Key Documents,* May 1995.
14. Martin Harwit, letter to Bill Constantine, Tom Alison, Gregg Herken, Don Lopez, Ken Robert, Steven Soter, April 26, 1994, *NASM/MH.*
15. J. Samuel Walker, "The Decision to Use the Bomb: A Historiographical Update," *Diplomatic History*, vol. 14, no. 1 (Winter 1990): 97.
16. Michael Neufeld to Martin Harwit, "Tiger Team" members, and exhibit team, Memorandum on "The 'decision to drop the bomb' and 'Crossroads,' " April 25, 1994, *NASM/MH.*
17. William M. Constantine, "Report of the National Air and Space Museum Review Team: Exhibit Script—Crossroads: The End of World War II, the Atomic Bomb and the Origins of the Cold War," May 25, 1994, *AFA, The Enola Gay Debate, Tiger Team*, May 1995.
18. Monroe W. Hatch, Jr., letter to Martin Harwit, July 19, 1994, *AFA The Enola Gay Debate, Key Documents*, May 1995.
19. John T. Correll, "Developments in the *Enola Gay* Controversy," August 22, 1994, *AFA The Enola Gay Debate, Script Analyses*, May 1995.

CHAPTER 22

1. Constance Berry Newman, letter to Bruce Thiesen, May [4?], 1994, *NASM/MH.*
2. Bob Roberts, "Veterans embroiled in *Enola Gay* Debate," *Journal Herald*, October 26, 1994, 1. Also, telephone conversation between Ben Nicks and the author, May 13, 1996.
3. National Executive Committee Meeting of the American Legion, Indianapolis, Indiana, Resolution No: 22, Smithsonian Exhibit of the *Enola Gay*, May 4–5, 1994, *NASM/MH*, and *AFA The Enola Gay Debate, Key Documents,* May 1995. Two earlier resolutions the American Legion also forwarded to us were Resolution concerning display of the *Enola Gay* in a patriotic manner in the National Air and Space Museum, Nashville, Tennessee, January 21, 1994, signed by Charles G. Norton, Tennessee Department; and Resolution 212 on "*Enola Gay*" referred to the Committee on Credentials and Internal Affairs, for the Seventh Annual National Convention, Louisville, Kentucky, September 6, 7, 8, 1988, submitted by adjutant (name illegible) Indiana Department, *NASM/MH.*
4. Hubert R. Dagley II, letter to Donald C. Rehl, June 10, 1994, *AFA, The Enola Gay Debate, Key Documents,* May 1995.

5. News Release, Airmen Memorial Museum, June 9, 1994, *AFA, The Enola Gay Debate, Key Documents*, May 1995. For an extensive excerpt see, *The American Legion* (November 1994), 29.
6. Ibid.
7. Martin Harwit, letter to Paul W. Tibbets, June 17, 1994, *NASM/MH.*
8. Notes of June 22 telephone conversation of Martin Harwit with Paul Tibbets, *MH.*
9. Martin Harwit, letter to Paul W. Tibbets, June 24, 1994, *NASM/MH.*
10. Paul W. Tibbets, letter to Martin Harwit, June 27, 1994, *NASM/MH.*
11. Martin Harwit, letter to Paul W. Tibbets, July 11, 1994. *NASM/MH.*
12. Martin Harwit, letter to Admiral T.J. Kilcline (USN Ret.) June 15, 1994, *NASM/MH.*

CHAPTER 23

1. Martin Harwit, letter to Monroe Hatch, June 22, 1994; and Martin Harwit to John Correll, June 22, 1994, *NASM/MH.*
2. Ken Goss to Military Coalition Members, Memorandum on "The *Enola Gay* Controversy or—Who Started World War II?" June 30, 1994, *AFA, The Enola Gay Debate, Key Documents*, May 1995.
3. Unsigned "Comments on 'The Last Act: The Atomic Bomb and the End of World War II,' " Office of Air Force History, July 12, 1994, *AFA The Enola Gay Debate, Script Analyses*, May 1995.
4. Luanne J. Smith, Memorandum for the Record on "Enola Gay Exhibit Meetings," July 26, 1994.
5. Tony Capaccio, "DOD Historians Lauded Revisions In A-Bomb Script," *Defense Week* (July 3, 1995): 1.
6. Michael J. Neufeld, cover letter addressed to nine members of the advisory board and four military historians, June 21, 1994, *AFA The Enola Gay Debate, Key Documents,* May 1995.
7. Luanne J. Smith, Memorandum, July 26, 1994.
8. Ibid.
9. Martin Harwit, letter to Gen. John Guthrie, Military Order of the World Wars, July 22 1994, *NASM/MH.* Identical letters dated the same day went to Lennox Gilmer, DAV; Bob Currieo, VFW; Chuck Cooper, TROA; and Phil Budahn, American Legion.
10. Luanne J. Smith, Memorandum, July 26, 1994.
11. Stephen P. Aubin, separate letters to Ed Simmons, Harold Nelson, and Dean Allard, July 26, 1994, *AFA The Enola Gay Debate, Key Documents*, May 1995.

12. Monroe Hatch, letter to Martin Harwit, July 19, 1994, *AFA The Enola Gay Debate, Key Documents*, May 1995.
13. Rowan Scarborough, "Museum director rebuked curators of A-bomb exhibit," *Washington Times,* August 29, 1994, p. 1.
14. Martin Harwit, "The *Enola Gay* and the 'last act' of World War II," Letters section, *Washington Times*, September 4, 1994, p. B2.
15. L.J. Smith, Memorandum for the Executive Director on "*Enola Gay* Exhibit"—Input for Mr. Hamilton's memo, August 26, 1994, *MH.*
16. See note 5.
17. L.J. Smith, Memorandum for Record on "Comments on Enola Gay Script," Sept 22, 1994, recorded this conversation with Wayne Dzwonchyk, *MH.*
18. Monroe Hatch, letter to Martin Harwit, August 24, 1994, *AFA The Enola Gay Debate, Key Documents,* May 1995.
19. Ibid.
20. See note 5.

Chapter 24

1. Arthur Hirsch, "Dismantled, a deadly courier holds on to its place in history." *Baltimore Sun*, March 24, 1994, see also *AFA Enola Gay Coverage, 1994,* May 1995.
2. David C. Morrison, "Airpower Über Alles," Focus Section, *National Journal,* April 2, 1994, p. 805; see also *AFA Enola Gay Coverage, 1994,* May 1995.
3. Tom Webb, "Grisly display for famed *Enola Gay* bomber angers vets," Knight-Ridder-Tribune News Service, *Wilmington Morning Star*, May 7, 1994.
4. J. Ivan Potts, Jr. Opinion Page, (Shelbyville) *Times-Gazette,* May 5, 1994. See also *AFA Enola Gay Coverage, 1994*. May 1995.
5. Eugene L. Meyer, "Dropping the Bomb," *Washington Post*, July 21, 1994, p. C2.
6. Tony Capaccio and Uday Mohan, "Missing the Target," *American Journalism Review* (July/August, 1995).
7. Martin Harwit, "The *Enola Gay*: A Nation's and a Museum's Dilemma," *Washington Post*, August 7, 1994, C7.
8. See chapter 14 or the following references cited there: Martilla & Kiley, Inc., Boston, Massachusetts, "A Survey of Public Attitudes in the U.S. and Japan," conducted in Japan, in the U.S., and in Detroit between November 2 and 5, 1991; Steven R. Weisman, "Japanese Think They Owe Apology and Are Owed One on War, Poll Shows," *New York Times,* December 8, 1991; ABC News Polling Unit and the Culture Research Institute of NHK Broadcasting, "U.S.-Japan Relations 50 Years After Pearl Harbor," for release after December 6, 1991.

9. Unsigned editorial, "Context and the *Enola Gay,*" *Washington Post,* August 14, 1994, p. C8.
10. "The Mission That Ended the War," five letters to the editor, *Washington Post*, August 14, 1994, p. C9.
11. Kai Bird and Martin Sherwin, United States Senate, Committee on Rules and Administration, 104th Congress, 1st Session, Hearings on the Smithsonian Institution—Management Guidelines for the Future, May 11 and 18, pp. 133, 138.
12. Garner, cartoon, *Washington Times,* August 31, 1994.
13. Harry Paine, cartoon, *Washington Times*, September 2, 1994.
14. Harry Paine, cartoon, *Washington Times*, October 7, 1994.
15. Cullum, cartoon *Washington Times*, November 7, 1994.
16. Bruce Tinsley, comic strip, *Washington Times*, November 28, 1994.
17. Unsigned editorial, "The Smithsonian and the Bomb," *New York Times*, September 5, 1994.
18. Letters to the editor by W.E. Cooper and John T. Correll, "Hiroshima Bomb Display Still Distorts History," *The New York Times*, September 10, 1994, 18.
19. Unsigned editorial, "War and the Smithsonian," *Wall Street Journal*, Review and Outlook section, August 29, 1994, p. A10.
20. See note 6.
21. Margaret Gaynor, letter to Martin Harwit, June 22, 1988, *NASM/MH.*
22. Hubert R. Dagley II, letter to Martin Harwit, May 7, 1996, *MH.*
23. Jack Giese, in an interview with Tony Capaccio, editor of *Defense Week*, April 10, 1995, *MH.*
24. Jack Giese, in a telephone conversation with Martin Harwit, April 27, 1996, *MH.*

Chapter 25

1. Constance B. Newman, letter to Herman G. Harrington, and an identical letter to Hubert R. Dagley II, June 21, 1994, *NASM/MH.*
2. Martin Harwit, letter to Hubert R. Dagley II, August 11, 1994, *NASM/MH.*
3. Bruce Thiesen, letter to The Hon. William Jefferson Clinton, August 12, 1994, *AFA, The Enola Gay Debate, Key Documents*, May 1995.
4. Jerry May, letter to Ken Goss, telefax, August 18, 1994, *AFA, The Enola Gay Debate, Key Documents*, May 1995.
5. Joanne Gernstein, informal note to the author, August 24, 1994. In the author's possession.
6. Martin Harwit, individual and identical letters to Ray Gallagher, Richard Nelson, Theodore Van Kirk, and Tom Ferebee, August 25, 1994, *NASM/MH.*

7. Martin Harwit, letter to Hubert R. Dagley II, August 26, 1994, *NASM/MH.*
8. Martin Harwit to Tom Crouch, Memorandum on "The latest Draft of '50 Years On,'" July 2, 1993, *NASM/MH.*
9. From an interview of Father John Dear by Nobile, cf. Philip Nobile, *Judgment at the Smithsonian* (New York: Marlowe & Co., 1995), xxxvii, xciv.
10. Philip Nobile, *Judgment at the Smithsonian* (New York: Marlowe & Co., 1995), xxxvii.
11. Martin Harwit, letter to Monroe Hatch, August 26, 1994, *NASM/MH.*
12. See note 8.
13. Martin Harwit, statement at press conference with the American Legion, September 22, 1994, in the author's possession; William Detweiler, statement at press conference with the Smithsonian, September 22, 1994, *AFA, The Enola Gay Debate, Key Documents*, May 1995.
14. National Air and Space Museum, exhibition script "The Last Act: The Atomic Bomb and the End of World War II," October 26, 1994, p. 200–50.
15. Martin Harwit, letter to Hubert R. Dagley II, October 25, 1994 and November 9, 1994, *NASM/MH.*
16. David Mannweiler, "Another look at A-bombs," *Indianapolis News,* September 27, 1994.
17. Eugene L. Meyer, "Smithsonian Bows to Critics, Revamps Atom Bomb Exhibit," *Washington Post*, September 30, 1994, 1.
18. Rowan Scarborough, "Legion in hot seat as exhibit is retooled," *Washington Times,* October 11, 1994.
19. Hubert R. Dagley II, "We're making sure that Smithsonian corrects *Enola Gay* exhibit," *Washington Times*, October 14, 1994, p. 20A.
20. Joyce Price, "Japanese threaten to cut ties with A-bomb exhibit," *Washington Times*, October 15, 1994, p. 1.
21. Charles D. Cooper, "The Nuking of NASM," *The Retired Officer* (November 1994): 4.
22. Brian D. Smith "Rewriting *Enola Gay*'s History," *The American Legion* (November 1994): 26.
23. "How the Legion Held Sway on *Enola Gay,*" Interview, *The American Legion* 138 (May 1995): 34, 36.
24. Richard H. Kohn, "History and the Culture Wars: The case of the Smithsonian Institution's *Enola Gay* Exhibition," *Journal of American History*, vol. 82, no. 3 (December 1995): pp. 1054–5.
25. Andrea Stone, "Wounds of war still color *Enola Gay*'s place in History," *USA Today*, October 5, 1994, 7a.

26. Same as note 14, but pages 100–33 and 100–34.
27. Hubert R. Dagley II, in *The American Legion* 138 (May 1995): 66.
28. John W. Dower, *War Without Mercy: Race and Power in the Pacific War* (New York: Pantheon, 1986).
29. Philip Nobile, *Judgment at the Smithsonian* (New York: Marlowe & Company, 1995), xxxvii.
30. Jo Becker, Letter to the Editor, *New York Times*, October 11, 1994, A20.
31. Robert K. Musil, letter to Martin Harwit, October 19, 1994, *NASM/MH*. Martin Harwit, letter to Robert K. Musil, October 26, 1994. *NASM/MH.*
32. I. Michael Heyman, quoted in Rowan Scarborough, "Veterans persuade museum to revise WWII exhibit again," *Washington Times*, October 21, 1994, p. A3.
33. Letter from forty-eight historians to Ira Michael Heyman, November 16, 1994, *NASM/MH* and *AFA, The Enola Gay Debate, Historians*, May 1995.
34. Organization of American Historians, Executive Committee Resolution, October 22, 1994, Organization of American Historians, Bloomington, Indiana.
35. Philip Nobile, *Judgment at the Smithsonian* (New York: Marlowe & Company, 1995), xxxix.
36. Barton J. Bernstein, "A postwar myth: 500,000 U.S. lives saved," *Bulletin of the Atomic Scientists* (June/July 1986): 38.
37. Robert Jay Lifton and Greg Mitchell, Letters to the Editor, *New York Times,* October 16, 1994, p. 14E.
38. Gar Alperovitz, "Beyond the Smithsonian Flap: Historians' New Consensus," *Washington Post*, October 16, 1994, p. C3.
39. Kai Bird, "The Curators Cave In," *New York Times,* October 9, 1994.
40. Julie A. Rhoad, "The Proposed *Enola Gay* Exhibit, Is It An Accurate Portrayal of History?" American Legion Auxiliary *National News* (January/February 1995): 15.
41. Monroe Hatch, letter to Martin Harwit, September 12, 1993, *NASM/MH* and *AFA, The Enola Gay Debate, Key Documents*, May 1995.
42. William D. Leahy, *I Was There* (New York: Whittlesey House, 1950), 441.
43. Dwight D. Eisenhower, *Crusade in Europe* (Garden City, New York: Doubleday, 1948): 443.
44. Henry Stimson to President Truman, "Memo Discussed with the President," April 25, 1945, Harry S Truman Library, Independence, Missouri.

Chapter 26

1. Jacqueline Trescott, "Smithsonian Search Narrows—Committee Head Resigns to Apply for Top Job," *Washington Post,* March 17, 1994, p. D1.
2. Jacqueline Trescott and Elisabeth Kastor, "Berkeley Ex-Chancellor Top Choice at Smithsonian—Regents Set to Vote on I. Michael Heyman," *Washington Post,* May 7, 1994, p. A1.
3. Jeffrey Staggs, "Searching for Himself," *Washington Times,* March 17, page C15.
4. Brett Moss, "Regents choose Berkeley lawyer—Ira Heyman will lead Smithsonian," *Washington Times,* May 26, 1994, p A3.
5. Martin Harwit, letter to Preble Stolz, September 13, 1995, *NASM/MH.*
6. Jeffrey Staggs, "Big Man on Mall campus," *Washington Times*, September 19, 1994, Metropolitan Times Section, p. C11.
7. Jacqueline Trescott, "Michael Heyman, Airing the Nation's Attic," *Washington Post*, September 20, 1994, p. D20.
8. Minutes of the National Air and Space Museum staff meeting for October 11, 1994, *NASM/MH.*
9. National Air and Space Museum Planning Document, "The Crossroads: The end of World War II, the Atomic Bomb and the Onset of the Cold War," July 1993; *AFA The Enola Gay Debate, Key Documents,* May 1995.
10. Heyman also repeated this story to a reporter: Stephen Budiansky et al. "A museum in crisis," *U.S. News & World Report*, February 13, 1995.
11. Michael Fetters to Martin Harwit, Memorandum on *"Enola Gay,"* October 17, 1994, *NASM/MH.*
12. Eugene L. Meyer, "Target Smithsonian—The Man Who Dropped the Bomb on Hiroshima Wants Exhibit Scuttled," *Washington Post,* January 30, 1995. Section D1.
13. Robert Manhan, U.S. Congress, Senate Committee on Rules and Administration, *The Smithsonian Institution: Management Guidelines for the Future*, 104th Cong., 1st sess., May 11, 18, 1995, 33.
14. Martin Harwit, "Comments on *Crossroads*—April 16, 1994, *NASM/MH,* AFA, *The Enola Gay Debate, Key Documents*, May 1995.
15. Rowan Scarborough, "Veterans persuade museum to revise WWII exhibit again," *Washington Times*, October 21, 1994, page A3.
16. I. Michael Heyman, letter to George A. Mannes, January 17, 1995, *NASM/MH.*

Chapter 27

1. Kenji Ohara, letter to Tom Crouch, August 23, 1994, *NASM/MH.*

2. Kenji Ohara, letter to Tom Crouch, August 30, 1994, *NASM/MH.*
3. Tom Crouch, letter to Kenji Ohara, August 30, 1994, *NASM/MH.*
4. Martin Harwit, letter to Takakazu Kuriyama, September 2, 1994, *NASM/MH.*
5. Martin Harwit, letter to Hitoshi Motoshima, September 6, 1994, *NASM/MH.*
6. Yoshio Saito, letter to President of the Smithsonian Institution, October 13, 1994, *NASM/MH.*
7. Takeshi Ito, Sakae Ito, and Senji Yamaguchi, letter to President of the United States, October 4, 1994, *NASM/MH.*
8. Hitoshi Motoshima, letter to Martin Harwit, November 17, 1994 *NASM/MH.*
9. Senji Yamaguchi, letter, translated by Wakako Takeuji, and forwarded to Martin Harwit by Hitoshi Motoshima on November 17, 1994, *NASM/MH.*
10. Brian Burke-Gaffney, letter to Martin Harwit, November 28, 1994, *NASM/MH.*
11. Martin Harwit, letter to Hitoshi Motoshima, December 17, 1994, *NASM/MH.*
12. Tatsuya Itoh, telefax to Martin Harwit January 7, 1995, *MH.*
13. Notes on phone call with Dr. Okajima, January 9, 1995, *MH.*

CHAPTER 28

1. Mark Rodgers, conversation with Martin Harwit, notes, September 19, 1995, *MH.*
2. Sam Johnson, Peter Blute, Henry Bonilla, Ralph Hall, Randy "Duke" Cunningham, Robert K. Dornan, and Duncan Hunter, letter to I. Michael Heyman, December 13, 1994, *AFA, The Enola Gay Debate, Congress*, May 1995. Edwin Bearss in an interview with Uday Mohan, April 11, 1995, *MH.*
3. Julie A. Rhoad, "The Proposed *Enola Gay* Exhibit, Is It An Accurate Portrayal of History?" American Legion Auxiliary *National News* (January/February 1995): 12.
4. Hubert Dagley II, letter to National Commander's Advisory Committee, December 2, 1994. I thank Mr. Dagley for making a copy of this document available to me with an explanatory letter on May 7, 1996. *MH.*
5. William M. Detweiler, to National Commander's Advisory Committee, Memorandum on "Recommended Final Position on NASM Display," January 4, 1995, *AFA, The Enola Gay Debate, Key Documents*, May 1995.
6. William M. Detweiler to Department Commanders and Adjutants, Memorandum on "National Air and Space Museum Exhibit,"

January 18, 1995, *AFA, The Enola Gay Debate, Key Documents,* May 1995.

7. I. Michael Heyman, letter to Sam Johnson, January 4, 1995, *AFA, The Enola Gay Debate, Congress*, May 1995.
8. Mike Heyman, Memorandum on "Exhibition and Public Programming Review," January 9, 1995. Martin Harwit to Mike Heyman, Memorandum on "Letter on Exhibition and Public Programming Review," January 9, 1995, *NASM/MH.*
9. Martin Harwit to Mike Heyman, Memorandum on "Sen. Hatfield's Exhibition on Hiroshima," January 9, 1995, *NASM/MH.*
10. Martin Harwit, letter to Hubert R. Dagley II, January 9, 1995, *NASM/MH.*
11. William M. Detweiler, "Position Statement of The American Legion Concerning The National Air and Space Museum Exhibit Entitled "The Last Act: The Atomic Bomb and the End of World War II," January 18, 1995, *AFA, The Enola Gay Debate, Key Documents,* May 1995; William M. Detweiler, "Background and Status: The American Legion Position on "The Last Act: The Atomic Bomb and the End of World War II," January 18, 1995, *AFA, The Enola Gay Debate, Key Documents,* May 1995.
12. See note 3.
13. See note 11.
14. See note 11.
15. I. Michael Heyman, quoted in Rowan Scarborough, "Veterans persuade museum to revise WWII exhibit again," *Washington Times*, October 21, 1994, p. A3.
16. See note 11.
17. James Kuhnhenn, "Museum scales back *Enola Gay* exhibit," *Kansas City Star,* January 31, 1995, p. A1; Andrea Stone, "Hiroshima display ends in rancor," *USA Today,* January 31, 1995, p. 1A.
18. Michael Kilian, "Smithsonian puts *Enola Gay* display on hold," *Chicago Tribune,* January 21, 1995.
19. William M. Detweiler, letter to President William Clinton, January 19, 1995, *AFA, The Enola Gay Debate, Key Documents*, May 1995.
20. Congressman Sam Johnson, press release, January 19, 1995, *AFA, The Enola Gay Debate, Key Documents*, May 1995.
21. Air Force Association press release, January 20, 1995, *AFA, The Enola Gay Debate, Releases,* May 1995.
22. Unsigned editorial, "The *Enola Gay* Explosion," *Washington Post*, January 20, 1995, p. A20.
23. Rowan Scarborough, "Hearings planned on WWII exhibit," *Washington Times*, January 20, 1995, p. A1
24. Congressman Gerald B. Solomon, letter to Martin Harwit, January 20, 1995, *NASM/MH.*

25. Letter from 81 Members of the House of Representative to I. Michael Heyman, January 24, 1995, *AFA, The Enola Gay Debate, Congress*, May 1995.
26. I. Michael Heyman, conversation with Martin Harwit, notes, January 28, 1995, *MH.*
27. Laurie Kellman, "Clinton firm on backing use of A-bomb," *Washington Times*, Wednesday, April 19, 1995, p. A10.
28. Wendell Phillippi, "Japan was anticipating annihilation," *World War II Times* (December 1994/January 1995).
29. Greg Pierce, "Under siege, Postal Service agrees to cancel A-bomb stamp," *Washington Times*, December 9, 1994, p. A10.
30. R.E. Smith, letter to Sen. Thad Cochran, January 25, 1995, *AFA Key Documents*, May 1995.
31. Roy K. Flint, letter to Chief Justice Rehnquist, January 26, 1995, *NASM/MH.*
32. Eric Foner, Gary B. Nash, and Michael Kammen, letter to Chief Justice Rehnquist, January 27, 1995, *NASM/MH.*
33. George F. Will, "The Real State of the Union," *Washington Post*, January 26, 1995, p. A25.
34. Oliphant, Cartoon in *International Herald Tribune,* January 26, 1995, p. A25.
35. "A Pared-Down *Enola Gay,*" *Washington Post,* January 28, 1995 p. D3; Joyce Price, "Smithsonian slant to stop, Gingrich says," *Washington Times,* January 28, 1995, p. A1.
36. "Hijacking History," *New York Times*, January 30, 1995, p. A18.
37. "Politics has no place in *Enola Gay* exhibit," *USA Today*, January 30, 1995, p. 14A.
38. I. Michael Heyman, Press release of statement made at press conference following the regents' meeting, Monday January 30, 1995, *NASM/MH.*
39. James Kuhnhenn, "Museum scales back *Enola Gay* exhibit," *Kansas City Star*, January 31, 1995, p. A1; Andrea Stone, "Hiroshima display ends in rancor," *USA Today,* January 31, 1995, p. 1A.
40. See note 5.
41. See note 3.
42. Preble Stolz, "Report to Secretary I. Michael Heyman concerning the *Enola Gay* Exhibition," January 26, 1995, page 18. In author's possession.
43. Paul W. Tibbets, Jr., "Our Job was to Win," *American Legion* Magazine, November 1994, p. 29. (See also news release, Airmen Memorial Museum, June 8, 1994).
44. Eugene L. Meyer, "Smithsonian Stands Firm On A-Bomb Exhibit," *Washington Post*, January 19, 1995, p. C3.

CHAPTER 29

1. Martin Harwit, letter to Frank Rabbitt, January 30, 1995, *NASM/MH.*
2. Martin Harwit, annotated "Statement 2, 01-31-95 Used at Town Meeting for staff Jan 31, 1995," *MH.*
3. Takashi Hiraoka, press release, January 31, 1995.
4. Hiroshi Harada, letter to Martin Harwit, February 7, 1995.
5. Shinichiro Asao, letter to Martin Harwit, February 1, 1995, *NASM/MH.*
6. J. Lynn Lunsford and George E. Hicks, "Interview—Paul W. Tibbets," *Dallas Morning News,* February 5, 1995, p. 1J.
7. Vago Muradian, "This is what we wanted," *Air Force Times,* February 13, 1995, 17.
8. Michael E. Ruane, "Flying into a web of historical dispute," *Philadelphia Inquirer,* February 5, 1995, p. C1.
9. Dell Herndon in the *Whittier Daily News,* February 20, 1995.
10. Murayama quotation: Kyodo News Service, as cited by the Washington Post service in "Japan criticizes decision to revamp *Enola Gay* exhibit," *San Jose Mercury News,* February 1, 1995; Clinton and Gore quotation from "Exhibit Torpedoed—Smithsonian Yields to Critics," Associated Press, *Tulsa World,* January 31, 1995, p. 1, and Don Kirkman, "Exhibit nuked—Smithsonian cuts controversial text from museum display on *Enola Gay,*" Scripps Howard News Service, *Knoxville News-Sentinel,* January 31, 1995, p. A1.
11. "The Smithsonian Changes Course," editorial, *Washington Post,* February 1, 1995.
12. Coleman McCarthy, "Glory-Seekers and the Bomb," *Washington Post,* February 7, 1995.
13. Benson, Cartoon in the *Arizona Republic,* January 25, 1995.
14. *Philadelphia Inquirer* editorial, "Bombing history," February 1, 1995, p. A14.
15. Stahler, Cartoon, *Cincinnati Post,* February 1, 1995.
16. Tony Auth, Cartoon, *Philadelphia Inquirer,* February 1, 1995.
17. Sargent, Cartoon, *Austin American Statesman,* February 1, 1995.
18. Wasserman, Cartoon, *Boston Globe,* February 2, 1995.
19. Stephen Budianski et al., "A Museum in Crisis" *U.S. News & World Report,* February 13, 1995, p. 73.
20. Ibid.
21. "American U. to Host Hiroshima Exhibit," *Washington Times,* March 31, 1995.
22. Eugene L. Meyer, "AU May Exhibit Artifacts from Hiroshima Bomb," *Washington Post,* April 6, 1995, p. D1.

23. Brian Blomquist, "New Hiroshima exhibit eyed," *Washington Times,* April 7, 1995, p. C4.

Chapter 30

1. Martin Harwit to Mike Heyman, Memorandum on Ann Arbor Forum, February 13, 1995, *NASM/MH.*
2. I. Michael Heyman, National Press Club speech, "The Future of the Smithsonian," and questions and answer session following the speech, February 23, 1995, transcript in author's possession, *MH.*
3. Mike [Heyman], letter to Martin [Harwit], March 3, 1995, in author's possession. *MH.*
4. Mike Feinsilber, "The full story of *Enola Gay* has to wait," *Cleveland Plain Dealer,* February 21, 1995, p. 2A.
5. Preble Stolz, "Report to Secretary I. Michael Heyman concerning the *Enola Gay* Exhibition," January 26, 1995, p. 8, in the author's possession.
6. I. Michael Heyman, letter to Gen. Monroe W. Hatch, Jr., March 31, 1995, *AFA, The Enola Gay Debate, Key Documents,* May 1995.
7. Thomas J. Kilcline, remarks at the symposium on "Presenting History—Museums in a Democratic Society," Ann Arbor, Michigan, April 19, 1995; Charles D. Cooper, in U.S. Congress, Senate Committee on Rules and Administration, *The Smithsonian Institution: Management Guidelines for the Future,* 104 Cong., 1 sess., May 11, 18, 1995, p. 15. Robert Manhan, ibid., pp. 33–34.
8. William M. Detweiler, letter to William F. Clinger, Jr., February 10, 1995, *AFA, The Enola Gay Debate, Key Documents,* May 1995.
9. Sam Johnson, letter to I. Michael Heyman, March 22, 1995, *AFA, The Enola Gay Debate, Key Documents,* March 1995.
10. I. Michael Heyman, letter to Sam Johnson, April 4, 1995, in U.S. Congress, Senate Committee on Rules and Administration, *The Smithsonian Institution: Management Guidelines for the Future,* 104th Cong., 1st sess., May 11, 18, 1995, p. 161.
11. Martin Harwit to Mike Heyman, Memorandum on "Handling requests for the current *Last Act* script." February 13, 1995, *NASM/MH.*
12. See notes 9 and 10.
13. Ibid.
14. National Academy of Public Administration, "The National Air & Space Museum—A Review of the Organization and Management," September 1995.
15. I. Michael Heyman, "The Future of the Smithsonian," National Press Club speech, February 23, 1995, transcript in the author's possession.

16. Robert McC. Adams, letter to William A. Rooney, June 13, 1986, *Carmen Turner File, tab 5, NASM/MH.* (See chapter 2.)
17. Joel Achenbach, "*Enola Gay* Exhibit: Plane and Simple," *Washington Post,* June 28, 1995, p. A1.
18. Martin Harwit, letter to I. Michael Heyman, May 2, 1995, *MH.*
19. Martin Harwit, in U.S. Congress, Senate Committee on Rules and Administration, *The Smithsonian Institution: Management Guidelines for the Future,* 104th Cong., 1st sess., May 11, 18, 1995, p. 111.
20. Senator Stevens, in U.S. Congress, Senate Committee on Rules and Administration, *The Smithsonian Institution: Management Guidelines for the Future,* 104th Cong., 1st sess., May 11, 18, 1995, pp. 64, 100.
21. Peter G. Powers to I. Michael Heyman, Memorandum on "Effect of National Armed Forces Museum Advisory Board Legislation on The National Air and Space Museum," January 24, 1995.

Index